Combustion Technology

Contributors

Philip M. Becker
J. M. Beér
Dennis J. Brown
Robert H. Essenhigh
I. H. Farag
Robert J. Heinsohn
H. C. Hottel
David W. Locklin
G. Malhouitre
Howard B. Palmer
Gunter J. Penzias
Abbott A. Putnam
William T. Reid
A. F. Sarofim
J. Swithenbank
Arvind C. Thekdi
Yih-Wan Tsai
Stewart Way

COMBUSTION TECHNOLOGY: Some Modern Developments

Edited by

HOWARD B. PALMER

Fuel Science Section
Department of Material Sciences
The Pennsylvania State University
University Park, Pennsylvania

J. M. BEÉR

Department of Chemical Engineering and Fuel Technology
The University of Sheffield
Sheffield, England

1974

ACADEMIC PRESS New York and London

A Subsidiary of Harcourt Brace Jovanovich, Publishers

ACADEMIC PRESS, INC.
111 Fifth Avenue, New York, New York 10003

United Kingdom Edition published by
ACADEMIC PRESS, INC. (LONDON) LTD.
24/28 Oval Road, London NW1

Library of Congress Cataloging in Publication Data

Main entry under title:

Combustion technology.

Includes bibliographies.
1. Combustion engineering. I. Palmer, Howard Benedict, Date ed. II. Beér, János Miklós, Date ed.
TJ254.5.C65 621.4′02 73-2066
ISBN 0–12–544750–7

PRINTED IN THE UNITED STATES OF AMERICA

Contents

IV Flame Stabilization in High Velocity Flow

J. Swithenbank

V Combustion Noise: Problems and Potentials

Abbott A. Putnam and Dennis J. Brown

VI Heat Transfer from Nonluminous Flames in Furnaces

H. C. Hottel, A. F. Sarofim, and I. H. Farag

VII Radiative Exchange in Combustion Chambers

H. C. Hottel, A. F. Sarofim, and I. H. Farag

XI Combustion Aspects of MHD Power Generation

Stewart Way

XII Temperature Measurements and Gas Analysis in Flames and Plasmas Using Spectroscopic Methods

Gunter J. Penzias

XIII Furnace Analysis: A Comparative Study

Robert H. Essenhigh, Arvind C. Thekdi, G. Malhouitre, and Yih-Wan Tsai

XIV An Introduction to Stirred Reactor Theory Applied to Design of Combustion Chambers

Robert H. Essenhigh

List of Contributors

Numbers in parentheses indicate the pages on which the authors' contributions begin.

PHILIP M. BECKER (239), Fuel Science Section, Department of Material Sciences, The Pennsylvania State University, University Park, Pennsylvania

J. M. BEÉR (61, 213), Department of Chemical Engineering and Fuel Technology, The University of Sheffield, Sheffield, England

DENNIS J. BROWN (127), Department of Chemical Engineering and Fuel Technology, The University of Sheffield, Sheffield, England

ROBERT H. ESSENHIGH (349, 373), Fuel Science Section, Department of Material Sciences, The Pennsylvania State University, University Park, Pennsylvania

I. H. FARAG (163, 189), Department of Chemical Engineering, Massachusetts Institute of Technology, Cambridge, Massachusetts

ROBERT J. HEINSOHN (239), Department of Mechanical Engineering, The Pennsylvania State University, University Park, Pennsylvania

H. C. HOTTEL (163, 189), Department of Chemical Engineering, Massachusetts Institute of Technology, Cambridge, Massachusetts

DAVID W. LOCKLIN (417), Battelle Memorial Institute, Columbus, Ohio

G. MALHOUITRE (349), Groupe Aerodynamique et Physique des Écoulements, Électricité de France, Direction des Études et Recherches, Chatou (Near Paris), France

HOWARD B. PALMER (1), Fuel Science Section, Department of Material Sciences, The Pennsylvania State University, University Park, Pennsylvania

GUNTER J. PENZIAS (321), Norcon Instruments, Inc., South Norwalk, Connecticut

ABBOTT A. PUTNAM (127), Battelle Memorial Institute, Columbus, Ohio

William T. Reid (35), Battelle Memorial Institute, Columbus Laboratories, Columbus, Ohio

A. F. Sarofim (163, 189), Department of Chemical Engineering, Massachusetts Institute of Technology, Cambridge, Massachusetts

J. Swithenbank (91, 275), Department of Chemical Engineering and Fuel Technology, The University of Sheffield, Sheffield, England

Arvind C. Thekdi (349), Research and Development Department, Surface Combustion Division, Midland-Ross Corporation, Toledo, Ohio

Yih-Wan Tsai (349), Melting and Forming Division, Pittsburgh Plate Glass Industries, Inc., Creighton, Pennsylvania

Stewart Way (291), Westinghouse Research Laboratories, Pittsburgh, Pennsylvania

Preface

Like many technical books, this volume is the outgrowth of a teaching experience. For a number of years we have participated in an intensive one-week course on modern developments in combustion engineering, taught at The Pennsylvania State University. Registrants in the course range from neophyte combustion technologists to mature professionals with years of experience. Fortunately, to meet the varied needs of this group, we have been able to assemble a distinguished and enthusiastic group of lecturers who are at the forefront of their specialties. The essentials of their most recent lectures are collected here. Furnace flames receive the most attention in this volume. However, we would stress that most matters treated have broad applicability. Certainly the discussions in the course have indicated this to be the case.

The book begins with a survey by H. B. Palmer of some aspects of the chemistry of flames. It includes an extensive bibliography directed especially to the reader who is relatively new to the field of combustion. This chapter is followed by W. T. Reid's discussion of an important area in applied combustion chemistry, the problem of corrosion and deposits. We then move to a group of chapters oriented toward aerodynamics and heat transfer in combustors, beginning with a summary of combustion aerodynamics by J. M. Beér that is followed by the application of aerodynamic principles to flame stabilization in high-speed flow as discussed by J. Swithenbank.

Another facet of combustion stability, and one of much practical significance, is treated by A. A. Putnam and D. Brown, in their chapter on combustion noise. Here the interaction between aerodynamics and other aspects of combustion emerges as crucial.

The three succeeding chapters, two by H. C. Hottel, A. F. Sarofim, and I. H. Farag, and one by J. M. Beér, treat problems of radiative heat transfer in combustion chambers. The first of these sets out the fundamentals, the second considers several possible models for calculations of radiative transfer, and the third is an overview of the whole problem, including effects of particulates.

We then present three chapters related to the electrical properties of flames. The chapter by R. J. Heinsohn and P. M. Becker describes the basic matter of flame–field interactions and goes on to discuss several practical applications. J. Swithenbank treats a specific application, generation of electricity by magnetohydrodynamic methods. This is followed by a chapter on critical considerations in the practical aspects of MHD power generation, by Stewart Way.

Although experimental measurements are touched upon in many of the chapters, one chapter by G. J. Penzias, devoted solely to the problem of temperature measurements and gas analysis, is included because of the special significance and potential of spectroscopic methods in studying high temperature flames and plasmas.

The final three chapters in the book are directed toward improved efficiency of combustors. R. H. Essenhigh, A. C. Thekdi, G. Malhouitre, and Y.-W. Tsai extend the concept of furnace analysis and test it against data from operating combustors. The influence of stirred reactor theory on design principles for high-performance combustion chambers is then considered by R. H. Essenhigh. The last chapter, contributed by D. W. Locklin, is a summary of developments in the design and utilization of oil burners.

Combustion technology continues to evolve in significant and increasingly powerful ways. It is reaching the point at which it does indeed provide that bridge, often discussed by M. W. Thring, which spans the "river of ignorance and prejudice" separating pure science and industrial processes. As D. B. Spalding has said, we are entering a period in which the engineer will be able to make real use of the results of research. It is an exciting period, to which we hope this book will make a contribution.

I

Equilibria and Chemical Kinetics in Flames

HOWARD B. PALMER

FUEL SCIENCE SECTION
DEPARTMENT OF MATERIAL SCIENCES
THE PENNSYLVANIA STATE UNIVERSITY
UNIVERSITY PARK, PENNSYLVANIA

If it's mixed, it's burnt
—Adage of the combustion engineer

There is just enough truth in that maxim to provide reason for examining its limitations. Indeed much of this book is devoted to the question of mixedness and its influence on combustion. For the most part, the emphasis is on mixing and burning in rather large flames. In this chapter we mainly consider the minutiae of combustion—the microscopic processes of chemical kinetics and the special kinetic balance that is chemical equilibrium.

The relation between mixing and burning is obvious. Experiment and theory both establish that mixing of a hydrocarbon with air at high temperature does not lead immediately to the stoichiometric quantities of CO_2 and H_2O, nor even to an equilibrium mixture containing these plus other species. The chemical conversion processes occur at finite rates. Some of the processes may be extremely fast; others may be so slow that they can be neglected—that is, considered to be "frozen." Some processes may be very sensitive to temperature (e.g., NO formation); others (e.g., atom recombination) may not be.

An engineer who is accustomed to thinking about large, practical flames may sometimes question the relevance of complex chemical equilibria and chemical kinetics. He wonders how small-scale research on chemical minutiae can help to solve his problems. To respond we cite again Professor Spalding's comments (1971) quoted in the Preface, that we are at the point where engineers can make significant use of the results of basic research. He is referring to advances in mathematical modeling using computers, in understanding of turbulent flow and transport, and in knowledge of reaction rates. Many of these advances have come through fundamental small-scale research, often on small flames, frequently on related phenomena such as reactions in shock waves and detonations, or in discharge-flow systems.

In principle it would be appropriate to summarize here what is now known about combustion in small premixed and diffusion flames and explosions, and perhaps include detonations as well. An attempt to present such a summary in a single chapter would be superficial and fruitless. This book is not intended as a general introduction to combustion; rather, it emphasizes recent developments related to practical flames. Hence, we summarize the knowledge of small-scale flames by presenting a bibliography. It appears at the end of this chapter. A few journals are also mentioned as sources of current research on flames.

However, an effort is made to indicate a few rather recent advances in fundamental studies of flame processes that are, or may be, of importance to a combustion engineer. The topics discussed reflect the writer's bias toward gas-phase kinetics and accordingly do not include all significant new work.

I. COMBUSTION EQUILIBRIUM

Although combustion is a process that starts from a nonequilibrium condition, we prefer to start with some comments on the final condition, that of chemical equilibrium. Equilibrium may not be achieved in a flame; that depends on the kinetics. Equilibrium might be termed the goal of the combustion chemistry. Frequently it is achieved, at least to a good approximation. In such cases, if combustion processes are fast relative to the rate of heat loss by conduction, convection, or radiation, one can calculate the maximum temperature of the combustion products and the composition of the products at that temperature, using only thermodynamic properties. This is the adiabatic flame temperature calculation. It is important to realize that this calculation is a practical one. The adiabatic assumption is often very good. Of course, the products eventually cool; but if heat losses are not severe, the adiabatic calculation is useful.

Methods of calculation are described in several of the books listed in the Bibliography [e.g., Lewis and von Elbe (1961) or Gaydon and Wolfhard (1970)]. The points to be made here are that:

(a) For hot flames—roughly for flames hotter than 2000°K—assuming simple burning to CO_2 and H_2O leads to significant errors. The errors get worse as the flame gets hotter. Assumption of simple burning in a stoichiometric CH_4–air ($CH_4 : O_2 = 1 : 2$) flame yields a flame temperature of about 2250°K. Including the appropriate dissociation equilibria lowers the calculated temperature by about 20°K, and of course reveals the formation of modest amounts of minor species. The effect in hotter flames is illustrated by the stoichiometric CH_4–O_2 flame. Here production of undissociated CO_2 and H_2O would yield a temperature of 5320°K. Including the equilibria lowers this to 3010°K and reveals, for example, that the products will contain more CO than CO_2.

(b) There is no longer any reason to oversimplify flame equilibrium calculations because they are becoming very easy with the aid of high-speed computers. They also are becoming highly reliable because of the continuing scrutiny and critical revision of thermodynamic properties being carried out by several organizations. Furthermore, the number of species for which thermodynamic data are available has increased enormously in the past decade. Probably the best-known and most used critical compilation is the JANAF Tables (Stull and Prophet *et al.*, 1970). Currently the U.S. National Bureau of Standards is undertaking a thorough revision of the famous Circular 500. Several reports (Wagman *et al.*, 1968, 1969) on that work have been issued at the time of writing. The reports provide a useful check on the 0 and 298°K values listed in the JANAF Tables and other tabulations. The engineer can now calculate high-temperature equilibria involving a remarkable variety of species, including a vast array of gaseous species and some liquids [e.g., $B_2O_3(1)$] and solids [e.g., C(s)].

Considerable attention has been given over the years to the methods and results of equilibrium calculations. Beyond the general discussions in the books listed in the Bibliography, the interested reader can find much of value in the proceedings of two conferences (Bahn and Zukoski, 1960; Bahn, 1963) held by the Western States Section of the Combustion Institute. For those concerned with equilibrium in flames containing species made up only of the elements, C, H, O, and N, two short papers by Harker (1967) and Harker and Allen (1969), and one by Smith (1969) will be useful. For more elaborate computations, the comprehensive NASA program (Gordon and McBride, 1971) is now available in report form.

As remarked earlier, equilibrium calculations are often valuable because they describe the ultimate kinetic destination of a reacting system, given sufficient time. A nice illustration of this point is presented in the work of Jeffers and Bauer (1971), who examined the high-temperature equilibrium compositions of various mixtures of SO_2, CO, oxides of nitrogen, and hydrocarbons. In the S–C–O system, 13 chemical species were considered. The results are significant in showing the composition patterns toward which the mixtures will move as they react chemically. Among other interesting features, the calculations show that removal of SO_2 by CO at high temperatures is very inefficient because the stability of SO_2 is comparable to that of CO_2. At low temperatures (below 1000°K), CO will remove SO_2, but here the homogeneous removal is very inefficient because of slow kinetics [as revealed by shock tube kinetic studies (Bauer *et al.*, 1971)].

Perhaps a word is in order about "partial equilibrium." It is possible in some chemical systems to have a collection of extremely fast reactions essentially equilibrate among themselves, so to speak, long before final equilibrium is reached. There is good evidence (Schott and Bird, 1964; Getzinger and Schott, 1965) that this happens in the shock-initiated combustion of H_2 and O_2. The reactions

$$H + O_2 \rightleftarrows OH + O, \tag{1}$$

$$O + H_2 \rightleftarrows OH + H, \tag{2}$$

$$OH + H_2 \rightleftarrows H_2O + H, \tag{3}$$

are intrinsically fast in both forward and reverse directions. They are much faster, at normal pressures, than the third-order reactions of atom–atom or atom–radical combination ($O + O + M \rightarrow O_2 + M$, etc., where M is any energy-removing species). The consequence is that the species H, O, OH, H_2, O_2, and H_2O adopt equilibrium relationships with respect to one another as required by reactions (1), (2), and (3) long before equilibria that include $H_2 \rightleftarrows 2H$, and so on, are established. There are indications that some analogous partial equilibria may occur in other flames, for example, O_2-supported ammonia flames (Kaskan and Nadler, 1972). From the standpoint of the engineer, partial equilibrium may be quite important in the formation of nitric oxide in combustion. Many hydrocarbon–air flames come close to equilibrating all other species before much NO is formed. Thus, for some purposes, a total H/C/O equilibrium model (Marteney, 1970) provides a useful starting point for the understanding of NO formation rates. However, Bowman (1973) has shown that a partial equilibrium model is in better agreement with detailed observations.

The freezing of CO levels in the exhaust of an internal combustion engine has been modeled rather successfully by Keck and Gillespie (1971) using a partial-equilibrium model in which the kinetic aspects of the system are controlled by three-body processes.

II. CHEMICAL KINETICS AND COMBUSTION

Flames in which the reactants are not mixed before entering the combustion region are aptly termed diffusion flames because the rate of burning is controlled by interdiffusion of the fuel and oxidant. Introduction of turbulence increases the burning rate, as expected, and indeed for such flames the adage at the head of this chapter is essentially true. Does that mean that chemical reaction rates in such systems can be ignored? For considerations of heat release rates, the answer is a cautious *yes*. But what we are really saying is that chemical kinetics are usually fast relative to mixing rates. Under some circumstances they may not be. Consider the limit of very high intensity turbulence, which would be a perfectly stirred system. Combustion in such a system is discussed in Chapter XIV by R. H. Essenhigh. For now, we merely point out that in a perfectly stirred system, transport limitations are removed and chemical kinetics become the controlling factor. Studies of combustion in strongly stirred systems were pioneered by Longwell and Weiss (1955) and have been continued more recently by Hottel *et al.* (1965), Wright (1969), and others. Clearly, combustion in very turbulent regimes that are strongly but not perfectly stirred will be governed by some combination of turbulent diffusion, molecular diffusion, and chemical kinetics. This complex problem has recently been tackled by Spalding (1972).

Even in poorly stirred systems—laminar diffusion flames, for example—one may have to consider chemical reaction rates. Suppose the concern is with promoting radiative heat transfer, that is, suppose we want a luminous flame, but we do not want soot emitted from the combustion chamber. Chemical kinetic factors, together with mixing rates, will determine where soot is formed in the flame, how much will be formed, what chemical composition and particle size distribution it will have, and whether it will be ultimately consumed.

There are many other examples of the significance of chemical kinetics in combustion. Here are a few.

(1) Premixed flames. Rates of burning are determined by the balance between chemical reaction rates, the associated production of heat and reactive intermediates (atoms and free radicals), and the rate of feedback of energy and reactive species to unburned gas through molecular transport processes. Propagation of premixed flames is discussed extensively in several books in the Bibliography.

(2) Flames in which there are strong fluctuations or oscillations in density and temperature. The mean composition of the burnt gases of such flames will reflect the ability or inability of the reactions (which would, let us say, be equilibrated in the absence of fluctuations) to follow the changes in gas concentration and temperature. Reactions with large energies of activation, such as $O + N_2 \rightarrow NO + O$, may not be very successful in following the fluctuations.

(3) Formation and disappearance of pollutants such as soot, NO, CO, and SO_2. These minor species, which paradoxically have assumed large importance, are challenges to combustion kineticists. Equilibrium considerations frequently fail to describe their behavior. We return to them later.

(4) Ignition and explosion behavior. When energy is fed into a combustible mixture (whether by electrical sparks, hot wires, mixing with hot gases, or heat conduction from walls), the consequences will depend on the rates at which chemical processes then occur, together with the rates of diffusion of the species and the energy created by those processes. A vast literature exists on ignition, explosions, quenching, and related matters. Valuable information on combustion reactions has issued from these studies. Ignition delay is of course a vital matter in the design and functioning of many combustion devices such as internal combustion and jet engines.

(5) Ion production in flames. Ion concentrations in many flames are orders of magnitude greater than concentrations calculated on the assumption of chemical equilibrium. Ions are formed by chemical kinetic processes that have high rates in the region of a flame where atoms and free radicals such as O and CH have relatively high concentrations. The residence time of gases in this region is normally too short to permit appreciable progress toward equilibration of ions with other species, and processes causing decay of ions later in the flame are rather slow. Hence, ions often overshoot equilibrium by a large margin in the reaction zone, and stay at high levels for a long time afterward. Some of the significance of ions in flames and

their potential to the engineer is pointed out in Chapter IX by R. J. Heinsohn and P. M. Becker, in Chapter X by J. Swithenbank, and in Chapter XI by S. Way.

(6) Emission of radiation by flames. Aside from emission by soot in flames, which is thermal (i.e., determined by the gas temperature and the concentration and optical characteristics of the emitter), one frequently finds peculiarities in the distribution of emission intensity with wavelength. This is especially true for the spectra of species such as CH, C_2, and OH. The peculiarities are a direct consequence of the chemical rates and mechanisms of their formation. Gaydon (1957) has written in detail about these spectroscopic matters. It should be mentioned, however, that CO_2 and H_2O, which are the principal infrared emitters in common flames and which emit the bulk of the radiant energy (in the absence of soot), exhibit essentially thermal (equilibrium) excitation. Further discussion of infrared emission can be found in Chapter VIII by J. M. Beér and in Chapter XII by G. Penzias. A detailed treatment is contained in the monograph by Tourin (1966).

III. SOME RECENT DEVELOPMENTS RELATED TO COMBUSTION KINETICS

In this section are discussed several areas of flame kinetics in which impressive progress has occurred during the past few years. In the writer's opinion, they are all areas that will have an impact on engineering design and practice—if not immediately, then soon. It is very significant to note that in all cases, the progress is occurring through the mutual stimulation of good experiments and theory. The power of the theoretical work arises in large part from the evolution and wise use of computers, while the advances in experimental findings arise through the creative use of the most modern instrumentation.

A. Profiles in One-Dimensional Flames

Of course no real flame is one dimensional. However, approximately one-dimensional flames can be created in the laboratory, and they are most amenable to theoretical treatment. Once a full match between theory and experiment is achieved for such flames, extension to three-dimensional flame propagation becomes a problem in mathematical modeling that can be set up in straightforward fashion. Spalding (1971) has noted that the three-dimensional problem is on the verge of becoming soluble with the aid of computers.

Interest in flame "structure" (i.e., profiles of temperature and species concentrations) is of long standing, but one can broadly mark the beginning of a modern era dating from about the early 1940s. Comprehensive theories of premixed flame propagation arose in that period, and one began to see experiments that revealed something of the microstructure of flames. Development of this effort during the succeeding two decades, particularly the experimental side, is summarized in the fine book of Fristrom and Westenberg (1965). Figure 1, adapted from that book, illustrates the excellent precision and detail with which probing of premixed flames can be accomplished to yield profiles of temperature and stable species. Sophisticated experiments have, in recent years, also yielded profiles of the labile

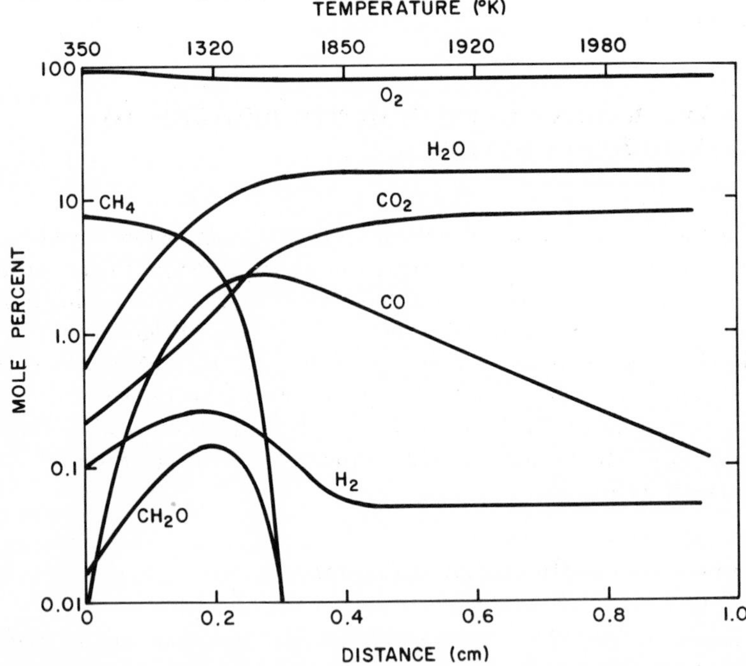

Fig. 1 Profiles of stable species (except N_2 and Ar) and temperature in a premixed, laminar CH_4–O_2 flame burning at a pressure of 0.1 atm. Input composition was, in mole percent: CH_4, 7.8; O_2, 91.5; balance, N_2, Ar, and CO_2. (Note the logarithmic scale for the concentrations.) [After Fristrom *et al.* (1960). *J. Phys. Chem.* **64,** 1386 (1960). Copyright 1960 by the American Chemical Society. Reprinted by permission of the copyright owner.]

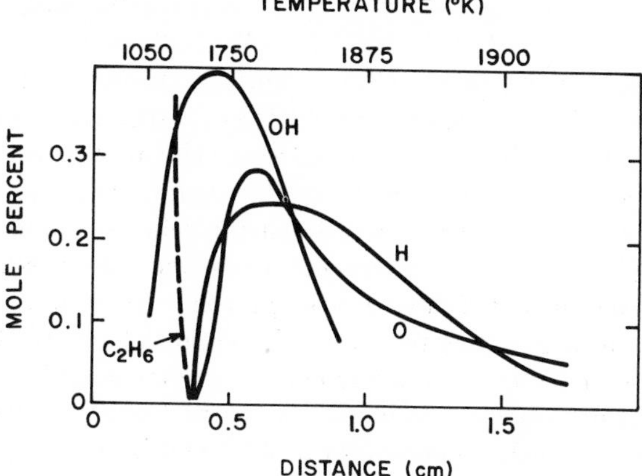

Fig. 2 Profiles of temperature, H, O, and OH in a premixed, laminar C_2H_6–O_2 flame at a pressure of 0.1 atm. A portion of the C_2H_6 profile (dashed line) is shown for reference. Input composition was, in mole percent: C_2H_6, 4.3; O_2, 95.0; balance, Ar, H_2O, and CO_2. [After Westenberg and Fristrom (1965).]

species H, O, and OH (Westenberg and Fristrom, 1965) in several flames. An example is shown in Fig. 2. The most spectacularly detailed probing of premixed flames has been accomplished by Peeters and Mahnen (1973). They have obtained profiles in CH_4–O_2 flames for the species O_2, CH_4, CO_2, H_2O, H_2, CH_2O, CH_3, HO_2, CH_3OH, CHO, CH_3O_2, O, H, and OH.

The precision of such profiles permits one to extract from them information on species fluxes, heat generation rates, and, in some cases, rates of elementary chemical reactions. However, the most important function of the profiles seems now to be to serve as a target for flame modeling. That is, one obtains from whatever source the best possible information on the rates of all pertinent elementary chemical reactions and on all transport coefficients; one then writes down all the relevant equations (i.e., the differential equations for conservation of mass, momentum, and energy, and for the continuity of species, plus the equation of state) applicable to the flame [the book by Strehlow (see Bibliography) presents the subject well]; finally one introduces the appropriate boundary conditions and, using a computer, numerically integrates the differential equations. [Even with a powerful computer this is no simple matter; see, for example, Wilde (1972).] If the result matches the experimental data, the proper conclusion may be that the model was correct. Of course, the danger is that there may be offsetting errors, so the model will normally be subjected to parametric study to assess the sensitivity of the output to variations in, let us say, rate constants of several crucial reactions.

It is perhaps slightly too early to speak of *complete* matching of models with experiments. Nevertheless, very great progress is being made, for example, by Dixon-Lewis (1967, 1970). The H_2–O_2 flame has been successfully modeled by Eberius *et al.* (1971). Work by Becker and co-workers (1972) suggests that a good model of the CH_4–O_2 flame may soon be available, including rates of chemi-ionization. One of the most interesting studies has been published in two papers by Porter *et al.* (1967) and by Browne *et al.* (1969a). The first paper reported detailed profiles for premixed flames using CH_4 and C_2H_2 as fuels. The measurements were noteworthy for their completeness. They included not only temperature and stable species but also the concentrations of positive ions and several radicals (OH, CH, and O atoms in ground states, plus electronically excited C_2, CH, and OH). In the second paper, kinetic modeling was carried out on several of the C_2H_2–O_2 flames for which experimental profiles had been obtained. A system of 16 reversible chemical reactions was adopted to describe the kinetics. The match between model and experiments was far from perfect, but was certainly very promising.

Reliable data on reaction rate constants of elementary reactions are essential to the success of kinetic modeling of flames. Despite the very great advances in knowledge resulting from rate measurements using (especially) shock tubes and discharge-flow systems, there are some serious deficiencies. A continuing critical assessment of the state of knowledge of rate constants has recently been undertaken by four different groups, one each in the U.S., the U.K., and the U.S.S.R., and one international. They are as follows:

(1) Chemical Kinetics Information Center, Dr. D. Garvin, Director, National Standard Reference Data System (NSRDS), NBS Office of Standard Reference Data, Washington, D.C.

(2) High Temperature Reaction Rate Data Group, Dr. D. L. Baulch, Director, The University, Leeds, England.

(3) Commission on Compilation of Rate Data (GSSSD), Professor V. N. Kondratiev, Chairman, Academy of Sciences, Moscow, U.S.R.R.

(4) Task Group on Data for Chemical Kinetics, Professor S. W. Benson, Chairman (Stanford Research Institute, Menlo Park, California), Committee on Data for Science and Technology (CODATA). Membership includes representatives from the U.K., France, West Germany, the U.S.S.R., Japan, and the U.S.

A number of valuable reports have already issued from these groups. New research on key reactions has been undertaken as a result, which means that one can expect continuing improvement in kinetic modeling of flames. Indeed, modeling of the H_2–air flame by Dixon-Lewis (1970) is now so successful that one can almost say that merely trivial details remain to be elucidated.

B. Ignition Delays and Induction Periods

Knowledge of the time to ignition upon sudden heating of a fuel–air mixture has immediate practical significance relative to the performance of some air-breathing engines and other high-intensity combustion systems. Measurements of ignition delays (or induction periods) in the laboratory have recently been carried out mainly with shock tubes, which provide virtually instantaneous heating of a gaseous fuel–oxidant mixture. Shock tube data further provide an important experimental reference for kinetic modeling, very analogous to flame profile data but possessing the great advantage from the chemical standpoint that transport of species and energy is negligible because of the very short times (microseconds to milliseconds) involved in the shock tube observations. Thus the equations to be solved are considerably simplified.

There are some difficulties in this approach. First there is the experimental problem of defining the moment of ignition, or the end of the induction period. In a typical study of ignition (e.g., of a hydrocarbon–oxygen–diluent mixture) in shock waves, the features followed may be the pressure, the concentration of one or more radical species (e.g., OH) as measured by light adsoprtion, the emission from excited OH, CH, or C_2, or a combination of these. The end of the induction period is associated with rapid but not discontinuous increases in these features; hence the definition of the moment of ignition is necessarily somewhat arbitrary. Investigators usually choose to define it in a way that will yield consistent results and permit comparisons with theory. Thus one might for example take it as the moment when the OH concentration reaches some particular value.

Assuming that the delay has been well defined, one measures it in shocks having a wide range of compositions, temperatures, and pressures. A correlation is then sought among the variables that will produce a straight line on an Arrhenius-type plot. [Thus Seery and Bowman (1970) found the most satisfactory correlation of their data on methane ignition to be given by

$$\tau[O_2]^{1.6}[CH_4]^{-0.4} = A \exp(E/RT),$$

where τ is the ignition delay time and A and E are constants.] This plot alone may well serve as a useful empirical correlation that will permit predictions of ignition delays. Its value becomes greater if one proceeds to kinetic modeling of the ignition process. The objective is, of course, to reproduce the empirical correlation. The modeling consists of a computer program by which the coupled differential equations of chemical kinetics and the gas dynamic and state equations are numerically integrated, yielding the time dependence of all pertinent variables: concentrations, pressure, and so on. The analytical results for the induction times can then be compared to the experimental values for various starting conditions.

The risk always is that one may succeed in matching the data with a seriously oversimplified or erroneous model, a danger recognized by most investigators. On the other hand, in the most careful work the procedure has led to very significant insight into combustion mechanisms and frequently to determination of previously unknown values of certain rate constants; for example, the concept of partial equilibrium in combustion reactions developed during the work by Schott and Kinsey (1958) on induction periods in the H_2–O_2 reaction in shock waves. The same work yielded a value for the rate constant of the reaction $H + O_2 \rightarrow OH + O$, at a higher temperature than had previously been available. This reaction is the crucial chain-branching step in hydrogen–oxygen flames.

Studies of shock-wave ignition have more recently been extended to complicated systems such as the methane–oxygen reaction (Skinner and Ruehrwein, 1959; Miyama and Takayama, 1965; Lifshitz *et al.*, 1971; Seery and Bowman, 1970; Higgin and Williams, 1969; Skinner *et al.*, 1972). The last four of these studies represent the most detailed experimental and, in the case of the last three, analytical work on ignition times in shock waves known to the writer. The experimental correlations found in the several studies differ somewhat, but there is rather complete agreement in two of them on the reactions required in a kinetic model that will match the absolute values of observed ignition delays. A comparison of the reaction schemes adopted by Seery and Bowman (1970) and Higgin and Williams (1969) is presented in Table I. For a comparison of rate constants and other features of the schemes, the original papers may be consulted. The only significant differences in the two schemes concern the best way to describe the route by which CH_3 radicals are oxidized to CO and H_2O. The reaction scheme of Skinner *et al.* (1972) contains most but not all of the reactions in Table I. It also contains about a dozen additional reactions that are introduced mainly with the aim of providing an improved correlation between low- and high-temperature regimes of ignition. The scheme involves too many uncertain rate constants to be entirely satisfying, but it may well point the way toward improving some details of the kinetic modeling of methane combustion.

The reaction schemes in Table I are likely to be essentially correct because they are based upon a wealth of evidence from probing of flames, from low-temperature oxidation studies in vessels, from reactions in discharge-flow systems, and from other sources. Much of this evidence has been reviewed in the book by Fristrom and Westenberg (1965), who were

TABLE I

PRINCIPAL REACTIONS IN THE HIGH-TEMPERATURE OXIDATION OF METHANE: COMPARISON OF TWO ANALYTICAL STUDIES

Reaction	Seery and Bowman (1970)	Higgin and Williams (1969)
$CH_4 + M \rightleftarrows CH_3 + H + M$	√	√
$CH_4 + O \rightleftarrows CH_3 + OH$	√	√
$CH_4 + H \rightleftarrows CH_3 + H_2$	√	√
$CH_4 + OH \rightleftarrows CH_3 + H_2O$	√	√
$CH_3 + O \rightleftarrows HCHO + H$	—	√
$CH_3 + O_2 \rightleftarrows HCHO + OH$	—	√
$HCO + OH \rightleftarrows CO + H_2O$	√	—
$HCHO + OH \rightleftarrows H + CO + H_2O$	—	√
$CO + OH \rightleftarrows CO_2 + H$	√	√
$H + O_2 \rightleftarrows OH + O$	√	√
$O + H_2 \rightleftarrows H + OH$	√	√
$O + H_2O \rightleftarrows 2OH$	√	√
$H + H_2O \rightleftarrows H_2 + OH$	√	√
$H + OH + M \rightleftarrows H_2O + M$	√	√
$CH_3 + O_2 \rightleftarrows HCO + H_2O$	√	—
$HCO + M \rightleftarrows H + CO + M$	√	—

themselves responsible for much of the flame-probing research and who outlined the main features of the methane combustion mechanism as a result of their work and that of others. It is unfortunately not possible here to cite all of the valuable work that led to establishment of the mechanism. We hasten to emphasize that the understanding of methane combustion is not yet complete. The kinetics of methane pyrolysis (Palmer and Hirt, 1963; Palmer *et al.*, 1968a; Hartig *et al.*, 1971) are not fully established, and the reactions of certain intermediates such as CH_3, HCHO, and HCO are in need of clarification. Even the much-studied reaction of CO with OH is still under investigation (Gardiner *et al.*, 1973) because of its elusive kinetics; however, this reaction is now reasonably well understood as a result of a simple theoretical observation by Dryer *et al.* (1972).

The investigator who has a powerful computer at hand for calculating reaction profiles behind shock waves would naturally like to match the computations with equally powerful experiments. He would prefer to examine the reaction profile in full detail. In recent work, some progress has been made toward this objective. Thus, Bowman (1970) measured "reaction times," defined as the times to reach 90% of the equilibrium concentrations, for the species CO, CO_2, and H_2O behind shocks in CH_4–O_4–Ar mixtures and C_2H_6–O_2–Ar mixtures. It was possible to match the methane data moderately well with a reaction scheme identical to that in Table I but with the addition of the reaction $CH_3 + O \rightarrow HCO + H_2$. The ethane modeling was similarly successful, but the reaction scheme was simplified by neglecting ethylene chemistry that might normally be expected to be significant.

The rates of growth of some transient species have been observed in shock tube studies of the induction periods of certain reactions in which there is significant chain branching, such as the H_2–O_2 and C_2H_2–O_2 reactions. At least three kinds of observations have been made, involving measurements of the time dependence of: (a) chemiluminescent emission from electronically excited species such as OH*, CO*, CH*, or C_2* (Kistiakowsky and Richards, 1962; Hand and Kistiakowsky, 1962; Belles and Lauver, 1964; Glass *et al.*, 1965; Gardiner *et al.*, 1968; Gutman *et al.*, 1968); (b) the concentrations of ions, using a time-of-flight mass spectrometer (Hand and Kistiakowsky, 1962; Glass *et al.*, 1965); or (c) the ground-state concentration of O atoms (Gutman and Schott, 1967; Gutman *et al.*, 1968; Schott, 1969), using added carbon monoxide to produce chemiluminescent emission from the reaction $O + CO \rightarrow CO_2 + h\nu$. From the rate of growth of these emissions or species concentrations, which is found to be exponential with time during a large fraction of its total growth, information has been obtained on certain rate constants and mechanistic steps in the reactions. Sophisticated as these studies are, they still have not reached the goal of complete matching of experimental profiles with theoretical models. The recent work of Brabbs *et al.* (1971) on the H_2–CO–O_2 system is somewhat nearer to this goal, while the study of density profiles in this same system by Browne *et al.* (1969b) is still closer. The study of NO–CO–Ar mixtures (which are not in the same chain-branching class of reactions) in shock waves by Sulzmann *et al.* (1971) comes very close indeed. This is an area where much useful work can be expected in the future. The successful application of shock wave profile-matching to studies of air pollutant formation is discussed in the next section.

C. The Rates of Formation of Gaseous Pollutants in Flames, Combustors, and Shock Waves

The importance of chemical kinetics to combustion engineering has become all too obvious with the emphasis upon controlling pollution from combustors of various types. Whereas three decades ago combustion was being considered as a potentially useful means of fixing atmospheric nitrogen for the production of fertilizer, the modest formation of oxides of nitrogen in flames is now viewed with dismay by many because of the role of NO_2 in smog-producing atmospheric photochemistry. Currently, some of the most interesting and useful research and development is centered on the production (and removal or prevention) of pollutants in combustion processes. Books are appearing on the subject (Starkman, 1971; Cornelius and Agnew, 1972), and a significant fraction of the papers in the *Fourteenth Symposium (International) on Combustion* (in press for 1973) deals with it. This makes the task of summarizing the state of knowledge very difficult. We shall attempt to present it only in a general way with several illustrative examples that may point the reader's way to further study of the subject.

1. Atoms and Free Radicals in Flames

Atoms and free radicals are not themselves important as species escaping from flames; but they are vital to the inner flame processes, all of which use or produce atoms or radicals. These processes of course include the reactions that produce or consume NO, SO_2, and CO.

There is adequate evidence now that several species, especially the stable molecules CO, H_2, and CH_2O and the free radical species H, O, and OH, as well as flame ions, normally go through concentration maxima in the reaction zone of hydrocarbon flames. These maxima may carry the atom and radical concentrations to levels of the order of 1 mole %, that is, considerably above their final equilibrium levels (see Fig. 2). It is the recombination of these radicals in the latter stages of the flame that is responsible, together with the conversion of CO to CO_2, for roughly half of the total heat release in the combustion process.

Atoms and radicals are the species that are kinetically responsible for producing intermediates and final combustion products. Thus, in the CH_4–O_2 system (Table I) one has

$$CH_4 + H \rightarrow CH_3 + \mathbf{H_2} \qquad \text{(intermediate)},$$

$$CH_4 + OH \rightarrow CH_3 + \mathbf{H_2O} \qquad \text{(product)},$$

$$CH_3 + O_2 \rightarrow HCO + \mathbf{H_2O} \qquad \text{(product)},$$

$$HCO + OH \rightarrow \mathbf{CO} + \mathbf{H_2O} \qquad \text{(CO may not be a \textit{final} product)},$$

$$CO + OH \rightarrow H + \mathbf{CO_2} \qquad \text{(product)}.$$

They also are responsible for ion formation, mainly through the reaction

$$CH + O \rightarrow HCO^+ + e^-$$

and for the formation of oxides of nitrogen and sulfur, to be discussed below.

They likewise probably play important roles in soot formation, as we shall see later.

2. Oxides of Carbon

The flame processes that produce CO are not well understood. Much of it may come from

$$HCO + OH \rightarrow CO + H_2O,$$

but the mechanism and rates of reactions leading to HCO are still obscure.

It is now well established that the main route for oxidation of CO to CO_2 is

$$CO + OH \rightarrow CO_2 + H.$$

This is intrinsically a very fast reaction. The direct oxidation of CO by O_2,

$$CO + O_2 \rightarrow CO_2 + O,$$

is relatively slow and is probably of minor importance in hydrocarbon- or H_2O-containing flames because they contain substantial concentrations of OH. Thus in typical flames, the rate of conversion of CO to CO_2 is controlled by the prevailing concentration of OH. The question is, why should appreciable CO escape from flames in which there is enough oxygen to (presumably) oxidize virtually all of it to CO_2?

First of all, since the concentration of CO goes through a maximum (Fig. 1) during the combustion process, it is clear that if flow rates were so fast that gas residence times were of the order of 10^{-3} sec, and if quenching were extremely abrupt, substantial CO could appear in the quenched products. This condition might be approached in a few air-breathing engines, but in general it cannot be the explanation; certainly not for typical furnace flames.

Inhomogeneities in the gas mixture are clearly one possible cause that is encountered in practice. The cure may not be simple, but at least it is obvious. However, there are some very well-mixed, or even premixed, flames of hydrocarbons from which CO escapes. Here there must be a kinetic explanation. The main clue is the fact that the CO occurs when the burnt gases of the flame are cooled very rapidly.

The final equilibrium condition (the "hot boundary") in a premixed flame at atmospheric pressure is approached very closely in a time of the order of a few milliseconds. In highly turbulent diffusion flames one also expects rapid attainment of equilibrium. At this condition, all individual elementary reactions are in equilibrium. The concentration of CO may be quite high. The equilibrium concentrations of CO, CO_2, H_2, and H_2O are related by the water–gas shift reaction,

$$CO + H_2O \rightleftarrows CO_2 + H_2,$$

$$K_{wg} = [CO_2][H_2]/[CO][H_2O].$$

The total mechanism by which the water–gas equilibrium is maintained in the flame is complex (Jost *et al.*, 1965), but the essential steps clearly must be two:

$$CO + OH \underset{-a}{\overset{a}{\rightleftarrows}} CO_2 + H,$$

$$H + H_2O \underset{-b}{\overset{b}{\rightleftarrows}} OH + H_2.$$

These constitute a pair of radical-shuttling reactions (for H and OH) not unlike those that are so significant in nitric oxide chemistry (see below). Note that the simultaneous equilibration of these yields the water–gas equilibrium.

Now if the temperature falls, the water–gas equilibrium generally tends to shift in such a way that CO_2 is produced from CO (Table II). But the process is sluggish, and above-equilibrium CO is found in the exhaust or stack gases [see Sawyer (1971)]. The kinetic reason may be qualitatively straightforward. Reaction a is fast, and its rate constant is quite insensitive to temperature. At 1500°K, reaction b is also fast, with a rate constant approximately equal to k_a. However, reaction b has an activation energy of about 20 kcal (Baulch *et al.*, 1972); the consequence is that at 1000°K, k_b is perhaps one-thirtieth of k_a; and at 800°K, $k_b \approx k_a/300$. Thus the oxidation of CO to CO_2 is quenched, but indirectly, through the fall in k_b. The same kind of conclusion is reached if one also considers the reaction $H + O_2 \rightarrow OH + O$ as a means of producing OH for reaction a. It, too, has a substantial activation energy and its rate constant decreases rapidly as the temperature falls.

We have described the explanation above as qualitative and have been cautious about it because we know that full understanding of the CO

TABLE II

A Short Table of the Water–Gas Equilibrium Constant

$$K_{wg} = P_{CO_2} \times P_{H_2}/P_{CO} \times P_{H_2O} = [CO_2][H_2]/[CO][H_2O]$$

T (°K)	298	500	1000	1500	2000	2500
K_{wg}	1.04×10^5	1.38×10^2	1.44	0.390	0.221	0.164

quenching phenomenon can come only through a comprehensive model of the system, including all of the chemistry. A model of postequilibrium reactions in an internal combustion engine cylinder has been published by Keck and Gillespie (1971). They invoke the assumption that the fast bimolecular reactions are in a state of partial equilibrium (discussed earlier) throughout the expansion stroke. Atom and radical recombination reactions are not assumed to be equilibrated. The model predicts above-equilibrium concentrations of CO in the cooled exhaust. One hopes that the Keck model will soon be extended to the case where partial equilibria are not assumed. This would provide a test of our qualitative explanation in the preceding paragraph. Clearly, good theoretical work is needed on this system. The inadequacy of knowledge of the kinetics of one of the most famous reactions of all, the water–gas reaction, is truly astonishing. On the practical side, more experimental work is needed on both a basic and an applied level. An example of the type of applied research that is needed appears in the recent work of Singh and Sawyer (1971) and Williams *et al.* (1972) on the conversion of CO to CO_2 in flames. Williams *et al.* showed that suitable cooling of flame gases by a heat exchanger quenches NO formation (see below) without freezing the CO-to-CO_2 conversion. This valuable research should be extended to establish the time-and-temperature history required for selective quenching in various practical combustors. Johnson (1970) studied the quenching of CO burnout in rapidly cooled gases from small propane–air flames. Again, further work is needed.

3. Oxides of Nitrogen

Studies related to the chemistry of heated air [see Camac and Feinberg (1967)] have shown that the atom-shuttle reactions of the Zeldovich (1946) mechanism:

$$O + N_2 \rightleftarrows NO + N, \tag{4}$$

$$N + O_2 \rightleftarrows NO + O, \tag{5}$$

account for most of the NO formation in hot air. The rate constants are given by (Baulch *et al.*, 1969):

$$k_4 = 1.36 \times 10^{14} \exp(-75.4 \text{ kcal}/RT) \quad \text{cm}^3/\text{mole sec},$$

$$k_5 = 6.4 \times 10^9 T \exp(-6.25 \text{ kcal}/RT) \quad \text{cm}^3/\text{mole sec}.$$

There is only a minor contribution from

$$N_2 + O_2 \rightarrow N_2O + O, \tag{6}$$

$$N_2O + O \rightarrow 2NO, \tag{7}$$

and virtually none from the four-center reaction

$$N_2 + O_2 \rightarrow 2NO. \tag{8}$$

Flame studies (Williams *et al.*, 1972; Shahed and Newhall, 1971; Heywood, 1971; Thompson *et al.*, 1972) suggest that the Zeldovich mechanism also dominates NO formation in combustion processes. In fuel-rich flames there may in addition be minor contributions from

$$N + OH \rightarrow NO + H \tag{9a}$$

or

$$N_2 + OH \rightarrow N_2O + H, \tag{9b}$$

the latter being followed by reaction (7).

Fenimore (1971) reported some observations that indicated a surprisingly large NO formation at early times in a flame, in the region before all the hydrocarbon is consumed. He suggested that in this region, N atoms might be created by reactions such as

$$N_2 + CH \rightarrow HCN + N. \tag{10}$$

Reaction (5) would then follow, making NO. But shock tube and kinetic modeling work by Bowman and Seery (1972) and by Bowman (1973) shows that it may not be necessary to invoke such a reaction. Superequilibrium concentrations of O atoms (and OH radicals) in the reaction zone apparently can account adequately for the fast NO formation rate, through reactions (4) and (5) [and (9)].

Considering then that reactions (4) and (5) are the main ones, the significant points are that:

(a) Reaction (4) has a very large activation energy.
(b) Reaction (4) is controlled by the O atom concentration.

Thus, at low temperatures NO will not form at an appreciable rate. In hot systems, NO will form fairly rapidly if there are O atoms present. It follows also that the NO will readily "freeze" on cooling because its consumption depends on reaction (−5), that is, on the consumption of NO by O atoms. The rate constant is given by

$$k_{-5} = 1.55 \times 10^9 T \exp(-38.6\ \text{kcal}/RT) \quad \text{cm}^3/\text{mole sec}.$$

The activation energy is such that the rate will be strongly temperature dependent. Furthermore, O atom recombination in the cooling gases will tend to slow the reaction.

From a practical point of view, one of the awkward points about NO formation is that the fuel-rich conditions that would keep the temperature low and the O atoms at a small concentration are the conditions under which CO formation and, especially, sooting become serious. Thus it is very difficult to create conditions that will simultaneously minimize all types of emissions. A great deal of engineering effort (Breen *et al.*, 1971; Breen, 1972; Bartok *et al.*, 1971; Bartok and Hall, 1971) is going into this problem. One quite successful procedure (greatly oversimplified here) has been to initially burn fuel-rich, extract some heat, then mix in sufficient air to bring the final stoichiometry to the desired value. The limitations on the method are, as expected, avoidance of CO emissions (especially in the case of natural gas firing) and soot emissions (especially with oil firing).

Finally, we note that the extreme temperature dependence of reaction (4) means that in many combustion systems the concentration of NO will not approach equilibrium ($\sim$3000 ppm typically). One observes levels around 50 ppm in most premixed hydrocarbon–air flames; higher, but still subequilibrium, levels are found in diffusion flames. Only in higher temperature systems such as gasoline engines does one encounter levels close to equilibrium values. In such systems one probably has to work toward removing NO from the exhaust rather than preventing its formation, unless one is willing to reduce peak temperatures.

4. Oxides of Sulfur

A comprehensive review of the kinetics of sulfur compounds in flames has recently been published by Cullis and Mulcahy (1972), who covered 425 references during their critical examination of the subject. Their paper is essential reading for anyone concerned with the origins and fate of oxides of sulfur in combustion processes. Because of the existence of this excellent review, only a few remarks are made here. They are directed toward a very simple view of SO_x pollutant formation in homogeneous flame processes.

The well-known oxides SO_2 and SO_3 are not the only significant SO_x species. The monoxide SO is very important kinetically in the oxidation of sulfur-containing compounds, and the species S_2O may also be, under suitable circumstances. A possible isomer of SO_2 has also been discussed in the literature. Both SO and S_2O are reactive and unstable at room temperature and hence are not observed as normal products of combustion. One finds SO_2 and SO_3. The equilibrium between these,

$$SO_2 + \tfrac{1}{2}O_2 = SO_3,$$

$$K_S = [SO_3]/[SO_2][O_2]^{1/2},$$

favors SO_2 at high temperatures and SO_3 at low temperatures (assuming in both cases that the oxygen concentration is that of typical fuel-lean combustion products). A few numbers are given in Table III. Hedley

TABLE III

A SHORT TABLE OF THE EQUILIBRIUM CONSTANT FOR THE SYSTEM $SO_2/O_2/SO_3$: $SO_2 + \frac{1}{2}O_2 = SO_3$

$K_p = P_{SO_3}/P_{SO_2}(P_{O_2})^{1/2}$

T (°K)	298	500	1000	2000
K_p ($atm^{-1/2}$)	2.6×10^{12}	2.6×10^{5}	1.8	5.6×10^{-3}

(1967) showed that in flame gases the SO_3 concentration normally *exceeds* the level given by the above equilibrium expression at that condition, while in the cooled exit gases SO_3 levels are far *below* equilibrium. (Heterogeneous catalysis can affect the exiting SO_3. This is discussed in Chapter II by W. T. Reid.)

This behavior of SO_3 is a kinetic phenomenon. To understand it qualitatively, we have to look at the elementary chemical steps responsible for forming and removing SO_3. Let us take for granted that in the reaction zone, sulfur-containing molecules will be oxidized to produce SO. The SO is converted to SO_2:

$$SO + O_2 \rightarrow SO_2 + O \tag{11}$$

and

$$SO + OH \rightarrow SO_2 + H. \tag{12}$$

The SO_2 is converted to SO_3 by the reaction

$$SO_2 + O + M \rightarrow SO_3 + M. \tag{13}$$

As a third-order process, this reaction is intrinsically rather slow; but in the reaction zone it will proceed at its highest possible rate because that is where O atoms achieve their maximum concentration.

The trioxide is removed by the steps

$$SO_3 + O \rightarrow SO_2 + O_2 \tag{14}$$

and

$$SO_3 + H \rightarrow SO_2 + OH, \tag{15}$$

as well as by thermal decomposition, the reverse of reaction (13). Reactions (14) and (15) have large rate constants. If, as an approximation, we assume that reactions (13) and (14) are the only significant steps for creating and removing SO_3, the maximum SO_3 concentration can be calculated from the condition $d[SO_3]/dt = 0$. One finds

$$[SO_3]_{max} = k_{13}[SO_2][M]/k_{14}.$$

Estimating k_{13} to be between 1 and 5×10^{15} cm^6/mole^2 sec [see Cullis and Mulcahy (1972)], [M] to be about 10^{-5} mole/cm^3, and k_{14} to be about 10^{12} cm^3/mole sec (Kondratiev, 1970), one finds that the maximum SO_3 is between 1 and 5% of the SO_2 concentration. This is of the order of Hedley's (1967) observations (Fig. 3). Note that the relationship between SO_3 and SO_2 is entirely different from the equilibrium expression K_S.

In hydrocarbon-rich flames, the O atom concentration is suppressed and the maximum concentration of SO_3 will be determined by the balance between steps (13) and (15):

$$[SO_3]_{max} = k_{13}[SO_2][M][O]/k_{15}[H];$$

the smallness of [O]/[H] will greatly depress the maximum conversion of SO_2 to SO_3.

Beyond the reaction zone, atom concentrations decay by recombination and, in practical systems, the temperature soon begins to fall. Both the formation and removal rates of SO_3 will decrease as atom concentrations are diminished. Before the temperature begins falling, one may then find some reduction in the SO_3 concentration because of the contribution of thermal decomposition to the total removal rate. The reaction

$$SO_3 + CO \rightarrow SO_2 + CO_2 \tag{16}$$

may also be significant (the rate is not well known). However, in this region, CO is also being converted to CO_2, so the contribution of the SO_3–CO reaction, whatever it may be, is presumably less than it was at the peak of the reaction zone. In fuel-rich systems, however, this reaction will continue to consume SO_3. Once the temperature falls appreciably, the thermal decomposition of SO_3 [reaction (−13)], which has a large activation energy, becomes insignificant.

The net effect of all this is that in fuel-lean systems one might expect a moderate drop in [SO_3] after its peak, but little change thereafter. Barring catalytic effects, the ratio of SO_2 to SO_3 in the cooled gases will remain at a few percent in fuel-lean combustion, as is observed in practice (Hedley, 1967); and of course one will find that, in terms of K_S, the conversion is subequilibrium (Fig. 3). In fuel-rich combustion, SO_3 will be almost totally absent.

Thus simple kinetic concepts can qualitatively rationalize the observations of SO_2–SO_3 relationships in sulfur-containing flames. As understanding of some of the equilibrium conditions (Jeffers and Bauer, 1971) and kinetics (Bauer *et al.*, 1971) in such flame systems improves, it will become possible to create kinetic models possessing validity comparable to that of current models of NO kinetics in flames.

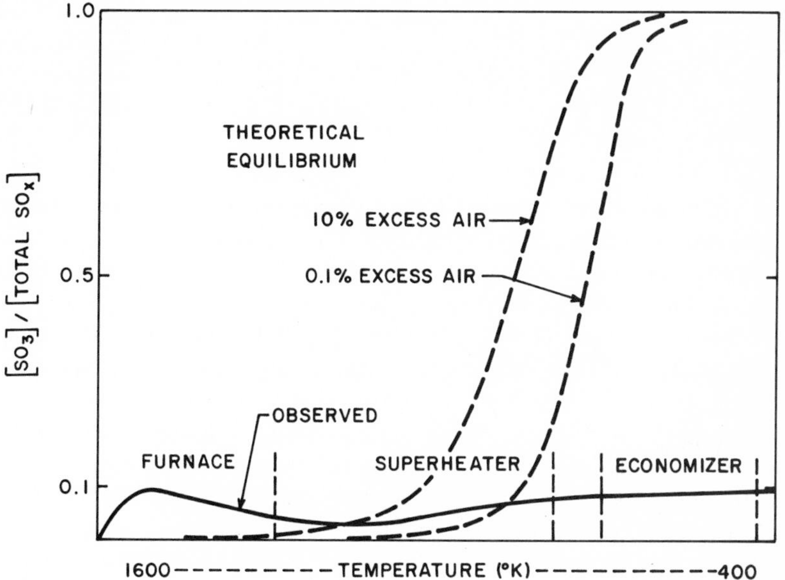

Fig. 3 Sketch of the theoretical ($SO_2/O_2/SO_3$ equilibrium) and observed fractional yield of SO_3 in a typical boiler. [After Hedley (1967).]

5. Sooting

We conclude this commentary on chemical kinetics and flames with some remarks on the awesomely complex matter of the formation and consumption of carbon in combustion processes. Rather comprehensive reviews of the subject of carbon formation in gases have been published by Palmer and Cullis (1965), Homann (1967), and in a chapter by Gaydon and Wolfhard (1970). Each review is an attempt to cover the whole question and objectively appraise the evidence; yet each carries somewhat different emphasis. For instance, the role of flame ions is barely mentioned in the first review, is discussed briefly in the second, and is considered rather carefully in the third. Hence the interested reader is advised to study all three in order to grasp fully the primitive state of knowledge of this subject. By this comment we are by no means dismissing the research that has been published. Important advances have been made in the past decade on the decomposition of hydrocarbons, the nature of intermediates in sooting flames, the rates of growth of carbon particles, the influences of additives or electric fields upon sooting, and the electric charges carried by carbon particles. Even so, the writer must agree with Miller (1973) when he says, "Clearly we are far from achieving a satisfactory picture of elementary carbon-forming chemistry." The same is true, unfortunately, of carbon-burning chemistry.

The understanding of sooting may still be primitive but research on it has been sophisticated, and has yielded many pieces of information that will eventually coalesce into a comprehensible pattern. Examples of recent work are as follows:

(1) The mass spectrometric sampling by Homann and Wagner (1967) of hydrocarbon species in sooting flames and comparison of these with mass spectra of species evaporated from soot collected early and late in the carbon-forming region of the flame. The early soots contain a vast array of compounds that are not present in the late soots and which presumably are intermediates. They seem to be aromatic molecules or radicals, many with side chains. The species evaporated from late soots are very similar to those found on the surfaces of carbon blacks (Palmer *et al.*, 1968b), which have been found to contain fused-ring aromatics having molecular weights as high as 646. These species are probably by-products of the process that leads predominantly to formation of a graphitic structure.

(2) The achievement by Ban and Hess (1969a,b) of resolution in electron microscopy sufficient to reveal the layer-plane structure of carbon blacks. The work shows that the particles contain graphitic layer planes in what is describable as an onion-like structure, but they are distorted and contain numerous dislocations. The observational technique may well produce important information on the clustering of soot particles and possibly on microscopic details of the interactions of growing particles with various gaseous species.

(3) The measurement by Jessen and Gaydon (1969) of the concentrations of the radicals C_2, C_3, and CH in fuel-rich C_2H_2–O_2 flames, using multiple-reflection spectroscopy. Their evidence suggests that C_3 is formed from soot, whereas it is possible that C_2 may serve as a starting point for the development of species that ultimately become soot particles.

(4) The use by Mayo and Weinberg (1970) of electric fields to show that all carbon particles in counterflow diffusion flames of hydrocarbons are charged and that each particle carries only one unit of charge [see also Wersborg *et al.* (1973)]. The study also indicates that carbon growth occurs both on neutral and on charged species; that is, some particles acquire a charge after substantial growth has occurred. It had previously been established (Place and Weinberg, 1967; Miller, 1967) that natural flame ions and certain ionizable additives such as cesium can serve as nuclei for carbon growth. At the time of writing, the relative importance of ions versus neutral species as centers for carbon growth has not been established. These valuable experiments have prompted Howard (1969)

to create a theoretical model for ion-nucleated soot formation that is supported by some of the observations and may well be modifiable to bring it into agreement with most of the experimental facts. However, it is of course known that soot can form under purely pyrolytic conditions, that is, in the absence of a flame and thus in the absence of flame-generated ions. Therefore, it is unlikely that focusing attention exclusively on ionic species will lead to an entirely comprehensive theory.

(5) Several ingenious studies of carbon formation under various conditions. These include Wright's (1969) use of a well-stirred flame reactor of the Longwell–Weiss type to observe C/O thresholds for sooting. The critical values were slightly different from those found in premixed flames. The results are significant with respect to high-intensity combustors. Lieb and Roblee (1970) used isotopically labeled (^{14}C) ethanol in a diffusion flame and were able to establish that the nonhydroxylated carbon atom in the molecule contributed twice as much carbon to the soot as the hydroxylated one. Dalzell *et al.* (1970) established that light scattering can be used successfully to study the kinetics of soot growth and burnout in flames [see also Kunugi and Jinno (1967)].

(6) Systematic examination of the reduction in sooting tendency of flames upon introduction of metallic additives. This includes the excellent work of Cotton *et al.* (1971), in which 40 metals were studied, and the work of Addecott and Nutt [as discussed by Miller (1973)]. Many metals are found to be effective, but the order in which they are ranked differs in the two studies. A detailed mechanism for the function of alkaline earths (Ba, Sr, Ca) and molybdenum based on catalysis of radical production is proposed by Cotton *et al.* (1971). Addecott and Nutt offer a possible alternative based on charge exchange between flame ions and metal atoms. The potential for productive work on this subject seems very large, and although a fair amount of research has been done [for a brief review see Salooja (1972)], it is somewhat surprising that there has not been more. One of the most useful studies has been reported by Friswell (1972). He studied the soot reduction in a kerosene-fired simulated gas turbine combustor using the additives barium, manganese, and iron. Barium and manganese were studied as a function of concentration for three air–fuel ratios (Fig. 4). Further support for the radical-production mechanism for soot suppression by Ba was found in the observation that Ba addition increases the rate of formation of NO at the same time that it inhibits sooting. Manganese and iron, in contrast, appear to operate by being incorporated into soot particles (Fenimore and Jones, 1967) and catalyzing soot burnout. Friswell found that the efficiency of Ba is approximately independent of the overall air–fuel ratio in his simulated gas turbine, whereas Mn and Fe

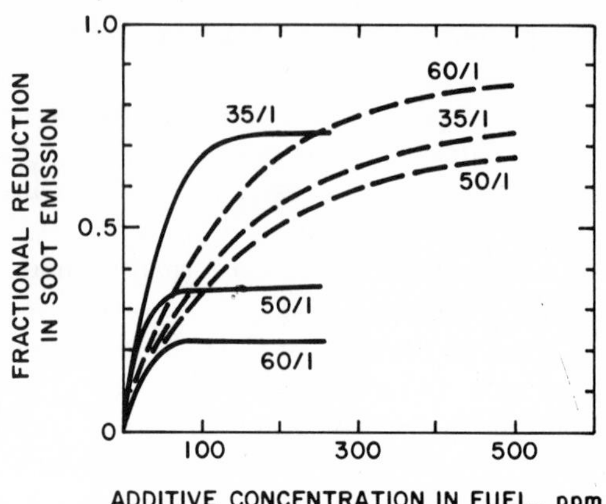

Fig. 4 Influence of added Ba and Mn on the fractional reduction in soot emitted from a simulated gas turbine combustor fired with kerosene. Solid lines: Mn; dashed lines: Ba. Air–fuel ratios are indicated. [After Friswell (1972, p. 167).]

function better as the exhaust temperature rises (i.e., as the air–fuel ratio moves toward stoichiometric).

(7) Strenuous efforts, only moderately successful to date, to understand and predict the rates of soot burnout. The basic point of reference in these studies is usually the work of Lee *et al.* (1962) on the rate of combustion of soot in a sooting laminar flame. Magnussen (1971) has extended the experiments to a turbulent flame and has presented a theoretical analysis that bears some resemblance to the qualitative model offered by Dalzell *et al.* (1970). It suggests that the laminar flame correlation of Lee *et al.* cannot be directly applied to turbulent flames. The burnout of residual carbon particles generated from heavy fuel oil is kinetically quite different from soot burnout (Kito *et al.*, 1971), mainly because the residual particles are orders of magnitude larger than soot and the rate is controlled by boundary layer diffusion rather than surface chemistry. Rosner and Allendorf (1968, 1970) have reported experimental rates of attack on graphite surfaces by O, O_2, OH, and H at a wide range of temperatures. Systematic data of this type, if extended to soot, may ultimately permit meaningful kinetic models of soot burnout to be constructed. An attempt to model the consumption of soot in gas turbines has been presented by Radcliffe and Appleton (1971). Considering the limitations of the available data, the model provides a worthwhile advance. One interesting conclusion in their work is that, contrary to the hypothesis of Fenimore and Jones (1967), OH is probably not the most significant oxidant of soot. The very high

efficiency of reaction of O atoms with graphite found by Rosner and Allendorf suggests that O atoms may be more significant oxidizers of soot than had previously been thought. If so, this offers a possible explanation for the extraordinary effect of SO_3 as an additive in promoting sooting. The reaction between SO_3 and O is undoubtedly extremely fast, forming the much less reactive species SO_2 and O_2. Thus a small amount of SO_3 could perhaps prevent the removal of incipient soot particles by O atoms and hence permit rapid growth. The fact that NO_2 does not have the same effect as SO_3 may perhaps be due to its lower thermal stability. It may act as a source of O atoms as well as a sink. As an odd-electron molecule, it may also attack active sites on the soot particles directly and more rapidly than SO_3.

IV. ON A NEW AREA OF COMBUSTION ENGINEERING

Some of the most exciting contemporary developments in science and technology are related to combustion reactions that produce molecules in excited states. The study of the energy distributions in radicals and molecules generated in flames has been of interest to combustion scientists for many years. Recently it has been found that in some flame systems the distribution of certain product species is actually inverted (fleetingly); that is, contrary to equilibrium conditions, one finds that the populations of the molecule in higher energy states are larger than in lower states. Chemical lasers may be created using such reaction systems as continuous sources of coherent radiant energy, provided one designs them so that the kinetics are favorable. The population inversion must not decay seriously in the direction of equilibrium before the radiant energy is extracted.

Perhaps the most straightforward of these combustion lasers is the one utilizing the H_2–F_2 flame (Meizner, 1970). A fluorine-rich flame burns, producing in the products a large concentration of F atoms. These are expanded through a nozzle to give them high velocity, and D_2 or H_2 is then added. The reaction

$$F + D_2 \rightarrow DF + D$$

produces DF having a vibrational population inversion. Laser energy in the infrared region can be extracted; or in principle one could mix in CO_2, which would absorb vibrational energy from the DF and in turn be produced in an inverted population distribution. A combustion-driven CO_2 laser could thus be created.

Interest in this new application of flames is high and, while no attempt is made here to cover the relevant literature, it is growing rapidly. Large-scale combustion lasers may well be considered an important new area of combustion engineering.

References

Addecott, K. S. B., and Nutt, C. W. (1969). Paper presented at 158th Nat. Meeting, Amer. Chem. Soc., New York, September 1969, as discussed by Miller (1973).

Bahn, G. S. (ed.) (1963). "Kinetics, Equilibria and Performance of High-Temperature Systems." Gordon & Breach, New York.

Bahn, G. S., and Zukoski, E. E. (eds.) (1960). "Kinetics, Equilibria and Performance of High Temperature Systems." Butterworths, London.

Ban, L. L., and Hess, W. M. (1969a). *Carbon* **7,** 723 (abstract only).

Ban, L. L., and Hess, W. M. (1969b). Private communication from Columbian Carbon Co., Princeton, New Jersey.

Bartok, W., and Hall, H. J. (1971). *Env. Sci. Tech.* **5,** No. 4.

Bartok, W., Crawford, A. R., and Skopp, A. (1971). *Chem. Eng. Progr.* **67,** 64.

Bauer, S. H., Jeffers, P., Lifshitz, A., and Yadava, B. P. (1971). *Int. Symp. Combust., 13th,* p. 713. Combust. Inst., Pittsburgh, Pennsylvania.

Baulch, D. L., Drysdale, D. D., Horne, D. G., and Lloyd, A. C. (1969). High Temperature Reaction Rate Data, No. 4. Dept. of Phys. Chem., The Univ. Leeds, England.

Baulch, D. L., Drysdale, D. D., Horne, D. G., and Lloyd, A. C. (1972). "Evaluated Kinetic Data for High Temperature Reactions," Vol. 1. Chem. Rubber Publ. Co., Cleveland, Ohio.

Becker, P. M. (1972). Private communication of work in progress. Fuel Sci. Sect., Dept. of Material Sci., Pennsylvania State Univ., University Park, Pennsylvania.

Belles, F. E., and Lauver, M. R. (1964). *J. Chem. Phys.* **40,** 415.

Bowman, C. T. (1970). *Comb. Sci. Tech.* **2,** 161.

Bowman, C. T. (1973). *Int. Symp. Combust., 14th.* p. 729. Combust. Inst., Pittsburgh, Pennsylvania.

Bowman, C. T., and Seery, D. J. (1972). *In* "Emissions from Continuous Combustion Systems" (W. Cornelius and W. G. Agnew, eds.), p. 123. Plenum, New York.

Brabbs, T. A., Belles, F. E., and Brokaw, R. S. (1971). *Int. Symp. Combust., 13th,* p. 129. Combust. Inst., Pittsburgh, Pennsylvania.

Breen, B. P. (1972). *In* "Emissions from Continuous Combustion Systems" (W. Cornelius and W. G. Agnew, eds.), p. 325. Plenum, New York.

Breen, B. P., Bell, A. W., de Volo, N. B., Bagwell, F. A., and Rosenthal, K. (1971). *Int. Symp. Combust., 13th,* p. 391. Combustion Inst., Pittsburgh, Pennsylvania.

Browne, W. G., Porter, R. P., Verlin, J. D., and Clark, A. H. (1969a). *Int. Symp. Combust., 12th,* p. 1035. Combust. Inst., Pittsburgh, Pennsylvania.

Browne, W. G., White, D. R., and Smookler, G. R. (1969b). *Int. Symp. Combust., 12th,* p. 557. Combust. Inst., Pittsburgh, Pennsylvania.

Camac, M., and Feinberg, R. M. (1967). *Int. Symp. Combust., 11th,* p. 137. Combust. Inst., Pittsburgh, Pennsylvania.

Cornelius, W., and Agnew, W. G. (eds.) (1972). "Emissions from Continuous Combustion Systems." Plenum, New York.

Cotton, D. H., Friswell, N. J., and Jenkins, D. R. (1971). *Combust. Flame* **17,** 87.

Cullis, C. F., and Mulcahy, M. F. R. (1972). *Combust. Flame* **18,** 225.

Dalzell, W. H., Williams, G. C., and Hottel, H. C. (1970). *Combust. Flame* **14,** 161.

Dixon-Lewis, G. (1967). *Proc. Roy. Soc.* (*London*) **A298,** 495.

Dixon-Lewis, G. (1970). *Proc. Roy. Soc.* (*London*) **A317,** 335.

Dryer, F., Naegeli, D., and Glassman, I. (1971). *Combust. Flame* **17,** 270.

Eberius, K. H., Hoyermann, K., and Wagner, H. G. (1971). *Int. Symp. Combust., 13th,* p. 713. Combust. Inst., Pittsburgh, Pennsylvania.

Fenimore, C. P. (1971). *Int. Symp. Combust., 13th,* p. 373. Combust. Inst., Pittsburgh, Pennsylvania.

Fenimore, C. P., and Jones, G. W. (1967). *J. Phys. Chem.* **71,** 593.

Fristrom, R. M., and Westenberg, A. A. (1965). "Flame Structure." McGraw-Hill, New York.

Fristrom, R. M., Grunfelder, C., and Favin, S. (1960). *J. Phys. Chem.* **64,** 1386.

Friswell, N. J. (1972). *In* "Emissions from Continuous Combustion Systems" (W. Cornelius and W. G. Agnew, eds.), p. 161. Plenum, New York.

Gardiner, W. C., Jr., Morinaga, K., Ripley, D. L., and Takeyama, T. (1968). *J. Chem. Phys.* **48,** 1665.

Gardiner, W. C., Jr., Mallard, W. G., McFarland, M., Morinaga, K., Owen, J. H., Takeyama, T., and Walker, B. F. (1973). *Int. Symp. Combust., 14th.,* p. 61. Combust. Inst., Pittsburgh, Pennsylvania.

Gaydon, A. G. (1957). "The Spectroscopy of Flames." Chapman & Hall, London. (A revised edition is in preparation at the time of writing.)

Gaydon, A. G., and Wolfhard, H. G. (1970). "Flames." Chapman & Hall, London.

Getzinger, R. W., and Schott, G. L. (1965). *J. Chem. Phys.* **43,** 3237.

Glass, G. P., Kistiakowsky, G. B., Michael, J. V., and Niki, H. (1965). *J. Chem. Phys.* **42,** 608.

Gordon, S., and McBride, B. J. (1971). NASA SP-273, Nat. Aero. Space Adm., Washington, D.C.

Gutman, D., and Schott, G. L. (1967). *J. Chem. Phys.* **46,** 4576.

Gutman, D., Hardwidge, E. A., Dougherty, F. A., and Lutz, R. W. (1967). *J. Chem. Phys.* **47,** 4400.

Gutman, D., Lutz, R. W., Jacobs, N. F., and Hardwidge, E. A. (1968). *J. Chem. Phys.* **48,** 5849.

Hand, C. W., and Kistiakowsky, G. B. (1962). *J. Chem. Phys.* **37,** 1239.

Harker, J. H. (1967). *J. Inst. Fuel* **40,** 206.

Harker, J. H., and Allen, D. A. (1969). *J. Inst. Fuel* **42,** 183.

Hartig, R., Troe, J., and Wagner, H. G. (1971). *Int. Symp. Combust., 13th,* p. 147. Combust. Inst., Pittsburgh, Pennsylvania.

Hedley, A. B. (1967). *J. Inst. Fuel* **40,** 142.

Heywood, J. B. (1971). *AIAA/AAE 7th Propulsion Joint Specialist Conf., Salt Lake City, Utah, June 14–18, 1971.* AIAA Paper No. 71-712.

Higgin, R. M. R. and Williams, A. (1969). *Int. Symp. Combust., 12th,* p. 579. Combust. Inst., Pittsburgh, Pennsylvania.

Homann, K. H. (1967). *Combust. Flame* **11,** 265.

Homann, K. H. and Wagner, H. G. (1967). *Int. Symp. Combust., 11th,* p. 371. Combust. Inst., Pittsburgh, Pennsylvania.

Hottel, H. C., Williams, G. C., Nerheim, N. M. (1965). *Ind. Eng. Chem.* **47,** 1634.

Howard, J. B. (1969). *Int. Symp. Combust., 12th,* p. 877. Combust. Inst., Pittsburgh, Pennsylvania.

Jeffers, P., and Bauer, S. H. (1971). *Combust. Flame* **17,** 432.

Jessen, P. F., and Gaydon, A. G. (1969). *Int. Symp. Combust., 12th,* p. 481. Combust. Inst., Pittsburgh, Pennsylvania.

Johnson, M. L. M. (1970). Doctoral dissertation, Pennsylvania State Univ., University Park, Pennsylvania.

Jost, W., Schecker, H. G., and Wagner, H. G. (1965). *Z. Phys. Chem. N.F.* **45,** 47.

Kaskan, W. E., and Nadler, M. P. (1972). *J. Chem. Phys.* **56,** 2220.

Keck, J. C., and Gillespie, D. (1971). *Combust. Flame* **17,** 237.

Kistiakowsky, G. B., and Richards, L. W. (1962). *J. Chem. Phys.* **36,** 1707.
Kito, M., Ishimaru, M., Kawahara, S., Sakai, T., and Sugiyama, S. (1971). *Combust. Flame* **17,** 391.
Kondratiev, V. N. (1970). "Rate Constants of Gaseous Reactions." Acad. Nauk S.S.S.R., Moscow.
Kunugi, M., and Jinno, H. (1967). *Int. Symp. Combust., 11th,* p. 257. Combust. Inst., Pittsburgh, Pennsylvania.
Lee, K. B., Thring, M. W., and Beér, J. M. (1962). *Combust. Flame* **6,** 137.
Lieb, D. F., and Roblee, L. H. S., Jr. (1970). *Combust. Flame* **16,** 385.
Lifshitz, A., Scheller, K., Burcat, A., and Skinner, G. B. (1971). *Combust. Flame* **16,** 311.
Longwell, J. P., and Weiss, M. A. (1955). *Ind. Eng. Chem.* **47,** 1634.
Magnussen, B. F. (1971). *Int. Symp. Combust., 13th,* p. 869. Combust. Inst., Pittsburgh, Pennsylvania.
Marteney, P. J. (1970). *Combust. Sci. Tech.* **1,** 461.
Mayo, P. J., and Weinberg, F. J. (1970). *Proc. Roy. Soc.* (*London*) **A319,** 351.
Meizner, R. A. (1970). *Int. J. Chem. Kinet.* **2,** 335.
Miller, W. J. (1967). *Int. Symp. Combust., 11th,* p. 252. Combust. Inst., Pittsburgh, Pennsylvania.
Miller, W. J. (1973). *Int. Symp. Combust., 14th.,* p. 307. Combustion Inst., Pittsburgh, Pennsylvania).
Miyama, H., and Takayama, T. (1965). *Bull. Chem. Soc. Japan* **38,** 37.
Palmer, H. B., and Cullis, C. F. (1965). *In* "Chemistry and Physics of Carbon " (P. L. Walker, Jr., ed.) Vol. I, p. 265. Dekker, New York.
Palmer, H. B., and Hirt, T. J. (1963). *J. Phys. Chem.* **67,** 709.
Palmer, H. B., Lahaye, J., and Hou, K. C. (1968a). *J. Phys. Chem.* **72,** 348.
Palmer, H. B., Voet, A., and Lahaye, J. (1968b). *Carbon* **6,** 65.
Peeters, J., and Mahnen, G. (1973). *Int. Symp. Combust., 14th,* p. 133 Combust. Inst., Pittsburgh, Pennsylvania.
Place, E. R., and Weinberg, F. J. (1967). *Int. Symp. Combust., 11th,* p. 245. Combust. Inst., Pittsburgh, Pennsylvania.
Porter, R. P., Clark, A. H., Kaskan, W. E., and Browne, W. G. (1967). *Int. Symp. Combust., 11th,* p. 907. Combust. Inst., Pittsburgh, Pennsylvania.
Radcliffe, S. W., and Appleton, J. P. (1971). *Combust. Sci. Tech.* **4,** 171.
Rosner, D. E., and Allendorf, H. D. (1968). *AIAA J.* **6,** 650.
Rosner, D. E., and Allendorf, H. D. (1970). *In* "Heterogeneous Kinetics at Elevated Temperatures" (G. R. Belton and W. L. Worrell, eds.), p. 231. Plenum, New York.
Salooja, K. C. (1972). *J. Inst. Fuel* **45,** 37.
Sawyer, R. F. (1971). *In* "Combustion-Generated Pollution" (E. S. Starkman, ed.), p. 60. Plenum, New York.
Schott, G. L. (1969). *Int. Symp. Combust., 12th,* p. 569. Combust. Inst., Pittsburgh, Pennsylvania.
Schott, G. L., and Bird, P. F. (1964). *J. Chem. Phys.* **43,** 3237.
Schott, G. L., and Kinsey, J. L. (1958). *J. Chem. Phys.* **29,** 1177.
Seery, D. J., and Bowman, C. T. (1970). *Combust. Flame* **14,** 37.
Shahed, S. M., and Newhall, H. K. (1971). *Combust. Flame* **17,** 131.
Singh, T., and Sawyer, R. F. (1971). *Int. Symp. Combust., 13th,* p. 403. Combust. Inst., Pittsburgh, Pennsylvania.
Skinner, G. B., and Ruehrwein, R. A. (1959). *J. Phys. Chem.* **63,** 1736.
Skinner, G. B., Lifshitz, A., Scheller, K., and Burcat, A. (1972). *J. Chem. Phys.* **56,** 3853.

Smith, M. Y. (1969). *J. Inst. Fuel* **42,** 248.
Spalding, D. B. (1971). *J. Inst. Fuel* **44,** 196.
Spalding, D. B. (1972). *In* "Emissions from Continuous Combustion Systems" (W. Cornelius and W. G. Agnew, eds.), p. 3. Plenum, New York.
Starkman, E. S. (ed.) (1971). "Combustion-Generated Pollution," p. 60. Plenum, New York.
Stull, D. R., and Prophet, H. *et al.* (1970). "JANAF Thermochemical Tables," 2nd ed. Nat. Std. Ref. Data Ser., Nat. Bur. Std. (U.S.), U.S. Govt. Printing Office, Washington, D.C.
Sulzmann, K. G. P., Leibowitz, L., and Penner, S. S. (1971). *Int. Symp. Combust., 13th,* p. 137. Combust. Inst., Pittsburgh, Pennsylvania.
Thompson, D., Brown, T. D., and Beér, J. M. (1972). *Combust. Flame* **19,** 69.
Tourin, R. H. (1966). "Spectroscopic Gas Temperature Measurement." Amer. Elsevier, New York.
Wagman, D. D., Evans, W. H., Parker, V. B., Halow, I., Bailey, S. M., and Schumm, R. H. (1968, 1969). Selected Values of Chemical Thermodynamic Properties, NBS Tech. Notes 270-3 (1968), 270-4 (1969), and subsequent reports. U.S. Govt. Printing Office, Washington, D.C.
Wersborg, B. L., Howard, J. B., and Williams, G. C. (1973). *Int. Symp. Combust., 14th.* p. 929. Combust. Inst., Pittsburgh, Pennsylvania.
Westenberg, A. A., and Fristrom, R. M. (1965). *Int. Symp. Combust., 10th,* p. 473. Combust. Inst., Pittsburgh, Pennsylvania.
Wilde, K. A. (1972). *Combust. Flame* **18,** 43.
Williams, G. C., Sarofim, A. F., and Lambert, N. (1972). *In* "Emissions from Continuous Combustion Systems" (W. Cornelius and W. G. Agnew, eds.), p. 141. Plenum, New York.
Wright, F. J. (1969). *Int. Symp. Combust., 12th,* p. 867. Combust. Inst., Pittsburgh, Pennsylvania.
Zeldovich, Ya. B. (1946). *Acta Physicochim. USSR* **21,** 577.

Bibliography

This bibliography, which is intended to be broad but not complete, includes a short list of general books on the basic aspects of combustion with brief comments on their contents, followed by a longer list of more specialized monographs and symposia, without descriptive comments. Most of the books presuppose a moderate or quite extensive background. For the completely uninitiated, a starting point might be one of the following:

Brame, J. S. S. and King, J. G. (1955). *Fuel,* Edward Arnold, London.
Smith, M. L. and Stinson, K. W. (1952). *Fuels and Combustion,* McGraw-Hill, New York.

General References

Bradley, J. N. (1969). "Flame and Combustion Phenomena." Methuen, London. A useful introduction to the subject; short (200 pages), well written, good breadth.
Strehlow, R. A. (1968). "Fundamentals of Combustion." Int. Textbook Co., Scranton,

Pennsylvania. More detail but less breadth than Bradley. Excellent on premixed flames, aerodynamics, and detonations. Unusually good illustrations including photographs.

Van Tiggelen, A., Burger, J., Clement, G., deSoete, G., Feugier, A., Kerr, C. and Monnot, G. (1968). "Oxydations et Combustions." Editions Technip, Paris. In French, in two volumes. Quite complete in coverage, remarkably thorough. One hopes it may soon be available in English.

Gaydon, A. G., and Wolfhard, H. G. (1970). "Flames," 3rd ed. Chapman & Hall, London. An excellent survey of flame behavior distinguished by pellucid writing and an absence of heavy mathematics.

Lewis, B. and von Elbe, G. (1961). "Combustion, Flames, and Explosions of Gases." Academic Press, New York. Probably the best-known book on combustion, possessing breadth and thoroughness. Particularly strong in discussions of mechanisms, explosion behavior, and characteristics of combustion waves.

Lewis, B., Pease, R. N., and Taylor, H. S. (eds.) (1956). "*Combustion Processes.*" Princeton Univ. Press, Princeton, New Jersey. Fifteen chapters by experts on topics ranging from chemical kinetics and combustion calculations to combustion of sprays, solid fuels, and propellants. Well conceived and well written.

Jost, W. (1946). "Explosions and Combustion Processes in Gases." McGraw-Hill, New York. Possibly the first "modern" book on flames and combustion, published originally in 1935, this book contains much that is still valuable.

Specialized Volumes and Symposia

Various volumes produced by AGARD under titles such as "Selected Combustion Problems" and published by Butterworths, London and Washington, D.C.

Int. Symp. Combust. sponsored by the Combust. Inst., Pittsburgh, Pennsylvania. Proceedings of thirteen symposia are in print at the time of writing, with a fourteenth in press for 1973.

"Oxidation and Combustion Reviews" (C. F. H. Tipper, ed.). Elsevier, Amsterdam. Two volumes appeared as books, the more recent ones as journals.

"Literature of the Combustion of Petroleum," ACS Advan. in Chem. Ser. No. 20. Amer. Chem. Soc., Washington, D.C. (1958).

Gaydon, A. G. (1957). "The Spectroscopy of Flames." Chapman & Hall, London. A new edition is in preparation.

Kondratiev, V. N. (1964). "Chemical Kinetics of Gas Reactions." Pergamon, Oxford. The last 200 pages of this book are on chain reactions and combustion processes.

Fristrom, R. M., and Westenberg, A. A. (1965). "Flame Structure." McGraw-Hill, New York.

Minkoff, G. J., and Tipper, C. F. H. (1962). "Chemistry of Combustion Reactions." Butterworths, London.

Williams, F. A. (1965). "Combustion Theory." Addison-Wesley, Reading, Massachusetts.

Penner, S. S. (1957). "Chemistry Problems in Jet Propulsion." Pergamon, Oxford.

Semenov, N. N. (1959). "Some Problems in Chemical Kinetics and Reactivity" (transl. by M. Boudart). Princeton Univ. Press, Princeton, New Jersey. A large part of this work, especially Vol. 2, is devoted to chain reactions and combustion.

Thring, M. W. (1962). "The Science of Flames and Furnaces." Wiley, New York. Not on small flames, but included here because of its basic importance to the combustion engineer.

Weinberg, F. J. (1963). "Optics of Flames." Butterworths, London.

Markstein, G. H. (1965). "Non-steady Flame Propagation." Macmillan, New York.
Fenimore, C. P. (1964). "Chemistry in Premixed Flames." Pergamon, Oxford.
Shchelkin, K. I., and Troshin, Ya. K. (1965). "Gas Dynamics of Combustion." Mono Book Corp., Baltimore, Maryland.
Soloukhin, R. I. (1966). "Shock Waves and Detonations in Gases." Mono Book Corp., Baltimore, Maryland.
Zeldovich, Ya. B. and Kompaneets, A. S. (1960). "Theory of Detonation." Academic Press, New York.
Lawton, J., and Weinberg, F. J. (1969). "Electrical Aspects of Combustion." Oxford Univ. Press, London and New York.
Holtzmann, R. T. (1969). "Chemical Rockets and Flame and Explosion Technology." Dekker, New York.
"Kinetics, Equilibria and Performance of High Temperature Systems," two volumes reporting proceedings of two conferences. The first, edited by G. S. Bahn and E. Zukoski, was published in 1960 by Butterworths, London; the second, edited by G. S. Bahn, was published by Gordon & Breach, New York, in 1963.
Vulis, L. A. (1961). "Thermal Regimes of Combustion." McGraw-Hill, New York.
Sokolik, A. S. (1963). "Self-Ignition, Flame and Detonation in Gases." Israel Program for Scientific Translations.
Beér, J. M., and Chigier, N. A. (1972). "Combustion Aerodynamics." Appl. Sci. Publ., London.
Chedaille, J., and Braud, Y. (1971). "Industrial Flames," Vol. 1, Arnolds, London.
Barrère, M., Jaumotte, A., Fraeys de Veubeke, B., and Vandenkerckhove, J. (1960). "Rocket Propulsion." Elsevier, Amsterdam.
Penner, S. S. (1962). "Chemical Rocket Propulsion and Combustion Research." Gordon & Breach, New York.
Starkman, E. S. (ed.) (1971). "Combustion-Generated Pollution." Plenum, New York.
Cornelius, W., and Agnew, W. G. (eds.) (1972). "Emission from Continuous Combustion Systems." Plenum, New York.
Spalding, D. B. (1955). "Some Fundamentals of Combustion." Academic Press, New York.

Journals

Unfortunately, research papers pertaining to combustion are to be found in a wide array of journals including most of the basic journals in chemistry, physics, and engineering. We mention here only those journals in English whose objectives include publication of research or reviews in combustion.

Combustion and Flame. The journal of the Combustion Institute, published by Elsevier, New York. Six issues per year.

Combustion Science and Technology. Published by Gordon & Breach, New York. Six issues per year.

Combustion, Explosion, and Shock Waves. Translation of a Russian journal; available in many libraries.

Oxidation and Combustion Reviews. Published by Elsevier, Amsterdam. The first two volumes appeared as books, more recent ones as journals.

Fire Research Abstracts and Reviews. Published by the Committee on Fire Research of the National Research Council. Three issues per year.

Journal of the Institute of Fuel. Published monthly by the Institute of Fuel, London.

High Temperature Science. Published by Academic Press, New York. Six issues per year.

II

Corrosion and Deposits in Combustion Systems

WILLIAM T. REID

BATTELLE MEMORIAL INSTITUTE
COLUMBUS LABORATORIES
COLUMBUS, OHIO

I. INTRODUCTION

Forecasts of our growing demand for energy lead to the inescapable conclusion that the fossil fuels coal, oil, and gas will supply our major needs for at least the next three decades. Electrical generation in particular will increase, to provide from about a fifth of the total energy consumed in the United States presently to about half of our energy demands at the turn of the century. Nuclear generation will soon begin supplying appreciable amounts of that electrical energy, but for these next 30 years the fossil fuels will be our mainstay for generating electricity. Natural gas, presently preferred for air pollution control, but in relatively short supply, probably will be diverted from power plants to residential and industrial consumers. Coal and oil, then, will be the major energy sources for central stations generating the bulk of our electricity. By 1980, this consumption should exceed 4×10^8 tons of coal and 4×10^7 tons of fuel oil annually.

With these fuels come troubles caused by the inorganic matter they contain. Coal is the worst offender with roughly 10% ash. Residual fuel oil generally contains less than 0.1% ash, but its ash constituents are more aggressive than those in coal. Both fuels lead to serious problems in external corrosion and deposits as these inorganic materials accumulate on heat-receiving surfaces. Such deposits of ash upset heat-transfer coefficients and the normal flow patterns of flue gases, but the main trouble is that rapid metal wastage can occur beneath deposits, forcing unscheduled outages for replacement of wall tubes or superheater elements. In gas turbines, combustor problems are not so severe, but deposits on turbine blading can be disastrous.

Deposits are objectionable in themselves, as thermal insulators or flow obstructors, but usually it is the corrosion conditions accompanying deposits that cause the greatest concern.

A typical result of metal wastage of superheaters and reheaters has been the trend toward lower steam temperature in boiler furnaces. Although steam temperature in central-station power plants increased an average of about 8°C per year between 1931 and 1955, so that steam temperature would now be higher than 704°C if that rate had been maintained, most boilers in 1960 were being designed for 566°C steam. In 1960, design steam temperature began to drop, with fewer boilers being built for 566°C and more for 538°C. In 1965, only one large unit was sold at the higher temperature. Since 1968, the steam temperature in all new units has been limited to 538°C. The reason for this is shown in Table I. Hence, with minimum trouble between 538 and 543°C, boiler practice has tended to stabilize at these levels.

A rough estimate a few years ago by the Corrosion and Deposits Committee of ASME placed the direct out-of-pocket costs of external corrosion and deposits in boiler furnaces at about $10 million a year. Replacing corroded wall tubes or superheater elements can be costly in itself, but the loss of availability during an unscheduled outage of a large steam generator

TABLE I

FORCED OUTAGES CAUSED BY SUPERHEATERS IN DRUM-TYPE UNITS DURING 1964[a]

Steam temperature (°C)	Forced outages (per unit)
482–513	1.5
538–543	0.6
560–571	1.4
593–649	1.9

[a] Data from Edison Electric Institute.

through a tube failure can be expensive indeed. Accurate cost figures are not available, but it is generally accepted that each day a 500-MW unit is out of service costs about $40,000. With the recent trend to larger and larger steam generators, even up to 1300 MW, the importance of eliminating such outages grows in proportion. This is the main reason why so much attention has been paid to investigating the causes of corrosion and deposits, and to seeking corrective measures. A more detailed account of the corrosion and deposit problem has been published recently (Reid, 1970b).

II. OCCURRENCE OF INORGANIC MATTER IN FUELS

All fuels used in boiler plants, except natural gas, contain appreciable amounts of noncombustible substances. The nature of the noncombustibles in coal is quite different from that of the noncombustibles in residual fuel oil. In the case of coal, usually only a small part of the inorganic matter was part of the original growing plant; the remainder was deposited with the plant material while coalification was taking place. In fuel oil, the major impurities are complex organic compounds originating in the life forms that became petroleum, plus inorganic material unavoidably included during transportation and refining.

A. Coal Ash

Inherent mineral matter in coal, seldom exceeding 2% of the weight of the coal, consists of the elements originally present in growing plants, such as iron, calcium and magnesium, phosphorus, potassium, and sulfur. Extraneous mineral matter, making up the remainder of the coal ash, is from sediment washed into the coal bed, or inorganic solids collected from percolating ground water.

Although as many as 100 minerals have been found in coal, most of the ash comes from four main groups—shale, clay, sulfur, and carbonates. Usually, 95% of the ash exists as kaolinite, pyrites, and calcite. Following combustion, these mineral forms convert to complex silicates, to unreacted oxides, and to partially or wholly sintered glasses of indeterminate chemical structure, depending upon the temperature to which they were heated and the time of exposure. Conventional chemical analyses of coal ash are reported as the oxides, typically SiO_2, Al_2O_3, Fe_2O_3, CaO, MgO, Na_2O, K_2O, and SO_3. In addition, TiO_2 and P_2O_5 may be present, and there will be trace amounts of many other elements.

Table II shows the usual range of composition of coal ash in the United States. The amount of SO_3 reported depends on the ashing procedure and the amount of CaO in the ash. Since the most stable sulfate, $CaSO_4$, dissociates in flue gas above 1232°C (Reid, 1970a), SO_3 is driven off completely when the ash is melted to form a slag. Hence analyses are usually recalculated on an SO_3-free basis in predicting slag characteristics.

Sulfur is an objectionable constituent in coal for reasons aside from air pollution considerations. It causes a great many problems in boiler furnaces, mainly because it leads to SO_3 and complex sulfates directly involved in corrosion. Coals contain varying amounts of sulfur, usually not exceeding 5%. Air pollution control ordinances now commonly limit sulfur to 1% with 0.4% maximum allowable sulfur possible in the future. There are

TABLE II

USUAL RANGE OF ASH COMPOSITION IN AMERICAN COALS

Component	Percentage
SiO_2	20–60
Al_2O_3	10–35
TiO_2	0.5–2.5
Fe_2O_3	5–35
CaO	1–20
MgO	0.3–4
$Na_2O + K_2O$	1–4
SO_3	0.1–12

very large reserves of low-sulfur coal in the United States, but most of these coals are low rank and are found in the midwestern states far from our coal-consuming centers.

Sulfur exists in coal in three forms: pyritic, organic, and as sulfates. Sulfate sulfur content usually is negligible. Figure 1 shows that pyritic and organic sulfur contents are about equal on the average, except with low-sulfur coals where organic sulfur predominates. Hence, although float-and-sink washing operations can separate large particles of pyrites from some coals, at best only about half the sulfur can be removed in this way. Pyrites often is present in particles too small to be liberated from the coal by crushing. Sulfur removal, therefore, depends not only on the quantity of pyrites in the coal but on the size of the pyrites particles as well. Organic sulfur, being a part of the coal substance, cannot be removed without breaking down the coal molecule, as by gasification.

Chlorine is frequently blamed for corrosion with English coals in which it occurs up to 1%; it seldom exceeds 0.3% in American coals, and it usually is less than 0.1%. Because less than 0.3% chlorine in coal does not cause problems through corrosion and deposits (Raask, 1963), chlorine in American coals generally may be neglected as a source of trouble. Phosphorus, which occurs up to about 1% as P_2O_5 in coal ash, was a frequent source of deposits when coal was burned on grates. With pulverized-coal firing, however, where highly reducing conditions do not occur as in fuel beds, phosphorus does not cause fouling.

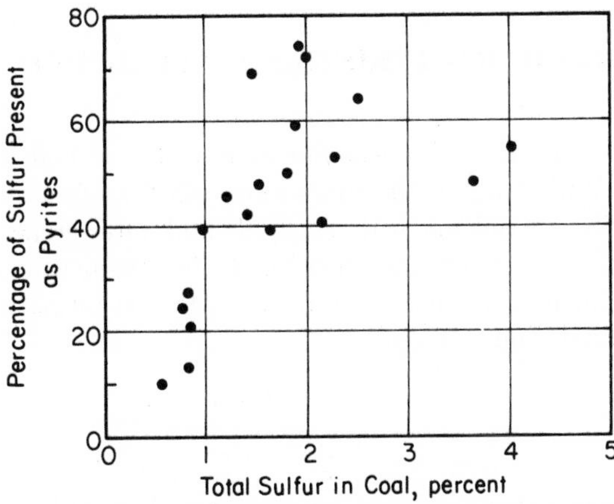

Fig. 1. Occurrence of pyrites in coal. [Data from Nicholls and Selvig (1932).]

TABLE III

TYPICAL RANGE OF MAJOR CONSTITUENTS IN RESIDUAL FUEL OIL LEADING TO CORROSION AND DEPOSITS

Constituent	In residual fuel oil (ppm)	In oil ash (%)
Vanadium	0–500 as V	0–40 as V_2O_5
Sodium	2–300 as Na	0.1–30 as Na_2O
Sulfur	1000–40,000 as S	0.1–40 as SO_3

B. Oil Ash

Extraneous ash occurs in residual fuel oil just as in coal, but the major impurity causing trouble through corrosion and deposits is vanadium, present as an oil-soluble porphyrin. Because this porphyrin is thermally stable, it is not removed during refining so that all the vanadium present initially in the crude oil appears in the residual fuel. The other elements in residual fuel oil contributing most to corrosion are sodium and sulfur.

Table III gives a typical range of composition of these objectionable constituents. Although such residual fuels are suited to boiler furnaces, they are much too dirty for gas turbines. Standards being proposed in 1970 for gas-turbine fuel call for a maximum of 1 ppm vanadium, 2 ppm sodium and potassium, 1 ppm calcium, and 1 ppm lead, with not more than 20 ppm total ash.

III. MELTING CHARACTERISTICS OF FUEL ASH

Much more is known of the behavior of coal ash at high temperature than of oil ash. With roughly 100 times more ash than fuel oil, coal poses much more serious problems with slagging and interference with heat transfer than does oil ash. On the other hand, the low-melting characteristics of oil ash can lead to special problems with the formation of liquid films of the vanadates on tube surfaces.

A. Coal Ash

Most of the earlier studies of coal ash were aimed at clinkering problems in fuel beds. Later, studies of ash were concerned with the unique problems involved with slag-tap pulverized-coal-fired boiler furnaces, where the viscosity of the coal-ash slag must be low to assure satisfactory removal of the slag. In these slag-tap furnaces, the slag is also fluid enough to provide essentially continuous liquid films on the lower sections of the heat-receiving wall tubes. In dry-bottom furnaces, wall deposits are made up largely of sticky particles that coalesce to cover the tubes in irregular patterns. As the gases cool while passing through superheaters and reheaters in either type of furnace, adherent ash deposits cause serious problems. In air heaters, ash accumulations again can be troublesome.

Those earliest attempts to relate ash characteristics to clinkering (Nicholls and Selvig, 1932) failed in part because no good criterion existed for measuring the fusibility of coal ash. The cone fusion test (ASTM, D 1857-68) provided some correlation between softening temperature and extent of clinkering, but the relationships were inexact. At best the cone-softening temperature divides coal ash into broad categories of fusibility—low fusion, 1000 to 1200°C; moderate fusion, 1200 to 1400°C; and high fusion, 1400 to 1650°C.

More useful data are available from measurements of the viscosity of molten coal-ash slags. Using a concentric cylinder viscometer of the Margules type, where viscosity is a function of the constant-speed torque required to rotate a 12.7-mm-diameter bob in a 25.4-mm-deep charge of slag contained in a 38.1-mm-diameter platinum crucible, the flow characteristics of a large number of slags were measured in 1940, over the temperature range 1100 to 1600°C (Nicholls and Reid, 1940). Other investigators extended these measurements (Hoy *et al.*, 1964) to include British coals. A good summary of the flow properties of coal ash slags has been prepared by the U.S. Bureau of Mines (Corey, 1964).

Figure 2 shows the flow behavior of three typical slags. Slag I is a Newtonian fluid over its entire viscosity range from below 10 Ns/m^2 to above 1000 Ns/m^2.* Whether heated or cooled between 1150 and 1510°C, this slag behaves as a glass. No devitrification occurs, and viscosity depends only on temperature, not on the previous thermal history of the slag.

Slag II typifies the larger group of slags where a solid phase separates abruptly at some given temperature as the slag is cooled. At A, the slag changes rapidly from a Newtonian fluid to a pseudoplastic solid, so that the slag is no longer fluid at temperatures below 1215°C. When reheated slowly, the solid phase gradually redissolves in the melt so that fluidity is reestablished at B. Between B and C, the last traces of solids disappear, this temperature range depending largely on the rate of heating. At temperatures above 1320°C for this slag, Newtonian behavior occurs again.

* 1 Ns/m^2 = 10 P (poise).

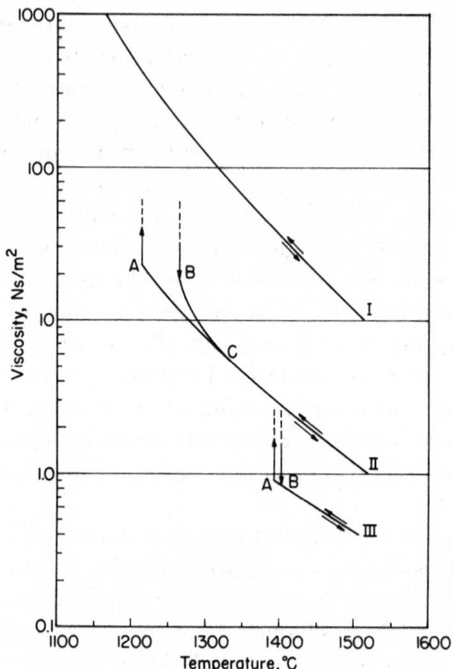

Fig. 2. Flow characteristics of three typical coal ash slags [after Nicholls and Reid (1940).]

	I	II	III
SiO_2	56.9	43.9	40.6
Al_2O_3	18.8	23.8	13.4
Fe_2O_3	12.9	12.1	34.1
CaO	8.9	20.2	9.2
MgO	1.2		1.2
Alkali	1.3	—	1.5

Slag III is similar to Slag II except that the range between A and B is much shorter and occurs at a much higher temperature. Such a slag converts abruptly from a fluid to a solid over a short cooling interval.

Although it may be inferred from these curves that viscosity is the main factor affecting this change from a Newtonian fluid to a pseudoplastic solid, such is not the case. Highly fluid slags may remain Newtonian, as well as more viscous ones. The separation of a solid phase is a function of the composition of the slag. The temperature at which this transition takes place is known as the "temperature of critical viscosity" (T_{cv}) (Reid and Cohen, 1944).

Relating these flow characteristics to composition is difficult because at least six variables are involved. Fortunately it has been found empirically that some of the variables have little significance and others can be combined.

For Newtonian viscosity, the relative proportions of silica (SiO_2) and the fluxes Fe_2O_3, CaO, and MgO give a good measure of viscosity. Expressed as the "silica percentage" (SP), calculated as $SiO_2/(SiO_2 + Fe_2O_3 + CaO + MgO) \times 100$ where these are the percentages of each component in the melt, the relationship is given approximately by

$$\log(\eta - 0.06) = (0.066)(SP) - 3.40, \tag{1}$$

where η is viscosity in Ns/m² at 1427°C, and SP is the silica percentage.

Figure 3 shows the original correlation from which this relationship was

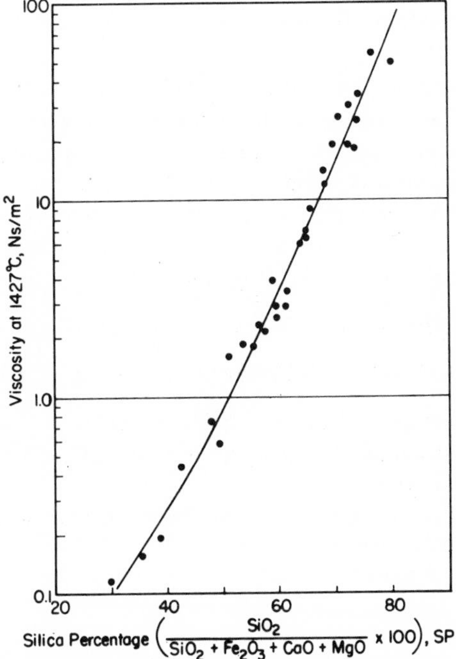

Fig. 3. Relationship between ash composition and viscosity at 1427°C.

derived, and the experimental points on which it is based. These cover a wide range of SiO_2/Al_2O_3 and $Fe_2O_3/(CaO + MgO)$ ratios.

A different type of correlation has been derived in England (Hoy *et al.*, 1964) to show the influence of temperature:

$$\log \eta = 4.468(SP/100)^2 + 1.265(10^4/T) - 8.44, \qquad (2)$$

where η is in Ns/m^2, SP is the silica percentage as defined above, and T is the absolute temperature in degrees Kelvin.

Still more refined correlations came later, based on a temperature-modified exponential Arrhenius equation for viscous flow (Watt and Fereday, 1969). This most recent and certainly the most precise relationship is

$$\log \eta = [10^7 M/(t - 150)^2] + C - 1, \qquad (3)$$

where η is in Ns/m^2, t is in degrees Celsius, $M = 0.00835SiO_2 + 0.00601 Al_2O_3 - 0.109$, and $C = 0.0415SiO_2 + 0.0192Al_2O_3 + 0.0276Fe_2O_3 + 0.0160CaO - 3.92$, with the percentage composition expressed as $SiO_2 + Al_2O_3 + Fe_2O_3 + CaO + MgO = 100$. These three correlations, of increasing complexity and precision, make it possible at present to predict the viscosity of essentially all coal ashes from their composition, making unnecessary any further difficult measurements in a high-temperature viscometer.

It must be recognized that these correlations are useful only when the slag behaves as a Newtonian fluid. Below T_{cv}, where solids are present, the condition of a slag depends upon much more complicated relationships, with time as an important variable. The T_{cv} for a slag must be known, then, to establish the minimum temperature at which these viscosity correlations can be used. The only useful relationship between T_{cv} and composition has come from BCURA (Hoy *et al.*, 1964):

$$T_{cv} = 2990 - 1470\ (SiO_2/Al_2O_3) + 360\ (SiO_2/Al_2O_3)^2 - 14.7\ (Fe_2O_3 + CaO + MgO) + 0.15\ (Fe_2O_3 + CaO + MgO)^2, \qquad (4)$$

where T_{cv} is in degrees Celsius and $SiO_2 + Al_2O_3 + Fe_2O_3 + CaO + MgO = 100$. The standard deviation here is high at $\pm 68°C$, but no better correlations are available.

The influence of deposits on external corrosion and on heat transfer is covered later.

B. Oil Ash

Relatively little information is available on the fusion characteristics of the inorganic matter in residual fuel oil. No systematic investigation has been made of the melting characteristics of oil–ash systems. Unlike the silicate glasses characterizing melted coal ash, a series of chemical compounds usually are formed from the residue left when fuel oil is burned. Generally these compounds involve vanadium, sodium, and sulfur, with vanadium predominating as was shown in Table III.

Table IV gives the melting points of some of the inorganic compounds likely to be present when residual fuel oil is burned. A great many other compounds may be formed, but these three classes of materials predominate.

Of the sulfates, only $Na_2S_2O_7$ melts at a temperature low enough to be molten on the wall tubes of boiler furnaces, and hence corrosive. Instability of $Na_2S_2O_7$ in air or in normal flue gas containing 25 to 50 ppm SO_3 eliminates it as a causative factor in corrosion at the high metal temperature of superheaters. But at the lower temperature of wall tubes, particularly

TABLE IV

MELTING POINT OF TYPICAL INORGANIC COMPOUNDS FORMED WHEN RESIDUAL FUEL OIL IS BURNED

Material	Melting point (°C)
Sulfates	
$Na_2S_2O_7$	399[a]
Na_2SO_4–$MgSO_4$ eutectic	671
Na_2SO_4–K_2SO_4 solid solution	832
Na_2SO_4	885
K_2SO_4	1077
Complex alkali metal sulfates	
$K_3Fe(SO_4)_3$	619[a]
$Na_3Fe(SO_4)_3$	624[a]
$NaFe(SO_4)_2$	690[a]
$KFe(SO_4)_2$	694[a]
Vanadium compounds	
$5Na_2O \cdot V_2O_4 \cdot 11V_2O_5$	578
$2Na_2O \cdot V_2O_5$	619
$Na_2O \cdot V_2O_5$	630
$Na_2O \cdot V_2O_4 \cdot 5V_2O_5$	658
V_2O_5	674
V_2O_4	1970
V_2O_3	1970

[a] Dissociates in air to lose SO_3.

beneath deposits where high SO_3 concentrations can exist, the pyrosulfates can contribute significantly to metal wastage. Both Na_2SO_4 and K_2SO_4 melt at temperatures too high to cause problems directly in boiler furnaces, but they lead to difficulties in gas turbines. The solid solution of these sulfates with a minimum melting point of 832°C can be particularly objectionable. The eutectic of Na_2SO_4–$MgSO_4$ melting at only 671°C is mainly of theoretical interest; it has not been found in deposits.

The complex alkali metal sulfates melt in a wide temperature range, depending upon the partial pressure of the SO_3 in the surrounding atmosphere. Beneath deposits where the SO_3 level can be high, as will be described later, the trisulfates are particularly objectionable. Enough alkalis and sulfur are present in essentially all residual fuel oils to allow for the formation of these complex sulfates in objectionable amounts.

Vanadium exists in boiler furnace deposits in a great many more forms than are illustrated in Table IV; the vanadates shown there are the most common. Since superheater metal temperatures may be as high as 650°C, it is evident why the vanadates are so troublesome. Only the lower oxides, V_2O_3 and V_2O_4, have melting points high enough to eliminate them as a cause of metal wastage. With the exception of these two oxides, essentially all the vanadates are potentially troublesome, particularly in gas turbines.

IV. EXTERNAL CORROSION

Loss of metal occurs throughout boilers on surfaces exposed under the proper conditions to combustion products. Generally, metal wastage of furnace wall tubes, superheaters, and reheaters is termed "high-temperature corrosion" because the metal temperature is greater than 325°C. In air heaters, where the metal temperature approaches the dew point of the flue gas, usually about 150°C, "low-temperature corrosion" takes place. Corrosion under both conditions usually results from the presence of a liquid film on the metal surface, but the nature and the formation of those liquid films are radically different.

A. High-Temperature Corrosion

1. Coal-Fired Boiler Furnaces

Tube wastage first posed serious problems in boiler maintenance beginning in about 1942, when a sudden rash of wall-tube failures in pulverized-coal-fired slag-tap furnaces was caused by external loss of metal (Reid *et al.*, 1945). In the worst cases, tubes failed within three months of installation. Measurements of tube-wall temperature showed that the tubes were not overheated, the maximum wall temperature being 370°C. Heat-transfer rates in the corrosion area were nominal. Examination of the tubes in the affected area, clearly related to the flame pattern in the furnaces, showed the presence of a tightly adherent porcelain-like coating on the tube beneath the overlying deposit of slag. This coating, with a greenish-blue to a pale-blue color, was moderately soluble in water, giving a solution with a pH as low as 3.0. The coating material was finally identified as an alkali iron trisulfate, principally $K_3Fe(SO_4)_3$, with $Na_3Fe(SO_4)_3$ also present as well as small and probably insignificant amounts of lithium compounds.

Although first related to corrosion at quite moderate metal temperatures, it was found later that the complex trisulfates also were involved with external corrosion at superheater metal temperatures (650°C). Superheater alloys are selected to be oxidation resistant at this temperature, and this resistance is not lowered in flue-gas atmospheres containing SO_2; gas-phase oxidation does not cause the observed increased rate of metal loss. In the presence of alkalis and appreciable amounts of SO_3, however, a liquid film of $K_3Fe(SO_4)_3$ and $Na_3Fe(SO_4)_3$ can cause catastrophic failure.

a. *Mechanism of High-Temperature Corrosion*

Alkali iron trisulfates are formed by reaction of SO_3 with Fe_2O_3 and either K_2SO_4 or Na_2SO_4, or with mixed alkali sulfates, according to the reaction (Corey *et al.*, 1945)

$$3K_2SO_4 + Fe_2O_3 + 3SO_3 \rightleftarrows 2K_3Fe(SO_4)_3. \quad (5)$$

Laboratory experiments showed that this reaction takes place at 538°C only when the SO_3 level is at least 250 ppm. At higher temperatures, the SO_3 concentration required is even greater. Hence, since flame-produced SO_3 normally present in furnace gases seldom exceeds 25 to 50 ppm, catalytic formation of additional SO_3 must occur to produce the trisulfates. This catalytic formation has been demonstrated by Cain and Nelson (1961) in the laboratory by exposing a "standard corrosion mixture" of Na_2SO_4, K_2SO_4, and Fe_2O_3 in the molar ratio 1.5 : 1.5 : 1 in a porcelain boat to a synthetic flue gas containing 3.6% O_2 and 2500 ppm SO_2. After 30 hr at temperatures from 550 to 700°C, large amounts of molten trisulfates were present because the SO_2 was converted largely to SO_3.

Two mechanisms have been proposed to explain how the alkali iron trisulfates lead to metal wastage:

Mechanism (1) A normal oxide forms on the tube metal surface and is covered with alkali sulfates from the furnace atmosphere. Particles of fly ash adhere to this alkali layer and accumulate to produce a gradually thickening layer of ash. As the ash layer becomes thicker, the thermal gradient through it increases and the exposed outer layer melts. Increased temperature within the deposits leads to dissociation of sulfates to release SO_3 which migrates to the cooler oxide-covered metal surface. This SO_3 reacts with the alkali sulfates and the iron oxide surface, thereby dissolving the protective film of Fe_2O_3 and forming alkali iron trisulfates. Later, after deslagging occurs normally with load change, the layer spalls off and the tube metal again oxidizes to reestablish its normal protective oxide film. This reoxidation causes irreversible loss of metal. The reactions are, in sequence:

$$K_2O + SO_2 + \tfrac{1}{2}O_2 \rightarrow K_2SO_4 \quad \text{(furnace atmosphere)}, \tag{6}$$

$$MSO_4 \rightarrow MO + SO_3 \quad \text{(dissociation of sulfates)}, \tag{7}$$

$$3K_2SO_4 + 3SO_3 + Fe_2O_3 \rightleftarrows 2K_3Fe(SO_4)_3 \quad \text{(formation of trisulfates at expense of } Fe_2O_3 \text{ on oxidized metal surface)}, \tag{8}$$

$$4Fe + 3O_2 \rightarrow 2Fe_2O_3 \quad \text{(reestablishment of oxide film on metal surface)}. \tag{9}$$

Mechanism (2) Loose, unbonded fly ash accumulates on the oxidized surface of the tube, and alkalis within this ash deposit form sulfates with SO_3 from the gas stream. These alkali sulfates, in turn, react with additional SO_3 and the iron oxide in the ash deposit to produce alkali iron trisulfates, which migrate as molten compounds through the ash deposit as a result of the normal thermal gradient in the deposit. The molten alkali iron trisulfates accumulating at the interface between the tube metal and the deposit react with the tube metal to cause metal wastage, since the liquid phase is rich in sulfate ion. The sequence of reactions is:

$$K_2O + SO_2 + \tfrac{1}{2}O_2 \rightarrow K_2SO_4 \quad \text{(reaction within the deposit with } SO_3 \text{ from gas stream)}, \tag{10}$$

$$3K_2SO_4 + 3SO_3 + Fe_2O_3 \rightarrow 2K_3Fe(SO_4)_3 \quad \text{(formation of trisulfates within ash deposit at expense of } Fe_2O_3 \text{ in the deposit)}, \tag{11}$$

$$2K_3Fe(SO_4)_3 + 10Fe \rightarrow 3Fe_3O_4 + 3FeS + 3K_2SO_4 \quad \text{(reaction of trisulfates with tube metal)}. \tag{12}$$

Figure 4 summarizes the reactions involved in these two chemical processes.

CORROSION REACTIONS

SCALE DESTROYING

$$3K_2SO_4 + \boxed{Fe_2O_3} + 3SO_3 \rightleftharpoons 2K_3Fe(SO_4)_3$$

FROM SURFACE OF TUBE

$$4Fe + 3O_2 \rightleftharpoons 2Fe_2O_3$$

SCALE FORMING

$$3K_2SO_4 + \boxed{Fe_2O_3} + 3SO_3 \rightleftharpoons 2K_3Fe(SO_4)_3$$

FROM FLY-ASH DEPOSIT

$$2K_3Fe(SO_4)_3 + 10\,Fe \rightarrow 3K_2SO_4 + 3Fe_3O_4 + 3FeS$$

Fig. 4. Possible mechanisms for metal wastage involving alkali iron trisulfates.

Both mechanisms depend critically on the availability of SO_3 at partial pressures greatly exceeding that in normal flue gas. As will be shown later, these high levels of SO_3 can be produced in a stagnant gas layer in the interface between the tube metal and the ash deposit, so that the SO_3 provided from dissociation of sulfates in mechanism (1) can be furnished instead by catalytic conversion of SO_2 to SO_3. A shortcoming of mechanism (2) is that sulfides are seldom found in corrosion areas, but this has been explained on the basis that rapid corrosion obscures sulfide penetration of the substrate metal. A much more serious shortcoming of mechanism (2) is that it calls for reaction between the alkali iron trisulfates and metallic iron. Because, at working temperature, formation of the oxide film occurs rapidly, it is much more likely that the tube surface is Fe_3O_4 or Fe_2O_3, rather than Fe. Since reaction of the trisulfates with iron oxides is unlikely, the only Fe that could be attacked would have to be dissolved through cracks and imperfections in the overlying oxide surface. Both mechanisms undoubtedly take place. Which one predominates is difficult to assess.

2. Oil-Fired Systems

Because alkalis and sulfur occur in residual fuel oil as well as in coal, alkali iron trisulfates also cause trouble in oil-fired boiler furnaces, as was noted earlier. In addition, high-vanadium fuel oils pose special problems.

As was shown in Table IV, several vanadates are molten at the temperature of superheater elements, and it is generally agreed that the sodium vanadates cause worse corrosion than V_2O_5. For gas turbines, maximum corrosion can be expected when the $V_2O_5/(V_2O_5 + Na_2SO_4)$ ratio falls between 0.78 and 0.85 (Buckland *et al.*, 1952). For boiler furnaces (Niles and Sanders, 1962), which of the sodium vanadates is formed is relatively important, $Na_2O \cdot 3V_2O_5$ being the most corrosive, with $Na_2O \cdot 6V_2O_5$ only slightly less troublesome; $Na_2O \cdot V_2O_5$ causes least trouble.

No assured mechanism has been derived for the way in which vanadium leads to metal wastage. Proposed mechanisms include: (1) vanadates serving as oxygen carriers, (2) vanadates distorting the normal crystal lattice of the protective oxide film on the metal, and (3) molten vanadates dissolving this protective film. It is generally agreed that a liquid film speeds up the corrosion rate, hence wastage will be most troublesome when the vanadate has a melting point lower than the temperature of the metal substrate.

As with most metal wastage problems, it has been difficult to translate small-scale laboratory experiments into meaningful full-scale operation. The most useful results have been obtained from moderately large experimental combustors burning residual fuel oil under more closely controlled conditions than is possible in a utility boiler.

B. Low-Temperature Corrosion

Corrosion of air heaters has been responsible mainly for keeping stack gas temperatures above 163°C. At lower temperatures the dew point of the flue gas is reached and a liquid film of sulfuric acid forms on the metal surfaces of the air heater, greatly hastening metal failure. Chlorides in the fuel, converted to hydrochloric acid during combustion, add to this problem. Chloride-aided corrosion is serious in England where coals contain appreciable amounts of this element, but it is not of much concern in the United States where most coals contain less than 0.3% chlorine.

Glass-coated air preheater elements have been used with good success to resist acid films, but the cost is generally considered excessive and the scheme has not caught on widely. The most obvious method of preventing low-temperature corrosion is to design the air heater so that the metal temperature never drops below the dew point of the flue gas.

C. Formation of SO_3

It is evident both in high- and low-temperature systems that SO_3 is a major factor in corrosion. Sulfur is always present in coal and in residual fuel oil, and although the amount varies, ample sulfur is generally available in the flue gas as SO_2 to present formidable problems if converted to SO_3.

As was shown in discussing the formation of the alkali iron trisulfates, 250 ppm SO_3 is required to form these salts at 538°C, while the partial pressure of SO_3 must be appreciably greater for the trisulfates to exist at higher temperatures.

Flame formation is inadequate to yield troublesome amounts of SO_3 as far as high-temperature corrosion is concerned. Depending on oxidation of SO_2 by oxygen atoms present in a flame, the amount of SO_3 so formed is generally only 1% of the amount of SO_2. Hence, in a combustion system with 2500 ppm SO_2 in the flue gas, the SO_3 level will be 25 to 35 ppm (Barrett *et al.*, 1966). This is ample to form sulfates such as K_2SO_4, but it does not account for the higher concentrations of SO_3 necessary to form $K_3Fe(SO_4)_3$.

Homogeneous catalysis contributes little to SO_3 in flue gas, so that the principal source of SO_3 to form the trisulfates must be by heterogeneously catalyzed oxidation reactions on available surfaces. A point of much disagreement many years ago, it is now generally conceded that the high concentrations of SO_3 required to form alkali iron trisulfates are the result of heterogeneous catalysis. Attempts to demonstrate catalytic oxidation

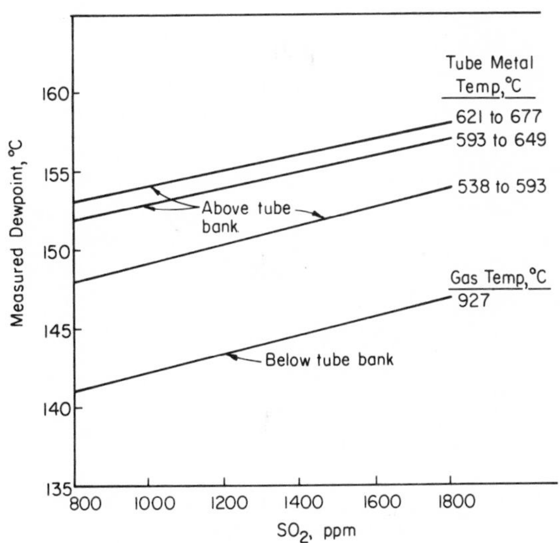

Fig. 5. Increase in dew point of flue gases passing over simulated superheater bank. [Data from Burnside *et al.* (1956).]

of SO_2 in flue gases passing over superheater surfaces have been only partially successful, but, beginning in 1944 (Harlow, 1944) with recognition of this effect, and continuing in 1956 with measurement in a pilot-scale unit by Burnside *et al.* (1956), it is now evident that a small but definite increase in SO_3 can occur in flue gas passing over boiler surfaces. Figure 5 illustrates this effect; increases in dew point of 12°C were observed at metal temperatures averaging 649°C, with markedly less effect at average metal temperatures of 621 and 566°C.

Although these relatively small changes in dew point represent much larger differences in SO_3 concentration, this bulk catalytic effect is inadequate to provide the high levels of SO_3 required to form the trisulfates. Also, with flue gas moving over surfaces at velocities approaching 25 m/sec, the residence time is short and reaction kinetics are still limiting. These questions led, in 1967, to probing the boundary layer in a dynamic gas system to show that a steep SO_3 gradient exists in a synthetic flue gas containing SO_2 and O_2 flowing at moderate velocities over catalytically active surfaces.

Figure 6 indicates the results obtained with a probe of small quartz tubing with an opening about 0.01 mm in diameter sampling the flowing gas stream at various distances above three surfaces at 593°C (Krause *et al.*, 1968). The gas stream velocity above the surface was 370 mm/sec, and the gas contained 5.8% SO_2, 19.4% O_2, and 74.8% N_2. It is evident that the SO_3 concentration near the surface is much greater than in the bulk gas stream. And although there is a significant difference in the SO_3

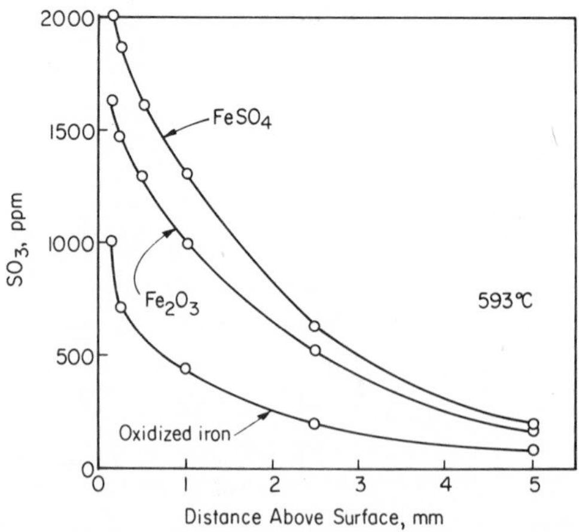

Fig. 6. Sulfur trioxide concentration above surfaces in flowing gas stream containing SO_2 and O_2. [Data from Krause *et al.* (1968).]

level 1 mm or more away from the surface with different materials, that difference becomes insignificant as the surface is approached more closely. The general conclusion here is that, beneath deposits where gas velocities are low, approaching stagnation, the SO_3 concentration can be 30 times or more greater than in the bulk, rapidly moving, gas stream. Thus enough SO_3 is easily available to permit the formation of the alkali iron trisulfates. This is not a surprising conclusion; these experiments simply confirm that conditions beneath deposits can be quite different from those in the bulk flue gas, and that an environment is provided beneath deposits in which corrosion can be rapid.

It is worth calling attention here to the usefulness of radioactive ^{35}S in corrosion studies. This isotope, with low-energy beta radiation (0.167 MeV) and a half-life of 86.7 days, is ideally suited to such tasks as differentiating between the activity of various sulfur oxides. By adding extremely small quantities of isotope-tagged SO_2 or SO_3 to gas mixtures, it has been shown by Krause *et al.* (1968) that SO_3 is thousands of times more reactive than SO_2 in most corrosion reactions.

D. Dew Point of SO_3

As was noted earlier, condensation of sulfuric acid causes serious low-temperature corrosion problems if the flue gas is cooled below its dew point. Many instruments based on electrical resistivity between electrodes on a glass surface have been devised to measure the dew point of flue gas, but all have shortcomings because the rate of acid buildup on the condensing surface influences the measurement. In addition, fly ash is captured by the dew-point meter and affects the results.

Because of these shortcomings of dew-point meters, direct measurements of dew point have largely been superseded by the analysis of the flue gas for SO_3, from which the dew point can be estimated. Figure 7 shows the acid dew point for a wide range of SO_3 concentrations, according to the best summary of many existing measurements (Reese *et al.*, 1964). The quantity of moisture in the flue gas influences the observed dew point, but the resulting variations are less than those in the results obtained by various investigators. Most significant here is the fact that even very small amounts of SO_3 elevate the dew point appreciably.

Of all the analytical schemes for determining the SO_3 concentration in flue gas, one of the simplest and most reliable is the technique of Goksøyr and Ross (1962). When hot flue gas is passed through a spiral glass condenser maintained at 80°C, SO_3 is separated from SO_2 by condensation. Washing out this condenser and titrating the washings gives the SO_3 con-

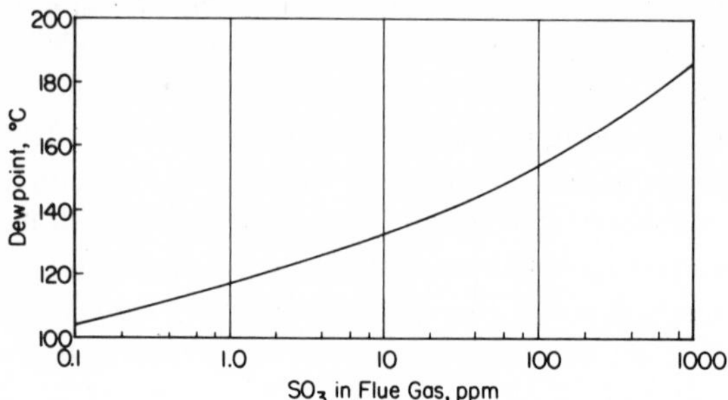

Fig. 7. Acid dew point of sulfur trioxide.

tent directly. Care must be taken to maintain sampling lines hotter than 250°C to prevent any possible condensation of SO_3 before reaching the spiral collector. Reproducibility is good, and less than 15 min is needed for an analysis.

V. FORMATION OF DEPOSITS

Although a great deal has been written about the way in which ash accumulates on heat-receiving surfaces in boiler furnaces and on turbine blades, little factual information is available. Most of the work done has been based on theoretical considerations. Experimental work in the field has been notably lacking in determining the influence of the many variables on the rate of accumulation of deposits under different conditions of gas velocity, tube distribution, ash characteristics, and temperature.

Figure 8 illustrates one cause for concern with deposits (Ely and Schueler, 1944)—their interference with heat transfer. Here the temperature gradient through the tube is a measure of heat flux. Immediately after deslagging, when the tube is relatively clean, heat flux on the face of the tube is a maximum and Δt is 209°C. After slag accumulates for 2 hr, Δt is down to 94°C, and after 23 hr it is only 50°C. Hence, within 23 hr, the layer of slag accumulating on the tube has decreased heat transfer to one-fourth of its initial value. The same sort of action, but less severe because the side of the tube is not so directly exposed to furnace radiation, occurs 85° away from the face of the tube. These data demonstrate forcibly how slag decreases heat transfer, but they give little information on how the deposits form. In a slag-tap furnace, such as this, where molten droplets of slag are present in the furnace cavity, accumulation of slag on the wall tubes is evidently a relatively simple process. Accumulation of fly ash on superheater and reheater tubes where the flue gas and metal temperature are below the temperature at which the ash particles are fluid or "sticky" is less easy to explain.

Four physical processes influence the accumulation of ash particles on boiler surfaces: (1) molecular diffusion, (2) Brownian motion, (3) turbulent diffusion, and (4) inertial impaction. The size of the particles determines which of these processes predominates in a specific case, molecular diffusion referring to particles having velocities like those of gas molecules, and inertial impaction describing the action of particles so large that their path is largely unaffected by changing gas direction and velocity.

Many excellent reviews are available on particle dynamics affecting the deposition of solids in boiler furnaces, such as that by Samms and Watt (1966). The main conclusions that can be drawn at present are that volatile components in flue gas are deposited mainly by molecular diffusion; small particles are captured by eddy diffusion; and the largest particles deposit by impaction. There are no hard and fast rules, and indeed all these processes probably often take place simultaneously because of the heterogeneous nature of most dust suspensions in flue gas. Further, thermal effects and electrostatic forces may influence deposition in boiler furnaces, but neither is likely to be significant except in unusual circumstances.

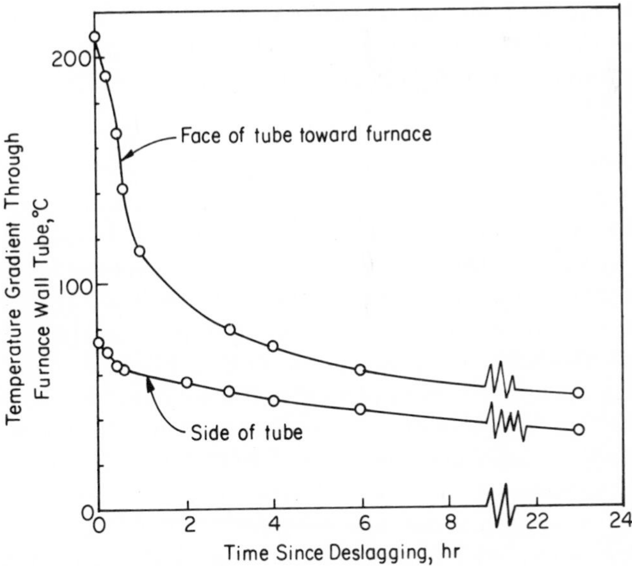

Fig. 8. Decrease in heat transfer caused by slag accumulation on furnace wall tube. [Data from Ely and Schueler (1944.)]

Deposition of solids, then, as on superheater and reheater elements, still poses many unsolved problems and, indeed, is poorly understood. Theoretical analyses have outstripped experimentation in boiler furnaces, and corrective measures are largely empirical.

VI. CORROSION CONTROL METHODS

Prevention of catastrophic failure of wall tubes in the 1940s was achieved by "ventilating" the tube surface. Most simply, this was done by modifying the air distribution through the burners so as to blanket the walls with a layer of air. The exact manner by which this simple expedient functioned was not understood at the time. It seems likely now that the air blanket over the wall tubes diluted the products of combustion reaching the interface between the slag layer and the tube metal, thereby lowering the SO_3 and alkali concentrations markedly and preventing the formation of the alkali iron trisulfates.

Such procedures are not possible with superheater tubes. In some cases, thin, loosely fitting, stainless steel sheaths have been clamped over superheater elements in particularly vulnerable positions. Because thermal conductivity is not high between the sheath and the superheater tubing, the sheaths are much hotter than the tube metal and the trisulfates do not form. These sheaths are relatively inexpensive. They can be replaced much more simply and at much lower cost than superheater elements.

A. Additives

Additives are useful in oil-fired boilers where the ash loading of the flue gas is low. Generally based on magnesia, corrosion-controlling additives function in two ways: (1) by neutralizing SO_3 through forming $MgSO_4$, and (2) by covering superheater surfaces with a layer of MgO to prevent the catalytic formation of large amounts of SO_3. Figure 9 shows how an MgO coating applied to a polished steel surface inhibited the development of a catalytically active surface, as occurred with unprotected bare steel (Barrett, 1967). In this case, after 5 hr at 643°C in the flue gas of an oil-fired laboratory furnace, a mild steel specimen was oxidized sufficiently to have increased the SO_3 in the flue gas from 47 ppm (the concentration when the test began) to 140 ppm. With an MgO coating on the specimen, the SO_3 concentration remained essentially unchanged.

When burning coal, the dust loading in the flue gas is so high that such

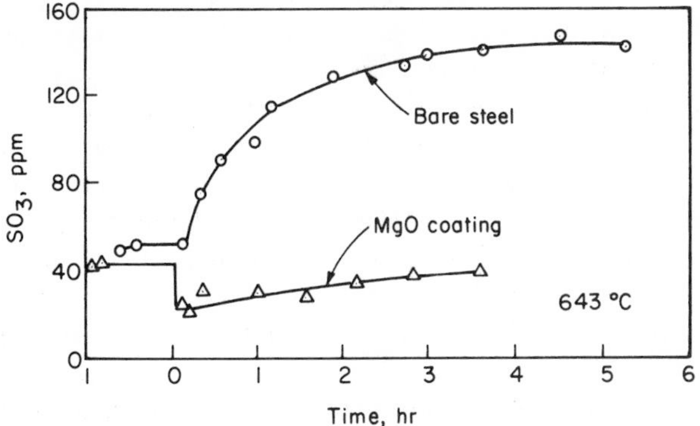

Fig. 9. Ability of MgO coating to prevent formation of SO_3 catalytically on iron surface. [After Barrett (1967).]

additives are impractical. However, air-pollution-control systems where pulverized limestone or dolomite is added to the boiler furnace to capture SO_2 are expected to prevent external corrosion. In such systems, the quantity of stone added is based on the sulfur content of the fuel; it often matches or exceeds the weight of ash in the coal. Hence the amount of additive can be very large. Because SO_3 is more reactive toward CaO and MgO compared with SO_2, it is likely that no SO_3 will exist and corrosion will cease. There is some indication that the accumulation of massive deposits of $CaSO_4$ on superheater surfaces may pose new problems.

B. Low Excess Air

A revolutionary approach has been taken over the past decade in Europe toward eliminating the formation of SO_3 in boiler furnaces fired with oil by limiting the excess air to an absolute minimum. Low excess air was proposed first in England as a means of decreasing corrosion and deposits when burning residual fuel. In 1960, highly favorable results were reported in Germany by Glaubitz (1960) who burned residual fuel oil with as little as 0.2% excess oxygen. By carefully metering fuel oil to each burner and properly adjusting air shutters, excess oxygen was lowered to as little as 0.1% before incomplete combustion became troublesome. By operating at these low levels of excess air, these boilers burned residual fuel oil for more than 30,000 hr without any corrosion and with no manual cleaning being required.

Low excess air in oil-fired equipment also has proven satisfactory in the United States and has been used successfully in large boiler plants. Precise metering of fuel and air to each burner has proven to be less troublesome than had been expected earlier, and in some instances, with high furnace turbulence, ordinary controls have been found satisfactory. In other cases, unburned combustibles have made low excess air undesirable. Sound principles guide the use of low excess air, but applying these principles gainfully is still largely a matter of judgment by boiler operators. It has been shown repeatedly, however, that SO_3 largely is eliminated, irrespective of the amount of sulfur in the fuel, when the products of combustion contain no more than about 0.2% oxygen. At this level, the dew point of the flue gas can be as low as 54°C when the dew point for the moisture in the flue gas is 41°C.

The benefits resulting from the use of low excess air include, in addition to a decrease in SO_3, a limitation on the oxidation of vanadium. Low excess air leads to the formation of V_2O_3 and V_2O_4, with melting points much higher than that of V_2O_5. These reduced forms of vanadium are less objectionable from the standpoint of corrosion. Opinion at present is that corrosion and deposits when burning residual fuel oil can be essentially eliminated by operating with low excess air.

VII. ULTRAHIGH-TEMPERATURE COMBUSTION

Ash-containing fuels are not basically suited for high-intensity combustion, where temperatures normally are much greater than in conventional combustion practice. In magnetohydrodynamic (MHD) combustors, for example, temperatures will be as high as 2800°C.

Large-scale magnetohydrodynamic systems certainly will demand conventional ash-containing fuels, probably bituminous coal. Natural gas, liquid distillate fuels, or de-ashed coal may be too expensive or too scarce for MHD generators operating in conjunction with a steam power plant. Hence, one of the problems with MHD systems is that the combustor, the MHD channel, and the secondary heat-recovery equipment must be able to handle ordinary fuels, with coal preferred. A further problem is that seed must be recovered in the presence of much larger quantities of coal ash. The development of an experimental MHD combustor in England beginning in 1964 demonstrated that control of slag was troublesome, and there was a "grievously difficult problem" in seed recovery (BCURA, 1966).

At the temperature of an MHD combustor, in a well-stirred reactor, or in any high-intensity system burning coal, a problem will be faced with volatilization of ash constituents. In the case of silica in the form of quartz, the volatilization of 40- to 50-μ particles decreases over the temperature range 2000 to 2150°C, but the rate of loss increases rapidly above 2200°C. Alumina and CaO with their vaporization temperature of 3500°C, and MgO with that of 3060°C, will not cause insuperable problems at 2800°C, but the other ash constituents may. The problem is not so much with these oxides in the gaseous phase as with their behavior as they cool and revert again to solids. Some preliminary evaluations have been made of ash behavior at these high temperatures, but much more experience will be necessary before the ash problem is settled. New concepts in high-intensity combustors burning coal will be judged largely on their ability to cope with ash.

References

ASTM, D 1857–68.

Barrett, R. E. (1967). *Trans. ASME J. Eng. Power Ser. A* **89,** 288.

Barrett, R. E., Hummell, J. D., and Reid, W. T. (1966). *Trans. ASME J. Eng. Power Ser A* **88,** 165.

BCURA Annu. Rep. (1966). p. 7.

Buckland, B. O., Gardiner, C. M., and Sanders, D. G. (1952). ASME Paper 52-A-161.

Burnside, W., Marskell, W. G., and Miller, J. M. (1956). *J. Inst. Fuel,* **29,** 261.

Cain, C., Jr., and Nelson, W. (1961). *Trans. ASME J. Eng. Power Ser. A* **83,** 468.

Corey, R. C. (1964). U. S. Bur. Mines Bull. 618.

Corey, R. C., Cross, B. J., and Reid, W. T. (1945). *Trans. ASME* **67,** 289.

Ely, F. G., and Schueler, L. B. (1944). Furnace Performance Factors. *Trans. ASME Suppl.* **66,** 23.

Glaubitz, F. (1963). *Combustion* **34,** No. 9, 25 (translation).

Goksøyr, H., and Ross, K. (1962). *J. Inst. Fuel* **35,** 177.

Harlow, W. F. (1944). *Proc. Inst. Mech. Eng.* **151,** 293.

Hoy, H. R., Roberts, A. G., and Wilkins, D. M. (1964). *Inst. Gas Eng.* Publ. 672.

Krause, H. H., Levy, A., and Reid, W. T. (1968). *Trans. ASME J. Eng. Power* **90,** Ser. A, 38.

Nicholls, P., and Reid, W. T. (1940). *Trans. ASME* **62,** 141.

Nicholls, P., and Selvig, W. A. (1932). U. S. Bur. of Mines Bull. 364.

Niles, W. D., and Sanders, H. R. (1962). *Trans. ASME J. Eng. Power* **84,** Ser. A, 178.

Raask, E. (1963). *In* "The Mechanism of Corrosion by Fuel Impurities" (H. R. Johnson and D. J. Littler, eds.), p. 150. Butterworths, London.

Reese, J. T., Jonakin, J., and Caracristi, V. Z. (1964). *Combustion* **36,** No. 5, 29.

Reid, W. T. (1970a). *Trans. ASME J. Eng. Power Ser. A* **92,** 11.

Reid, W. T. (1970b). "External Corrosion and Deposits, Boilers and Gas Turbines." Elsevier, New York.

Reid, W. T., and Cohen, P. (1944). Furnace Performance Factors. *Trans. ASME Suppl.* **66,** 83.

Reid, W. T., Corey, R. C., and Cross, B. J. (1945). *Trans. ASME* **67,** 279.

Samms, J. A. C., and Watt, J. D. (1966). *BCURA Mon. Bull.* **30,** 225.

Watt, J. D., and Fereday, F. (1969). *J. Inst. Fuel* **42,** 99.

III

Combustion Aerodynamics*

J. M. BEÉR

DEPARTMENT OF CHEMICAL ENGINEERING AND FUEL TECHNOLOGY
THE UNIVERSITY OF SHEFFIELD
SHEFFIELD, ENGLAND

Theoretical and experimental research on combustion aerodynamics is reviewed in this chapter. Procedures aimed at predicting distributions of velocity concentration and temperature in flames and local heat flux distribution to bounding surfaces require turbulent transport properties to be determined as a function of the local fluid state. Parallel with the solution of the mathematical problem, detailed measurements of the fluctuating properties in both isothermal flows and in flames are necessary for the development of a satisfactory effective viscosity hypothesis. Advances in experimental research include interpretation of hot wire anemometer signals in fields of high-intensity turbulence, the light scattering method, and measurement of the static pressure defect corresponding to the transverse component of turbulence intensity. In flames radical emission from the reaction zone, ionization of the fluid element, and both stagnation and

* The material in this chapter originally appeared in a paper by J. M. Beér, *VDI-Berichte* **146** (1969), under the title "Verbrennung und Feuerungen," and is used with permission of the copyright owner, VDI-Verlag, Dusseldorf, Germany.

static pressure measured by probes were used and the signals interpreted in terms of fluctuating components of velocity and concentration.

Flows with adverse pressure gradients leading to strong recirculation are of special interest to combustion engineers. Measurement data on turbulence characteristics in an isothermal stream in the wake of a bluff body are shown to indicate the strong spatial variation of eddy diffusivities and of turbulence scales. For rotating flows, it is shown that there are two significantly different flow types: rotating jets with recirculation in their central region in which very high turbulence shear stresses develop, and systems in which the rotation is imparted to the environment instead of the jet with the result of strong damping of turbulence and lengthening of the flame. Possible trends in future research are discussed, in particular the need for combining detailed experimental work on the spatial distribution of both time mean average and fluctuating components of flame properties with mathematical methods of prediction.

I. INTRODUCTION

In flame processes with flow, aerodynamic effects are significant. This is because important elements of the process, such as ignition, propagation, and stabilization of flames, depend greatly on the transfer of heat mass and momentum to which fluid flow, whether laminar or turbulent, is instrumental. Due to the considerable increase in rates of transport under turbulent flow conditions, aerodynamic effects are more prevalent, and when the fuel and oxidant are not premixed prior to ignition—the case of turbulent diffusion flames—aerodynamics become the rate-determining factor. This means that the most important characteristics of turbulent diffusion flames such as spatial distributions of concentration and temperature in the flame, its physical dimensions, can be determined from fluid flow considerations alone.

Early development of the theory and engineering of turbulent diffusion flames was helped by the fact that all turbulent jets can be considered to be similar, which in turn meant that the knowledge of the nozzle dimension, of the initial velocity of the jetting flow, and of the type of fuel burned was sufficient for predicting important parameters such as the flame length, velocity and concentration distributions, and so on. Because this simple but powerful generalization was valid only for the free axial constant density turbulent jet, efforts were made to extend the simple jet laws to more complex systems of nonconstant density compound or enclosed jet geometry, and so forth. The complicating factor was taken into account in the form of an "equivalent nozzle diameter," thus leading the problem back to that of the free constant density turbulent jet.

Modeling studies and the measurement of time mean average properties—velocities, static concentrations, and so on—in flames enabled descriptions of flames to be made in terms of flow patterns, streamline distributions. The importance of recirculating flows to flame stability and to combustion intensity has become evident and interest has turned to the practical ways in which, as a result of an adverse pressure gradient in the flow, recirculating flows can be obtained: enclosed jet flows, wakes behind bluff body and rotating flows.

In the first instance the experimental information available on these recirculating systems was also in the form of streamline patterns and static pressure distributions, spatial distributions of time mean values of flame properties. More recently development of computational techniques for predicting velocity and concentration fields in boundary-layer-type flows and improvements in methods of measurement of turbulent characteristics have given new directions to research in combustion aerodynamics.

II. THEORY

A description of a flow field in terms of the spatial distribution of velocities, concentrations, temperatures, and so on, has to be based on statements of conservation of mass, of momentum, and of energy. For the case of axisymmetric steady, incompressible flow and using cylindrical coordinates where r, ϕ, and x denote radial azimuthal and axial coordinates, and v, w, and u are radial, tangential, and axial components of the velocity, the continuity equation can be written as

$$\partial u/\partial x + \partial v/\partial r + v/r = 0 \tag{1}$$

and the force balance on a volume element yields (Navier–Stokes equation)

$$\begin{aligned}\rho[u(\partial u/\partial x) + v(\partial u/\partial r)]\\ = (-\partial p/\partial x) + \mu[(\partial^2 u/\partial x^2) + (1/r)(\partial u/\partial r)(\partial^2 u/\partial r^2)]\end{aligned} \tag{2}$$

$$\begin{aligned}\rho[u(\partial v/\partial x) + v(\partial v/\partial r)]\\ = (-\partial p/\partial r) + \mu[(\partial^2 v/\partial x^2) + (1/r)(\partial v/\partial r) - (v^2/r) + (\partial^2 v/\partial r^2)].\end{aligned} \tag{3}$$

After introducing the notation of $u = \bar{u} + u'$ and $v = \bar{v} + v'$, where the instantaneous velocity is given as a sum of the time average and the fluctuating component, and assuming that jet and wake flows are boundary-layer-type flows, that is, changes in flow properties along the flow are small compared with radial gradients, that is, $\partial/\partial r \gg \partial/\partial x$, the equation of continuity becomes

$$(\partial\bar{u}/\partial x) + (\partial\bar{v}/\partial r) + (\bar{v}/r) = 0 \tag{1a}$$

and Eq. (2) becomes

$$\bar{u}(\partial\bar{u}/\partial x) + \bar{v}(\partial\bar{u}/\partial r) = (1/r)(\partial/\partial r)[r\epsilon(\partial\bar{u}/\partial r)], \tag{2a}$$

where the turbulent shear stress term $u'v'$ has been substituted for by the product of the eddy viscosity and the radial gradient of the axial velocity.

It can be shown (Schlichting, 1968) that the momentum equation in the radial direction [Eq. (3)] can be reduced with good approximation to

$$\bar{p} - p_0 = -\rho v'^2. \tag{3a}$$

For the treatment of the problem of turbulent flows they can either be considered in their full complexity, which requires the solution of the partial differential equations for time-dependent motion, or they can be treated as though they were steady laminar flows with transport properties which varied in magnitude from place to place.

The solution of these differential equations for suitable boundary conditions yields the detailed velocity distribution. Although the equations describe turbulent flow there are only time mean average values in them. Because of turbulent flow conditions, however, the effective or turbulent viscosity ϵ is not a property of the fluid and will vary spatially. In order that the differential equations represent the correct physical description of turbulent fluid motion, it is necessary therefore that the effective viscosity is related to the local fluid state.

Recent publications by Spalding and co-workers indicate that there are computational procedures available for the approximate solution of the differential equations. It seems, however, that the success of the predictions depends greatly on the choice of an effective viscosity hypothesis. Clauser (1954), following measurements on shearing stresses, supposes that the effective viscosity is proportional to the product of the stream velocity and displacement thickness. The displacement thickness is defined as

$$\delta = \int_{r=0}^{r=\infty} [1 - (u/u_\infty)]\, dr.$$

The mixing length theory of Prandtl (1925) is based on the analogy between molecular and turbulent motion and assumes that the mixing length is proportional to the boundary layer thickness. According to a hypothesis by Kolmogorov (1942) and Prandtl (1945) the effective viscosity is proportional to the square root of the local kinetic energy of turbulence, while Townsend (1961) and Bradshaw *et al.* (1967) assume a direct proportionality between kinetic energy of turbulence and effective viscosity. None of these hypotheses can claim to be valid generally: For example, the displacement thickness hypothesis is not applicable when the velocity profile exhibits a maximum; the mixing length hypothesis breaks down in a flow region without a velocity gradient in which turbulence is generated by a grid. In more recent work on prediction procedure (Gossman *et al.*, 1968) the Kolmogorov–Prandtl hypothesis seemed to have satisfied conditions for a wide range of recirculating flows. It was assumed by Spalding and his group (Gossman *et al.*, 1969) after Kolmogorov (1942) and Prandtl (1945) that properties of turbulence can be characterized by two quantities, the kinetic energy of turbulence k and the length scale of the turbulence l. Thus the effective dynamic viscosity can be given as

$$\mu_{\mathrm{eff}} = C_{\mu}\rho k^{1/2} l, \tag{4}$$

where C_{μ} is a constant which depends on the local Reynolds number of turbulence $\mathrm{Re_t}$, where

$$\mathrm{Re_t} = \rho k^{1/2} l/\mu. \tag{5}$$

In fully turbulent flow C_{μ} attains a value of unity. As $R_{\mathrm{t}} \to 0$ the total fluid viscosity $\mu + \mu_{\mathrm{eff}}$ tends to μ. The l and k are calculated from balance equations and it was assumed that l and k diffuse at rates proportional to their respective gradients. Analogously to the formulation of the concept of effective viscosity, effective Prandtl and Schmidt numbers can also be determined. It follows from the line of argument above that for the case of very large values of R_{t} it should be expected that the value of these "effective" transport numbers tends toward a constant. There is considerable uncertainty and conflicting evidence about the values of the effective Schmidt and Prandtl numbers and even more so about values of transport coefficients for the kinetic energy of turbulence k and its length scale l. Further development of the hypothesis linking local fluid state and turbulent transport coefficients requires the comparison of predictions with detailed measurement data covering wide ranges of input conditions.

III. MEASUREMENT OF TURBULENCE CHARACTERISTICS

Measurements of time-averaged properties of flames have been widely reported in the literature: For small laboratory-scale flames, Fristrom and Westenberg (1965) developed both microsampling and optical methods; Weinberg (1963) reports on optical measurements and, in a recent paper, on the use of lasers (Schwar and Weinberg, 1969). Optical and probe

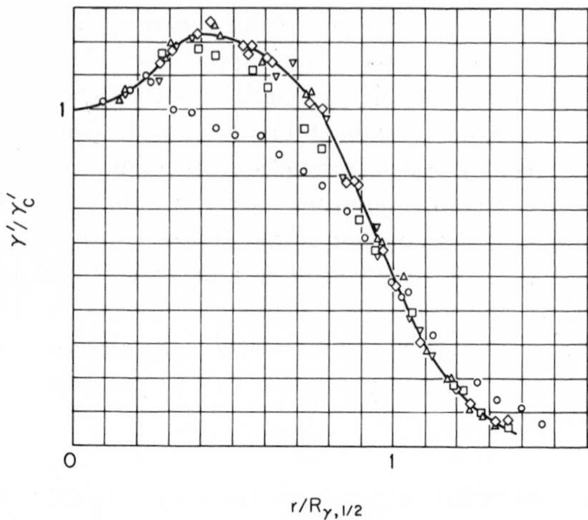

Fig. 1. Normalized profiles of concentration fluctuations. [After Becker *et al.* (1967).]

measurement techniques applicable to industrial-sized flames are described in several publications of the International Flame Research Foundation (Kissel, 1960; Thring and Beér, 1962; Pengelly, 1962).

In this chapter we intend to discuss the measurement of time-dependent properties in turbulent flow, in particular that of fluctuating velocity, temperature, and pressure.

The difficulties with such measurements are partly due to the high

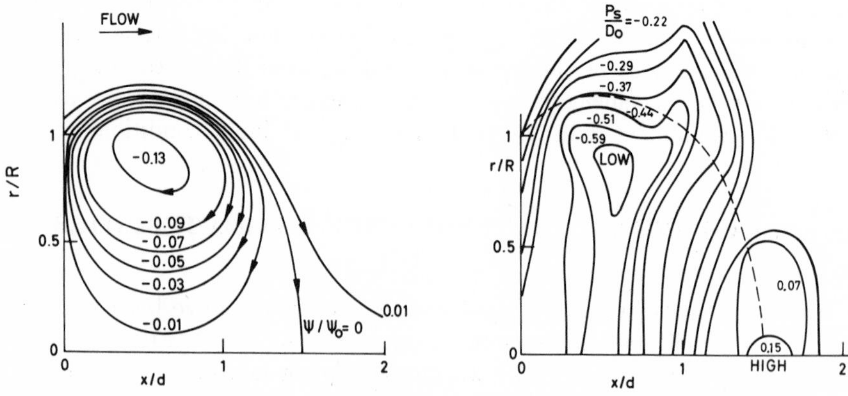

Fig. 2. Spatial distributions of stream function and static pressure in the wake of a disk. [After Davies and Beér (1973).]

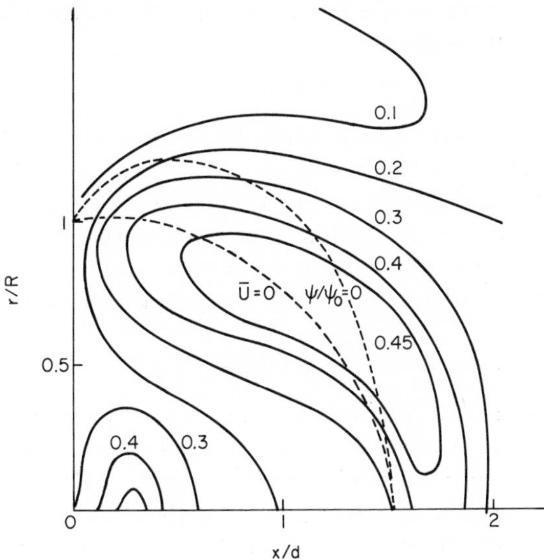

Fig. 3. Spatial distribution of $(\overline{u'^2})^{1/2}/\bar{u}$ in the wake of a disk. [After Davies and Beér (1973).]

intensity of turbulence, particularly in wakes and recirculating flows—and also due to the high temperature in flames.

In constant-temperature and constant-density flow conditions the most generally used research tool is the hot wire anemometer. The basic measurement is the rate of heat loss from the heated wire to the gas stream. Based on a heat transfer relationship derived by King (1914) the interpretation

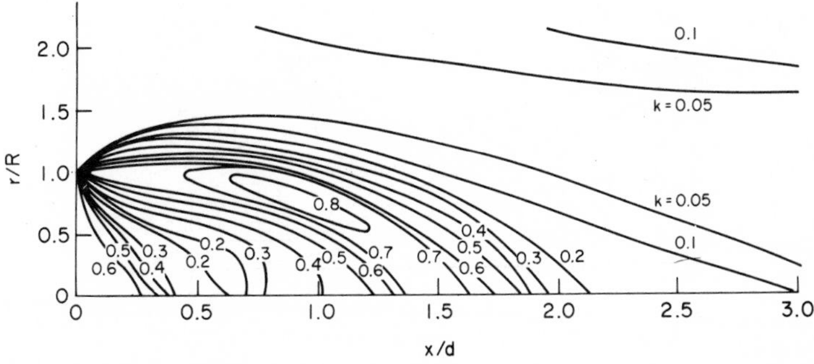

Fig. 4. Spatial distribution of local kinetic energy of turbulence (k) in the wake of a disk. [After Davies and Beér (1973).]

of electrical signs obtained in a bridge in which the wire is maintained at constant temperature is given by

$$E^2 = A + B(U_e)^n, \tag{6}$$

where E is the indicated wire voltage, U_e is the total effective cooling velocity, and A and B are constants.

Recently Siddall and Davies (in press) have developed a new method of analysis which is based on an improved form of the steady-state heat transfer relationship for a heated wire situated normally to a laminar flow. The recommended relationship between wire voltage and effective velocity is given as

$$E^2 = A + BU_e^{1/2} + CU_e. \tag{6a}$$

This relationship has been checked experimentally. In regions of turbulent flow with turbulence intensities smaller than 20% good agreement was found with results calculated from King's formula. In highly turbulent regions of flow, however, the conventional methods of analysis indicated turbulence intensities much higher than those computed from the new method of analysis. Although it is clear from the less restrictive assumptions underlying the derivation of this new method that it can be expected to yield more satisfactory results than the conventional formulas, for its full assessments, results obtained by its use would have to be checked against an independent type of measurement of turbulence characteristics, preferably an optical method.

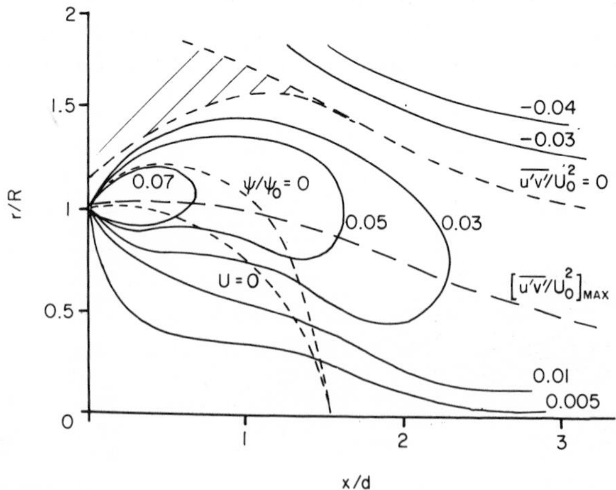

Fig. 5. Spatial distribution of $(u'v')/\bar{u}^2$ in the wake of a disk. [After Davies and Beér (1973).]

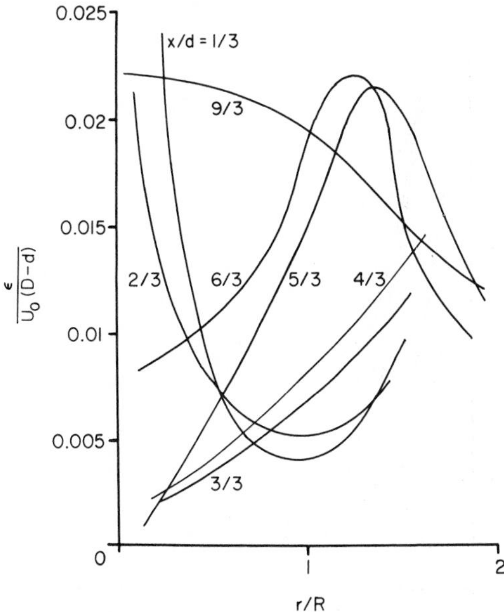

Fig. 6. Radial distributions of Boussinesq eddy viscosity in the wake of a disk. [After Davies and Beér (1973).]

Becker *et al.* (1963, 1965, 1967) in a series of publications reported on the use of a light scattering method for determining intermittency of concentration and nozzle fluid concentration fluctuations in a cold ducted jet system. The nozzle fluid was labeled by the introduction of an oil fog. Scattered light was picked up at right angles to a pencil beam of light traversing the jet. The multiplier phototube in the scattered light detection system produced a voltage signal proportional to the space average instantaneous concentration of oil fog in a small volume of 1 to 2 mm diameter under observation. Townsend (1949) named the fraction of time during which a point at the boundary of a turbulent jet is inside the turbulent fluid the "intermittency factor." In jets without recirculation the intermittency is the complement of the probability distribution function of the radial location of the jet boundary and can be given as

$$\beta = 1 - F(r) = \tfrac{1}{2}\{1 - \mathrm{erf}[(r - \bar{R})/\sigma\sqrt{2}]\}, \tag{7}$$

where r is the radial distance, $\bar{R}$ is the mean position of jet boundary, and σ is the standard deviation of R.

Since σ is a measure of the average size eddies at the jet boundary, scale of turbulent flows could be determined (Townsend, 1949) from inter-

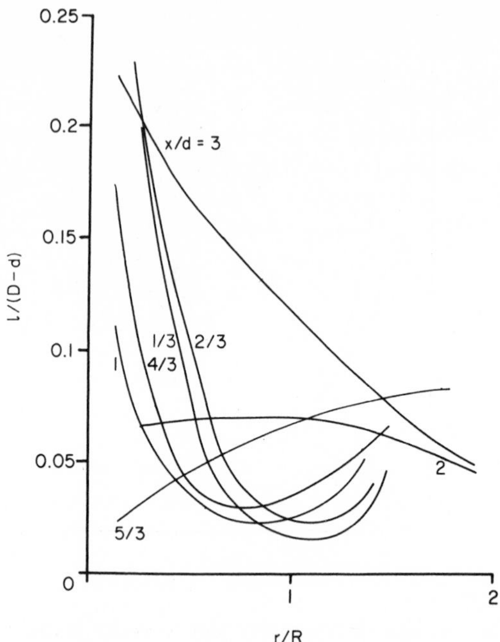

Fig. 7. Radial distributions of Prandtl mixing length in the wake of a disk. [After Davies and Beér (1973).]

mittency measurements. Becker *et al.* (1965) found for free isothermal jets that $\bar{R} = 1.57R_{\gamma/2}$ and $\sigma = 0.177\bar{R}$, where $R_{\gamma/2}$ is the radius at which the concentration is half of that on the jet axis. The normalized profiles of concentration fluctuations measured by the same technique are represented in Fig. 1. Here γ' is the rms value of the fluctuating component of nozzle fluid concentration, γ_c' is the rms value on the jet axis and the normalized jet radius, and $R_{\gamma,1/2}$ is the radius at which $\gamma' = \gamma_c'/2$. The light scattering technique also enables correlation coefficients to be determined by simultaneously observing two points at the same light beam (Becker *et al.*, 1967).

Using the improved method of analyzing hot wire anemometer signals in high-intensity turbulence flow Davies (1969) has carried out a comprehensive study of isothermal annular wake flows. Davies and Beér (1971) reported on measurements of time average and of turbulence characteristics in the wake of bluff-body stabilizers of varying forebody geometry and blockage ratio.

The spatial distribution of the mean stream function calculated as $\psi = \int_0^\infty \bar{u}r\,dr$ is represented for a disk in an annular flow in Fig. 2. Figure

3 represents the spatial distribution of the turbulence intensity $(\overline{u'^2})^{1/2}/\bar{u}$, and Fig. 4 shows the distribution of local kinetic energy of turbulence $k = (\overline{u'^2} + \overline{v'^2} + \overline{w'^2})/\bar{u}^2$. The normalized shear stress distribution $\overline{u'v'}/\bar{u}_0^2$ in the wake of a disk is given in Fig. 5.

Correlations between mean flow and turbulence characteristics were used to draw conclusions about various turbulent mixing hypotheses. Figures 6 and 7 were constructed to show the variation of the Boussinesq eddy viscosity, defined as $-\rho\overline{u'v'} = \epsilon(d\bar{u}/dr)$, and the Prandtl mixing length, defined as $l = (\overline{u'v'})^{1/2}/(d\bar{u}/dr)$.

Runchal (Gossman *et al.*, 1968), using Spalding's new model of turbulence based on theories by Prandtl and Kolmogorov, has predicted the correlation between stream function and local kinetic energy of turbulence

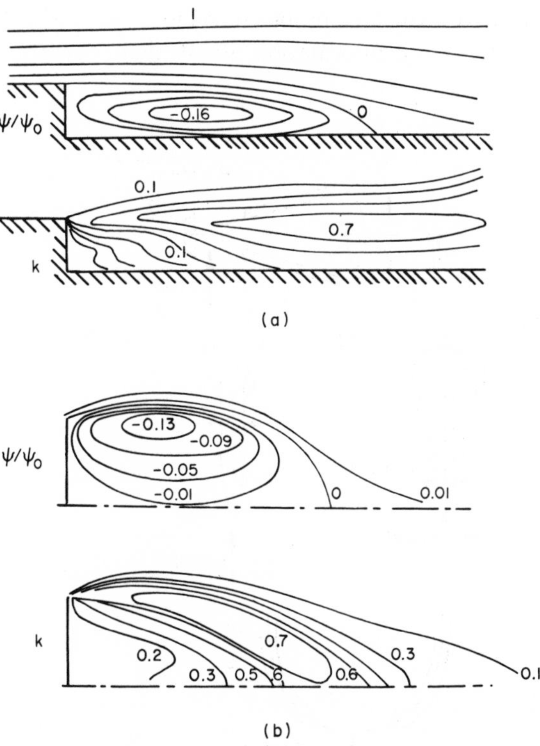

Fig. 8. (a) Correlation between *predicted* stream function and local turbulence kinetic energy fields downstream of a sudden pipe expansion. [After Davies and Beér (1971, 1973).] (b) Correlation between *measured* stream function and local turbulence kinetic energy fields downstream of a disk-in-annulus. [After Davies and Beér (1971).]

for a sudden enlargement in a pipe. These predicted correlations are shown together with those determined experimentally for a disk-in-nozzle system in Fig. 8. Although the two systems are somewhat different in geometry the good qualitative agreement between the stream function and kinetic energy of turbulence distributions as predicted and as measured holds out good hope for the application of this model of turbulence in prediction methods. On the other hand, Figs. 6 and 7 show the inadequacy of simple models of turbulence for prediction in the intermediate zone behind the disk.

A. Measurement of Turbulence Characteristics in Flames

A series of experimental studies of turbulent, reacting systems were carried out more recently at the University of Karlsruhe (Kremer, 1964; Günther and Simon, 1969; Eickhoff, 1969) and reported by Günther (1969). Schematic drawings of probes developed at Karlsruhe are shown in Fig. 9 and some basic information about their operation is given in Table I.

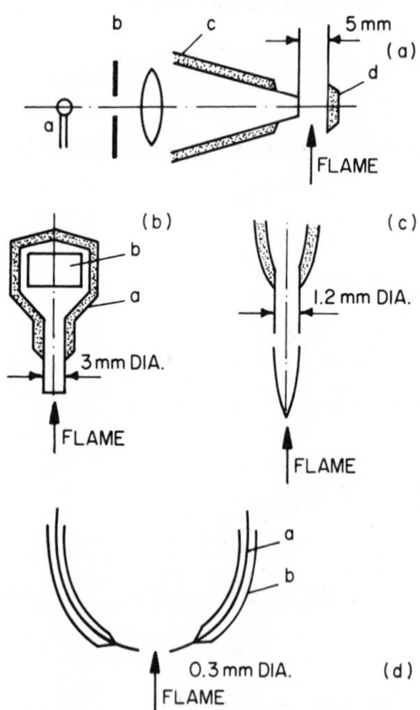

Fig. 9. Schematic drawings of probes for the measurement of turbulence in flames. [After Günther (1969).]

TABLE I

METHODS OF TURBULENCE MEASUREMENT DEVELOPED AT KARLSRUHE[a]

Method	Diagrammatic drawing in Fig. 9	Measured quantity	Size of observed volume (mm diameter)	Interference with flame
Emission probe (photodiode)	a	Radical emission 300 to 1200 nm	5	Significant
Microphone probe	b	Stagnation pressure giving axial velocity fluctuations (u')	2–3	Significant
Static pressure probe	c	Static pressure giving transverse velocity fluctuations (v')	2	Small
Ionization probe	d	Electrical current giving ion concentration	Very small	Very small

[a] From Günther (1969).

In connection with the interpretation of measurements based on radical emission from the reaction zone, Günther (1969) warns that the results are not unambiguous: Even if it is assumed that the flame front coincides with the loci of boundaries of fuel-rich and air-rich fluid particles and that the reaction is confined to the surface of fluid particles or eddies, it does not follow that each fluid particle is reacting. In the initial region of the jet neighboring particles can be both fuel rich, around the edge of the jet they can be both air rich, and toward the end of the flame they can be composed of fully burned combustion products. Emission measurements give therefore a picture of the reaction zone rather than the flow field.

An approximate analysis of the equation of motion yields for the radial component of the momentum equation the relationship

$$\bar{p} - p_0 = -\rho\overline{v'^2}, \tag{3a}$$

which shows that a static pressure defect exists in jet flows which in turn can be related to the transverse velocity fluctuations. Becker and Brown (1969) have used static pressure measurements for the determination of distribution of the Reynolds stresses both in a free jet and in a free propane–air diffusion flame. They found that profiles of $\rho v'^2$ and of $\rho w'^2$ in the diffusion flame are quite similar to those in the air jet but the levels of turbulence are as much as 50% lower. This statement is in qualitative agreement with some of the results by Günther and Simon (1969) who used the radical emission method and obtained spectral density distributions from flames. These latter could then be compared with spectral density distributions of fluctuating velocity and concentration measured in isothermal jets. They showed that in the frequency range above 3000 Hz the intensities in flames decay more rapidly than in isothermal jets, and have attributed this to the high viscosity of the flame gases. Günther and Simon have calculated longitudinal Eulerian scales of turbulence from their measurements of spectral density functions. These together with intensities of turbulence measured by Eickhoff (1969) are representative of macroscopic turbulence in flames. Figure 10 represents the axial distributions of intensities of turbulence $(\overline{v'^2})^{1/2}/\bar{u}$ determined in turbulent diffusion flames by Eickhoff using the static pressure measurement method compared with the distribution in an isothermal jet. The turbulence intensity increases at first in lifted town-gas flames due to the lower density of the nozzle fluid, but the turbulence rapidly drops in the reaction zone and then increases steadily

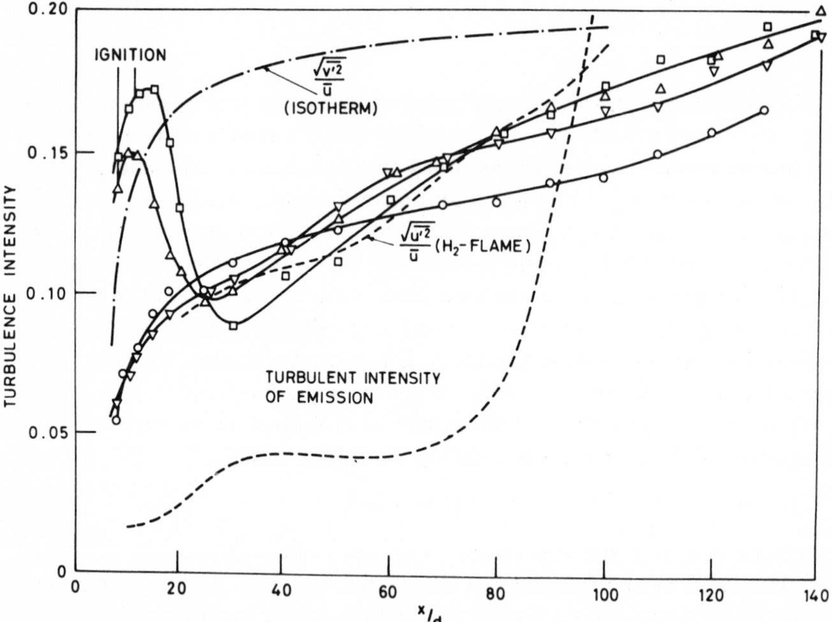

Fig. 10. Axial distributions of turbulence intensities determined in turbulent diffusion flames. Isotherm flames are shown by solid and dash–dot curves, flames of other methods by dashed curves. For lifted flames: □ $Re_0 = 7.14 \times 10^4$; △ $Re_0 = 5.05 \times 10^4$. For attached flames: ▽ $Re_0 = 5.05 \times 10^4$; ○ $Re_0 = 3.2 \times 10^4$. [After Eickhoff (1968).]

along the jet flame; the level of turbulence intensity is, however, generally lower than in isothermal jets.

IV. FLAMES IN ROTATING FLOWS

While swirl burners are not new to combustion, it is only recently that a concentrated effort was made to understand better how and why rotating flow has such an important influence on the stability and combustion intensity of flames.

The physical systems of interest can be divided into three groups: first, turbulent swirling jet diffusion flame with weak swirl; second, strongly swirling jet flame with torroidal recirculation vortex; and third, buoyant or turbulent jet diffusion flame in a rotating flow environment.

For the first case, the region of interest is the part of the jet boundary layer which lies between the edge of the jet and the stagnant surroundings, where the angular momentum decreases from a maximum near the edge of the jet to a zero in the surroundings. The boundary layer is here highly unstable, and the intensity of turbulence is high, which in turn increases the rates of mixing between fuel and oxidant in a pure diffusion flame, with the effect of shortening the flame length. The high radial transfer of angular momentum in this region has an effect also on premixed flames, the high-intensity turbulence makes a contribution to increasing the flame speed and thereby increasing the intensity of combustion, but the effects in this case will not be as significant as in pure diffusion flames.

Chigier and Chervinsky (1967) have shown that after integrating the turbulent equations of momentum with suitable boundary conditions, these could be reduced to expressions of the conservation of the axial fluxes of linear and angular momenta. By making further assumptions about the similarity of velocity, density, and temperature profiles, the decay of the maximum values of axial and swirl velocity and of temperature could be described. These expressions together with the semiempirical ones describing the radial spread of these properties could then be used for the detailed description of velocity, air temperature distributions in a jet with a weak swirl.

The case of great practical significance is that of swirling flow with recirculation. A detailed account of such flows was given in reports and in several publications of the I.F.R.F. (Beér and Chigier, 1963; Chigier and Beér, 1964; Chedaille *et al.*, 1966; Leuckel, 1967; Fricker and Leuckel, 1969). Both the theoretical description of the flow and experimentation are more difficult in this case, partly because the region of greatest interest is very close to the nozzle, the profiles of velocity cannot be considered to be similar, and the intensity of turbulence in the reverse flow region is very high, thus creating considerable experimental difficulties.

From the study of the swirling flows, two simple but rather significant generalizations could be made regarding the stability and combustion intensity of flames. It was shown that the strength and position of the recirculating vortex had a significant influence on the stability of flames. Also, it was postulated from the studies at Ijmuiden that for cases of non-premixed flames, high combustion intensities can be obtained by adjusting the fuel and airflow so that there are high fuel concentrations in the region of large velocity gradients, that is, in the regions where the intensity of turbulence is the highest. For the case of swirling jets with recirculation this is the region between the recirculating vortex flow and the nozzle fluid flow.

Recently Johanssen (1967) carried out a detailed experimental study at Trondheim University on swirl burning. In his experiments he has measured the turbulent stress and intensities of turbulence in swirling cold jets and then injected an oil spray into the swirling air jet, and measured combustion characteristics of the flame while varying the spatial distribution of the fuel relative to the air jet. He concluded that the criterion for high-intensity combustion as mentioned above was correct. Highest combustion intensi-

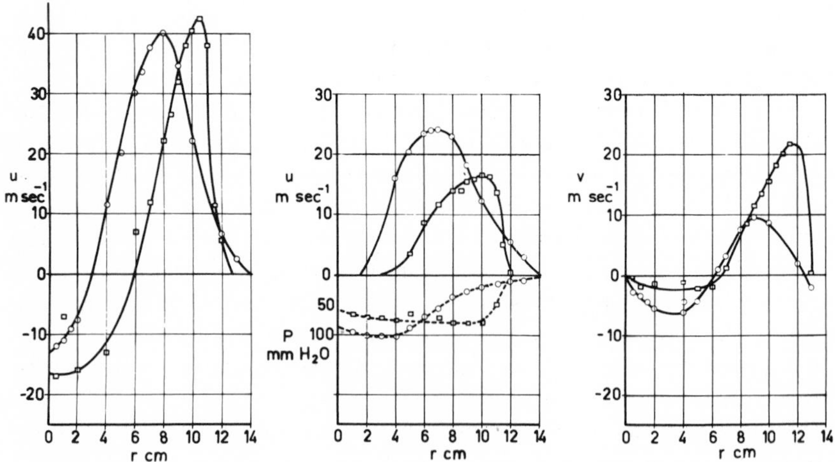

Fig. 11. Radial distributions of the axial, tangential, and radial components of the velocity in jets with strong swirl. ○ Convergent only; □ convergent–divergent. [After Beér and Chigier (1963).]

ties were achieved when high spray concentrations overlapped regions of the flow in which the turbulent stress was maximum.

A. The Effect of Nozzle Geometry

The Ijmuiden studies of swirling flames have also shown the significance of nozzle geometry in determining the strength and position of the recirculating vortex. Figures 11 and 12 show the radial distributions of the axial, tangential, and radial components of the velocity in the jet and the position of the reverse flow region, respectively, for two jets with the same swirl degree, but with different exit nozzle geometry. The use of the divergents has the combined effects of increasing both the radial distance of separation between the peaks of axial and tangential velocities, and the

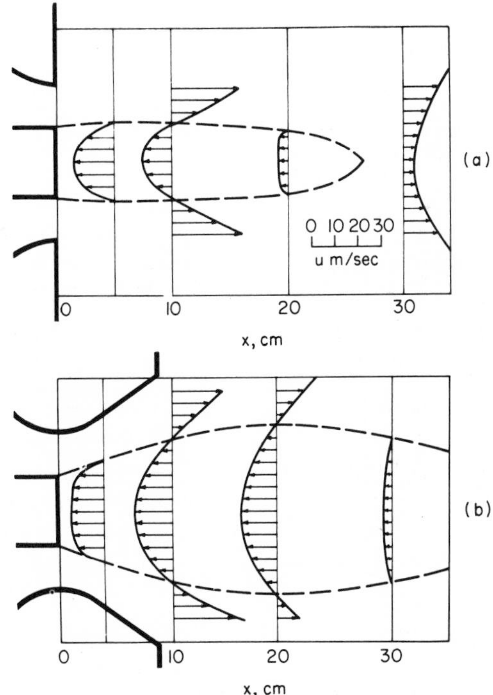

Fig. 12. The effect of a divergent nozzle upon the size and position of recirculating flow in strongly swirling jets. (a) Without divergent nozzle; (b) with divergent nozzle. [After Beér and Chigier (1963).]

intensity of turbulence and entrainment between the nozzle flow and the recirculating eddy, because of the wall jet character of the flow in the nozzle. A mathematical treatment of the effect of change of cross section of the tube on a stream of rotating fluid is given by Batchelor (1967). The qualitative changes in u and w, the axial and tangential velocity components, due to expansion of cross section of flow, can be explained in terms of the shape of the vortex lines: Because of conservation of linear and angular momentum, as one end of the vortex line passes into transition and moves outward in a divergent nozzle, the velocity of the material point on the vortex line changes according to the rule wr = constant. This causes a negative gradient in the vorticity, which in turn yields a positive value of the radial gradient of the axial velocity du/dr. Thus the expansion in the cross section produces a minimum of the axial velocity on the axis. A practical application of the high combustion intensity criterion is the multi-

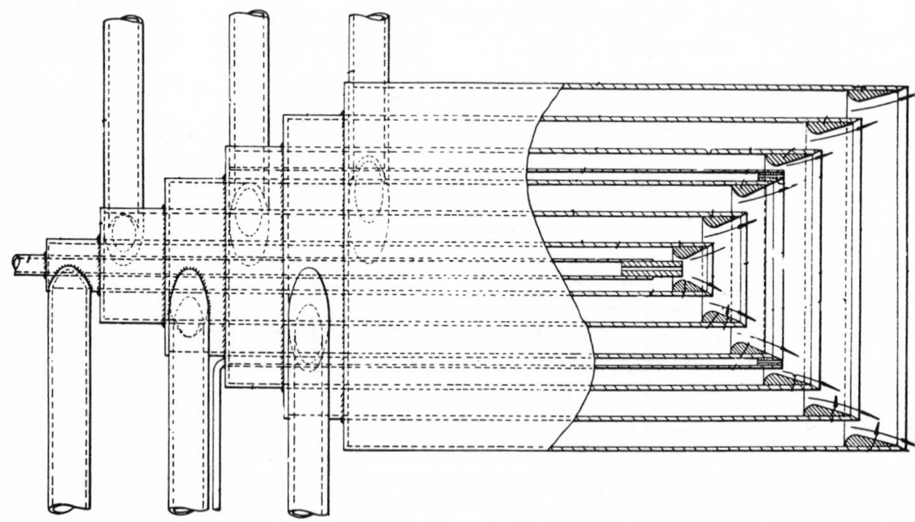

Fig. 13. Diagrammatic arrangement of Multiannular Divergent Nozzle Burner. [After Beér (1965).]

Fig. 14. The Multiannular Divergent Nozzle Burner in operation burning heavy fuel oil.

annular nozzle burner in which layers of high shear are produced in the form of a Swiss roll (Beér, 1965). This arrangement enables a better control of the velocity distribution in the boundary layer of a divergence, and enables also very high combustion intensities to be obtained within the confines of the divergent nozzle. Figure 13 represents the diagrammatic burner arrangement, and Fig. 14 shows the burner in operation with a combustion intensity of about 7×10^6 Btu/ft^3 hr obtained with a very crude atomization of a distillate fuel.

B. Buoyant or Turbulent Jet Diffusion Flame in a Rotating Flow Environment

The typical system of this type consists of a buoyant column of gas surrounded by a rotating environment. The conclusions reached in the study of swirling jet flames as summarized above are, almost without exception, invalid for this case. Emmons and Ying (1967), who have studied the fire whirl from a liquid fuel pool formed at the center of a rotating screen, gave a clear description of this type of flow. The phenomenon of fire whirl is produced by a concentrating mechanism which brings vorticity together into the vortex core of the rotating system. In the case of flames, the concentrating process is the rising gas in the buoyant column. In addition to this, the boundary between the flame and the surrounding air is stable because the centrifugal force opposes entrainment of the air into the rising column once it is above ground level. Vorticity is concentrated into the core, where again the fluid is stable with regard to radial interchanges of fluid as shown by Rayleigh (1916) (rotating fluid is stable if ρwr increases with r). The wall boundary layer on the ground also plays a significant role by slowing down the rotation of the fluid and causing thereby an imbalance between the centrifugal and pressure forces: Due to the radial pressure gradient, the flow attains a radial velocity component at the ground level. In the case of pure diffusion flames where only fuel is present in the center of the whirl, this is most important, as the rate of radial inflow of air determines the rate of burning of the fuel. As an extension of the Emmons investigation, an experimental study has been carried out at Sheffield (Chigier *et al.*, 1970) on a system where the pool burning in the center of the rotating screen is replaced by a turbulent jet. The experimental apparatus is shown in Fig. 15. A cylindrical wire mesh screen 2 ft in diameter and 4 ft high is driven by a variable-speed motor and can be rotated at speeds of 4 to 40 rpm. The burner is fitted centrally so that the jet fluid is introduced along the axis of the screen.

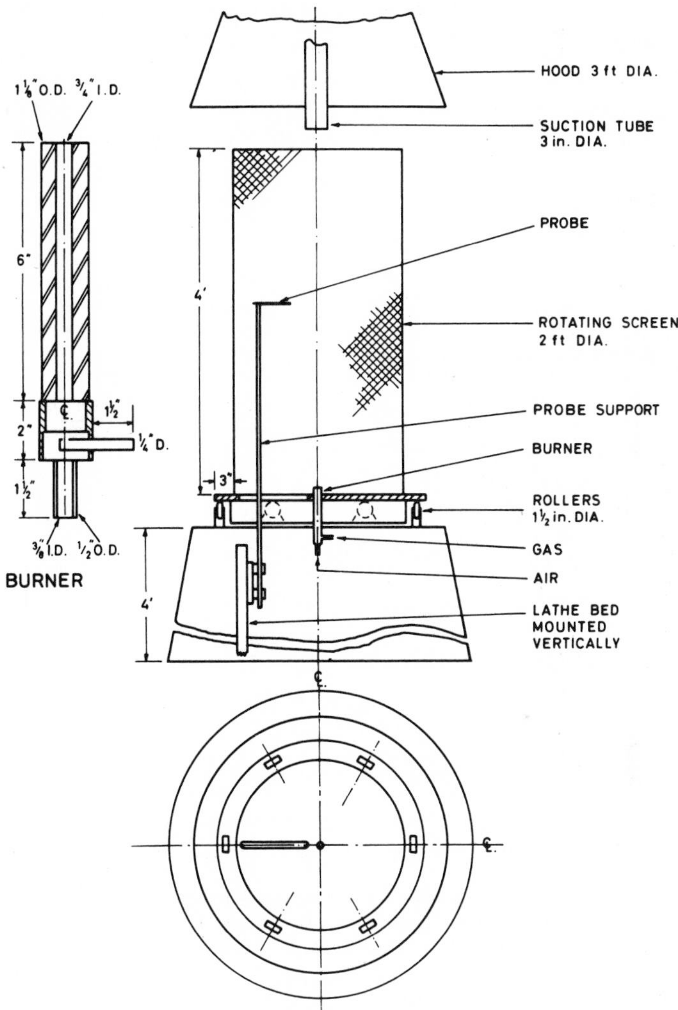

Fig. 15. Rotating screen apparatus. [After Chigier *et al.* (1970).]

1. Flame Length

Flame lengths for methane diffusion flames in the range of Reynolds numbers 400 to 5000 are shown in Fig. 16 for a burner exit diameter, $d = 3.17$ mm, where the ratio of flame length to nozzle diameter is plotted as a function of Reynolds number based on nozzle exit conditions. Flame length measurements for zero circulation were found to lie between the

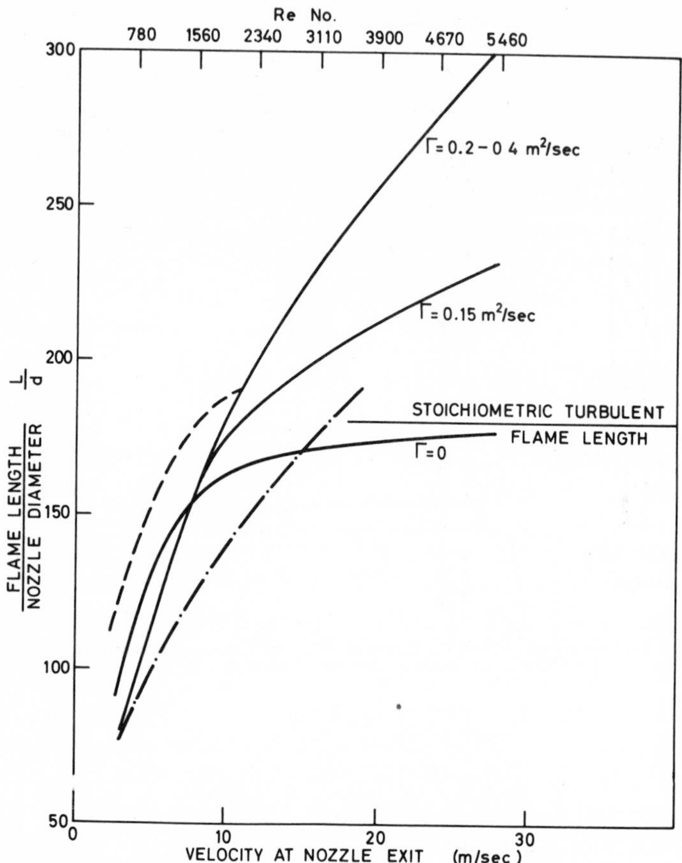

Fig. 16. Flame lengths for methane in a rotating atmosphere. Nozzle diameter d = 3.17 mm. Dashed curve is plotted from data of Hottel and Hawthorne (1951), dash–dot curve from Wohl *et al.* (1951). [After Chigier *et al.* (1970).]

values calculated from the empirical formulas of Hottel and Hawthorne (1951) and that of Wohl *et al.* (1951). These flames were laminar in the initial region near the nozzle exit, followed by a turbulent brush. Increases in flame length are mainly due to the effect of rotation upon the turbulent brush of the flame.

Hottel and Hawthorne (1951) made an analysis of combustion and diffusion in laminar flame jets and proposed a relation for flame length of the form

$$L = A \log(Q - \theta_f) + B, \tag{8}$$

where L is flame length; A and B are constants depending only on fuel gas and primary air–fuel ratio; θ_f is a generalized time, defined as

$$\theta_f = 4D_v t_f / d^2, \tag{9}$$

where t_f is the time for gas to flow from nozzle to flame tip, D_v is molecular diffusivity coefficient, and d is the nozzle diameter, or

$$\theta_f = 1 - 4 \ln(1 - C_f),$$

where C_f is the unaccomplished fractional change in concentration at the flame tip. For methane, $C_f = 0.095$, $\theta_f = 2.7$.

Equation (8) can be reduced to a nondimensional form

$$L/d = a \log(\text{Re}\, \theta_f) + b, \tag{8a}$$

where Re is the Reynolds number based on conditions at the nozzle exit.

For the results shown in Fig. 16 the flame length data for $\Gamma = 0$ agree in form with those of Hottel and Hawthorne in the laminar region with small differences in the constants a and b. At the higher Reynolds number values (Re > 4000) the results agree with those of Hawthorne *et al.* (1951) for stoichiometric turbulent flame lengths. It may be noted that flame length was found to increase continuously as flow rate was increased until the turbulent flame length was reached. The decrease in flame length found by Hottel and Hawthorne for city gas flames in the transition region was not found in the experiments of Chigier *et al.* (1970). The values of constants a and b in Eq. (8a) are listed in Table II.

It was found that as the circulation was increased flame lengths increased until a maximum value was reached. Further increase in rotational speed leads to decreased stability and ultimately to decreases in the flame length. Flame lengths were increased by a factor of 2 (Fig. 16) at the higher Reynolds number range. These increases in flame length are less than those reported by Emmons and Ying (1967).

In diffusion flames, flame length is essentially dependent on the rate of

TABLE II

VALUES OF a AND b IN EQUATION $L/d = a \log(\text{Re}\, \theta_f) + b$

Gas	m²/sec	a	b
Methane	0	100	−225
Methane	1.0	315	−950
Methane	1.5	470	−1575
Methane (Hottel and Hawthorne, 1951)	0	170	−337

mixing between fuel and air. Entrainment of air into the core is reduced and since fuel is introduced and remains within the vortex core, the consequent delay in mixing results is increased flame length.

2. Flame Stability

Flame stability measurements were made by setting the fuel flow rate to a fixed value, rotating the screen at the required speed, and after the vortex was allowed to stabilize the airflow rate was increased until blowoff occurred. Figure 17 shows the blowoff velocities for propane–air flames as a function of fuel concentration in the burner, for varying degrees of circulation. Rotation is seen to lead to very large increases in blowoff velocities, and blowoff curves rise almost asymptotically. There is a progressive

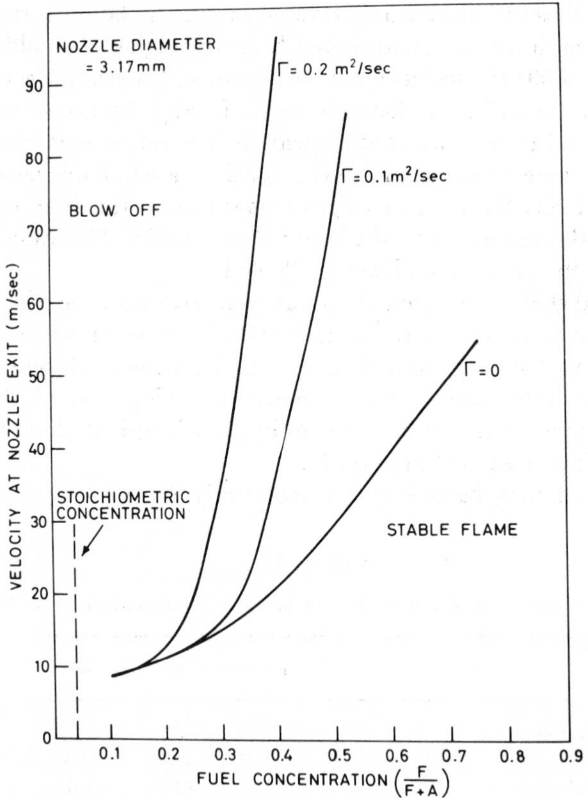

Fig. 17. Flame stability diagram for propane–air turbulent diffusion flame. [After Chigier *et al.* (1970).]

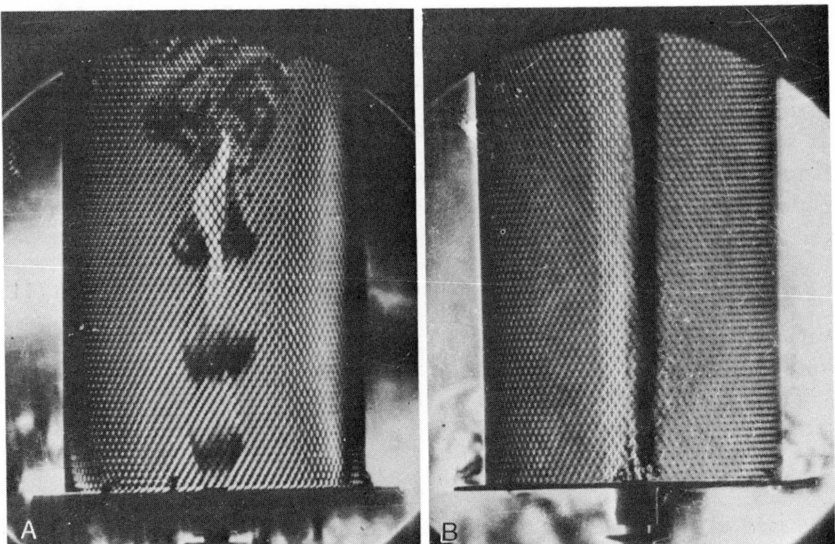

Fig. 18. Schlieren photographs of a turbulent methane–air flame with (a) stationary screen, and (b) rotating screen. [After Chigier *et al.* (1970).]

increase in blowoff velocity with increase in Γ from 0 to 0.2 m^2/sec. It can be seen for example that for a fuel concentration of 0.35 the blowoff velocity is increased from 15 to 70 m/sec. In cases where the flame was lifted off the burner with the screen rotating, it was found that the flame blew off when the screen rotation was reduced. Conversely, a flame lifted with the screen stationary could be brought back to the burner rim by rotating the screen. Flames could be maintained in a lifted position with considerable variation in burner exit velocity, while the screen was rotating.

The conditions of stability of the boundary layer at the interface between the vortex core and the free vortex region deserve consideration because this interface is the boundary of the cylindrical flame. Coupled with the centrifugal force field the radial density gradient in the flow causes stratification and the boundary to become even more stable. This is illustrated by the Schlieren photographs in Fig. 18.

One of the effects of turbulence is to make more uniformly distributed any property possessed by the fluid. If the density of fluid decreases toward the center, turbulence will tend to make it more uniformly distributed. The carrying of density inward requires work done against centrifugal forces in a rotating flow by the kinetic energy of turbulence.

The rate at which work is done by the turbulence for unit mass of fluid is equal to the rate at which density is transferred by the turbulence in unit density gradient (represented by the ϵ_ρ coefficient of eddy transfer of heat or density) multiplied by the radial density gradient and multiplied by the centrifugal force on unit mass. This can be written as

$$\epsilon_\rho(1/\rho)\ (\partial\rho/\partial r)(w^2/r), \tag{10}$$

the rate of work per unit mass. The source of turbulence in the fire whirl-jet system is of course the jet and vortex core of the rotating flow field. If ϵ is the coefficient of transfer of momentum by turbulence, the stress is

$$\tau = \epsilon\ (\partial U/\partial r), \tag{11}$$

the force per unit area for the case where the velocity U is vertical and varies only radially.

The rate at which work is done per unit mass against this stress is the stress multiplied by the velocity gradient, that is,

$$\epsilon\ (\partial U/\partial r)^2,$$

the rate of work per unit mass. The work done against these stresses is converted into energy of turbulence, and if this exceeds the work done against the centrifugal forces, the energy of turbulence can be expected to increase.

Thus if

$$\epsilon_\rho(1/\rho)\ (\partial\rho/\partial r)(w^2/r) < \epsilon(\partial U/\partial r)^2,$$

turbulence will prevail, otherwise the force field prevails with the effect of damping turbulence. A modified Richardson number for the case of the fire whirl can therefore be given as

$$\mathrm{Ri}^* = [(1/\rho)\ (\partial\rho/\partial r)(w^2/r)]/(\partial U/\partial r)^2.$$

V. FUTURE TRENDS

During the last decade, combustion scientists and engineers have turned increasingly toward the problems of predicting flame behavior in combustors, and furnace performance from design input variables. The progress made in the solution of the mathematical problem is quite significant, but this progress highlighted also the lack of information on physical–chemical input data necessary for the computations concerning turbulent flow systems. The most sophisticated methods for predicting furnace performance or the formation and emission of pollutants from combustion systems require the knowledge of the detailed flow and heat release patterns in flames, including fluctuating components of the velocity, species concentration, and temperature. Methods of determining these are therefore of great importance to the progress of any predicted procedure whether oriented toward heat flux distribution or pollutant emission calculations. For engineering design calculations, the combination of mathematical and physical modeling based on dimensionless groups derived from the governing differential equations of the system seems at present to provide the most economical route to take.

Experimental research is necessary partly for the further development of a "turbulence hypothesis" linking properties of turbulent fluid with local state of flow, and also for determining chemical kinetic data for reactions, the rate of which cannot be assumed to be "infinite" even in turbulent diffusion flames. Careful and rigorous testing of the prediction methods is then required before they can be used with sufficient confidence. A significant step in this direction has been taken by the International Flame Research Foundation when they decided to devote a proportion of their research effort to the testing of mathematical models in their experimental furnace at Ijmuiden.

List of Symbols

A, B	Constants in Eqs. (6) and (8)
C	Constant in Eq. (6a)
D_v	Coefficient of molecular diffusion
d	Nozzle diameter
E	Indicated wire voltage (hot wire anemometer)
erf	Error function
$F(r)$	Probability distribution function of radial location of the jet boundary
k	Kinetic energy of turbulence $= (\overline{u'^2} + \overline{v'^2} + \overline{w'^2})/\bar{u}^2$
L	Flame length
l	Prandtl mixing length, scale of turbulence
p	Pressure $\bar{p}$ time mean value of fluctuating pressure
$\bar{R}$	Mean radial position of jet boundary
Re_t	Reynolds number of turbulence $= \rho k^{1/2} l/\mu$
r	Radial space coordinate
t	Time for gas to flow from nozzle to a point in the flame
t_f	Time for gas to flow from nozzle to flame tip
U	Axial velocity component
u	$= \bar{u} + u' =$ time mean plus fluctuating velocity component for turbulent flow (Reynolds' notation)
u_∞	Stream velocity far from the wall
u'	Fluctuating component of the axial velocity in turbulent flow
$(\overline{u'^2})^{1/2}$	Root mean square (rms) value of the axial, fluctuating velocity component
v	Transverse velocity component; $v = \bar{v} + v'$
w	Tangential velocity component; $w = \bar{w} + w'$
x	Axial space coordinate

β	Intermittency
Γ	Circulation, $2\pi rw$
γ'	Root mean square value of nozzle fluid concentration
σ	Displacement thickness
ϵ	Turbulent or "eddy" viscosity
ϵ_ρ	Coefficient of eddy diffusivity of heat or matter
μ_{eff}	Effective or turbulent dynamic viscosity
$\emptyset$	Azimuthal space coordinate
ψ	$\int_0^\infty Ur\,dr$, the stream function
ρ	Density
σ	Standard deviation of R
θ	Generalized time, $\theta = 4D_v t/d$

References

Batchelor, G. (1967). "An Introduction to Fluid Dynamics." Cambridge Univ. Press, London and New York.

Becker, H. A., and Brown, P. G. (1969). *Int. Symp. Combust. 12th,* p. 1059. Combust. Inst., Pittsburgh, Pennsylvania.

Becker, H. A., Hottel, H. C., and Williams, G. C. (1963). *Int. Symp. Combust., 9th,* p. 7. Academic Press, New York.

Becker, H. A., Hottel, H. C., and Williams, G. C. (1965). *Int. Symp. Combust., 10th,* p. 1253. Combust. Inst., Pittsburgh, Pennsylvania.

Becker, H. A., Hottel, H. C., and Williams, G. C. (1967). *Int. Symp. Combust., 11th,* p. 791. Combust. Inst., Pittsburgh, Pennsylvania.

Beér, J. M. (1965). British Patent No. 45652/65.

Beér, J. M., and Chigier, N. A. (1963). *5e Journee d'Etudes sur les Flammes*. Int. French Flame Res. Foundation French Committee, Paris.

Bradshaw, P., Ferris, D. H., and Atwell, N. P. (1967). *J. Fluid Mech.* **28,** 593.

Chedaille, J., Leuckel, W., and Chesters, A. K. (1966). *J. Inst. Fuel* **39,** 506.

Chigier, N. A., and Beér, J. M. (1964). *Trans. ASME Ser. D: J. Basic Eng.* **86,** 788.

Chigier, N. A., and Chervinsky, A. (1967). *Int. Symp. Combust., 11th,* p. 489. Combust. Inst., Pittsburgh, Pennsylvania.

Chigier, N. A., Beér, J. M., Grecov, D., and Bassindale, K. (1970). *Combust. Flame* **14,** 171.

Clauser, F. H. (1954). *J. Aeron. Sci.* **21,** 91.

Davies, T. W. (1969). Ph.D. Thesis. Sheffield Univ., Sheffield, England.

Davies, T. W., and Beér, J. M. (1971). *Int. Symp. Combust., 13th,* p. 631. Combust. Inst., Pittsburgh, Pennsylvania.

Davies, T. W., and Beér, J. M. (1973). *Int. Seminar Heat Mass Transfer, Herceg Novi, 1969.* Int. Center Heat and Mass Transfer, Belgrade, Yugoslavia (in press).

Eickhoff, H. (1968). Doctoral Dissertation. Univ. of Karlsruhe, Germany.

Emmons, H. W., and Ying, S.-J. (1967). *Int. Symp. Combust., 11th,* p. 475. Combust. Inst., Pittsburgh, Pennsylvania.

Fricker, N., and Leuckel, W. (1969). Doc. No. G.02/a/18, Int. Flame Res. Found., Ijmuiden, Holland.

Fristrom, R. M., and Westenberg, A. A. (1965). "Flame Structure." McGraw-Hill, New York.

Gossman, A. D., Launder, B. E., Pun, W. M., Runchal, A. K., Spalding, D. B., Taylor, R. G., and Wolfstein, M. (1968). Heat transfer Sect. Res. Reps. Dept. of Mech. Eng., Imperial College, London.

Gossman, A. D., Pun, W. M., Runchal, A. K., Spalding, D. B., and Wolfstein, M. (1969). "Heat and Mass Transfer in Recirculating Flows." Academic Press, New York and London.

Günther, R. (1969). *Chem.-Ing.-Tech.* **41,** 315.

Günther, R. and Simon, H. (1969). *Int. Symp. Combust., 12th,* p. 1069. Combust. Inst., Pittsburgh, Pennsylvania.

Hawthorne, W. R., Weddell, D. S. and Hottel, H. C. (1951). *Int. Symp. Combust., 3rd,* p. 266. Williams & Wilkins, Baltimore, Maryland.

Hottel, H. C., and Hawthorne, W. R. (1951). *Int. Symp. Combust., 3rd,* p. 254. Williams & Wilkins, Baltimore, Maryland.

Johanssen, J. (1967). Doctoral Thesis. Univ. of Trondheim, Norway.

King, L. V. (1914). *Proc. Roy. Soc. (London) Ser. A* **214,** 373.

Kissel, R. R. (1960). *Brennstoff-Warme-Kraft* **12,** 340.

Kolmogorov, A. N. (1942). *Izv. Akad. Nauk SSSR Ser. Fiz.* **33,** 1–2.

Kremer, H. (1964). Doctoral Dissertation. Univ. of Karlsruhe, Germany.

Leuckel, W. (1967). Doc. No. G.02/a/16. Int. Flame Res. Found., Ijmuiden, Holland.

Pengelly, E. S. (1962). *J. Inst. Fuel* **35,** 210.

Prandtl, L. (1925). *Z. Angew. Math. Mech.* **5,** 136.

Prandtl, L. (1945). *Nachr. Akad. Wiss., Göttingen, Math-Phys. Ke II* **6.**

Rayleigh, Lord (1916). *Proc. Roy. Soc. (London) Ser. A* **93,** 148.

Schlichting, H. (1968). "Boundary Layer Theory." McGraw-Hill, New York.

Schwar, M. J. R., and Weinberg, F. J. (1969). *Combust. Flame* **13,** 335.

Siddall, R. G., and Davies, T. W. (1973). Communication to *Int. Seminar Heat Mass Transfer, Herceg Novi, 1969.* Int. Center Heat and Mass Transfer, Belgrade, Jugoslavia. (in press).

Thring, M. W., and Beér, J. M. (1962). *Symp. Temp. Measurement,* p. 3. Inst. Mech. Eng., London.

Townsend, A. A. (1949). *Aust. J. Sci. Res. Ser. A* **2,** 451.

Townsend, A. A. (1961). *J. Fluid Mech.* **11,** 97.

Weinberg, F. J. (1963). "Optics of Flames." Butterworths, London.

Wohl, K., Gazley, C., and Kapp, N. (1951). *Int. Symp. Combust., 3rd,* p. 288. Williams & Wilkins, Baltimore, Maryland.

IV

Flame Stabilization in High Velocity Flow

J. SWITHENBANK

DEPARTMENT OF CHEMICAL ENGINEERING AND FUEL TECHNOLOGY
THE UNIVERSITY OF SHEFFIELD
SHEFFIELD, ENGLAND

I. INVISCID COMPRESSIBLE FLOW EQUATIONS*

The kinetic theory of gases shows that temperature and pressure are due to the random motion of molecules. Flow is a mean motion superimposed on this random motion, and it is compressible when the flow velocity is comparable with the molecular velocity. The speed of sound is proportional to the molecular velocity and provides a convenient parameter. The flow is incompressible when the Mach number (M) is much less than unity.

$$V/c = \mathrm{M} \ll 1 \tag{1}$$

* See Shapiro (1954).

Applying the conservation equations to one-dimensional flow

$$W = \rho A V \qquad \text{(mass)}, \tag{2}$$

$$F = Ap + WV \qquad \text{(momentum)}, \tag{3}$$

$$H = h + (V^2/2) \qquad \text{(energy)}, \tag{4}$$

$$S = s \qquad \text{(2nd law)}, \tag{5}$$

$$p = \rho R t \qquad \text{(state equation)}. \tag{6}$$

From the energy equation (4),

$$T = t + (V^2/2C_p);$$

therefore

$$T/t = 1 + (\gamma R V^2/2C_p \gamma R t). \tag{7}$$

But $C_p - C_v = R$, that is,

$$R/C_p = (\gamma - 1)/\gamma \tag{8}$$

and

$$\gamma R t = c^2. \tag{9}$$

Substituting (8) and (9) in (7) gives

$$T/t = 1 + \tfrac{1}{2}(\gamma - 1)\mathrm{M}^2. \tag{10}$$

Equation (10) is plotted in Fig. 1, where it can be seen that the tempera-

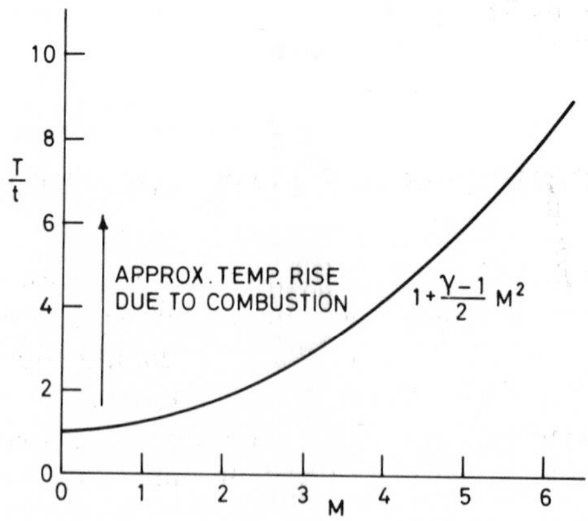

Fig. 1. Total to static temperature ratio (T/t) versus Mach number (M).

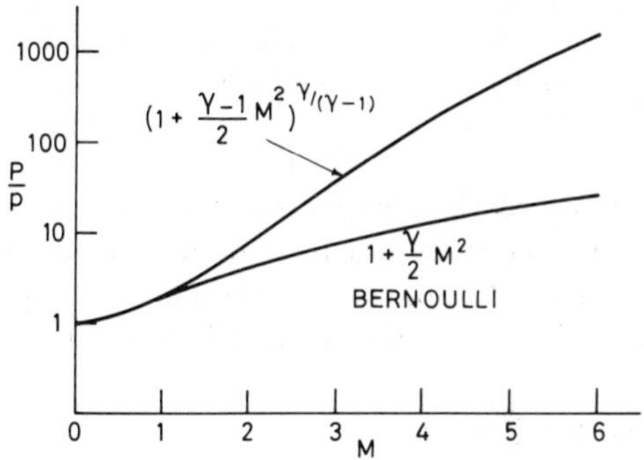

Fig. 2. Total to static pressure ratio (P/p) versus Mach number (M).

ture increment due to the flow velocity is comparable to that due to flame reactions at approximately $\mathrm{M} = 5$. For adiabatic flow

$$p/\rho^\gamma = \text{constant}, \tag{11}$$

hence

$$P/p = (T/t)^{\gamma/(\gamma-1)};$$

therefore

$$P/p = [1 + \tfrac{1}{2}(\gamma - 1)\mathrm{M}^2]^{\gamma/(\gamma-1)}. \tag{12}$$

When $\mathrm{M} \ll 1$ (i.e., incompressible flow) we can apply the binomial expansion to (12)

$$P/p = 1 + \tfrac{1}{2}\gamma\mathrm{M}^2, \tag{13}$$

but

$$\tfrac{1}{2}\gamma p\mathrm{M}^2 = \tfrac{1}{2}\rho V^2;$$

therefore

$$P = p + \tfrac{1}{2}\rho V^2. \tag{14}$$

Thus Bernoulli's equation (14) is a low-speed approximation. Equations (12) and (13) are plotted in Fig. 2.

II. HEAT ADDITION TO AN IDEAL GAS*

For simplicity, heat addition to an inviscid ideal gas (constant γ) in a constant-area duct will be considered. Since we wish to include the case

* See Shapiro (1954).

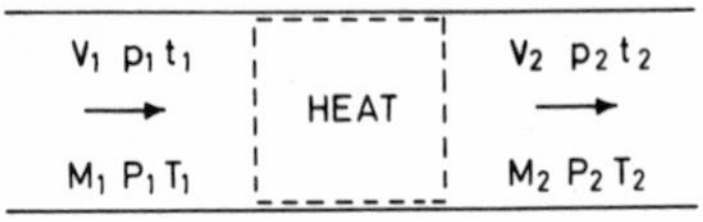

Fig. 3. Heat addition in a constant-area duct.

when the velocity energy can be comparable with the chemical energy, the compressible fluid flow relations must be used. We utilize a coordinate system in which the heat addition front is stationary (Fig. 3) and use the conservation equations to derive the relation between the initial and final Mach numbers.

Momentum conservation gives

$$p + \omega V = p + \rho V^2 = p + \gamma p \mathrm{M}^2 = p(1 + \gamma \mathrm{M}^2);$$

therefore

$$p_1/p_2 = (1 + \gamma \mathrm{M}_2^2)/(1 + \gamma \mathrm{M}_1^2). \tag{15}$$

Energy addition gives

$$\Delta H/C_{\mathrm{p}} = T_2 - T_1. \tag{16}$$

But from (10)

$$T_2/T_1 = (t_2/t_1)\{[1 + \tfrac{1}{2}(\gamma - 1)\mathrm{M}_2^2]/[1 + \tfrac{1}{2}(\gamma - 1)\mathrm{M}_1^2]\}. \tag{17}$$

Mass conservation and the equation of state give

$$t_2/t_1 = p_2\rho_1/p_1\rho_2 = p_2V_2/p_1V_1 = [p_2\mathrm{M}_2(t_2)^{1/2}]/[p_1\mathrm{M}_1(t_1)^{1/2}];$$

therefore

$$t_2/t_1 = p_2^2\mathrm{M}_2^2/p_1^2\mathrm{M}_1^2 = [(1 + \gamma\mathrm{M}_1^2)^2\mathrm{M}_2^2]/[(1 + \gamma\mathrm{M}_2^2)^2\mathrm{M}_1^2]. \tag{18}$$

We can therefore relate T_2/T_1 to M_1 and M_2:

$$\frac{T_2}{T_1} = \frac{\{\mathrm{M}_2^2(1 + \gamma\mathrm{M}_1^2)^2[1 + \tfrac{1}{2}(\gamma - 1)\mathrm{M}_2^2]\}}{\{\mathrm{M}_1^2(1 + \gamma\mathrm{M}_2^2)^2[1 + \tfrac{1}{2}(\gamma - 1)\mathrm{M}_1^2]\}}, \tag{19}$$

and from (12) and (15)

$$\frac{P_2}{P_1} = \frac{\{(1 + \gamma\mathrm{M}_1^2)[1 + \tfrac{1}{2}(\gamma - 1)\mathrm{M}_2^2]^{\gamma/(\gamma-1)}\}}{\{(1 + \gamma\mathrm{M}_2^2)[1 + \tfrac{1}{2}(\gamma - 1)\mathrm{M}_1^2]^{\gamma/(\gamma-1)}\}}. \tag{20}$$

Equation (19) is plotted in Fig. 4 and shows that the gas dynamic equations can be satisfied for two ranges of M_1 separated by a region in the vicinity of $\mathrm{M}_1 = 1$. This forbidden region becomes larger for greater heat release. On the subsonic leg of the curve, a free flame normal to the flow direction would be stabilized at the laminar or turbulent flame speed. If, however, the flame front is inclined or limited by mixing, any other higher velocity on this leg can also be obtained. With stoichiometric hydrocarbon flames the maximum initial subsonic velocity in a constant-area burner is about 100 m/sec. The duct then becomes thermally choked and no further increase in initial velocity is possible.

The free flame speed on the supersonic leg of the curve is given by the Chapman–Jouguet rule as $M_2 = 1$ (relative to a stationary flame front) and represents a detonation wave. By the use of turbulent mixing control of the effective flame speed, flames on the upper supersonic leg of the curve can be produced. To produce flames on the lower supersonic leg of the curve, a strong shock can be used, resulting in an overdriven detonation wave. In this latter case the Mach number after the heat addition is subsonic with respect to the wave, while in the former case it is supersonic.

The associated changes in total pressure are plotted from Eq. (20) in Fig. 5. This graph shows that there is a fundamental pressure loss associated with heat addition which increases with the Mach number (i.e., velocity) of the stream. This total pressure loss amounts to about 20% at the thermal choking point for subsonic heat addition, and becomes extremely large for supersonic heat addition. We can therefore conclude that this fundamental pressure loss can be minimized by adding heat at the lowest possible stream velocity. If necessary, the flow may be accelerated subsequently by converging the duct.

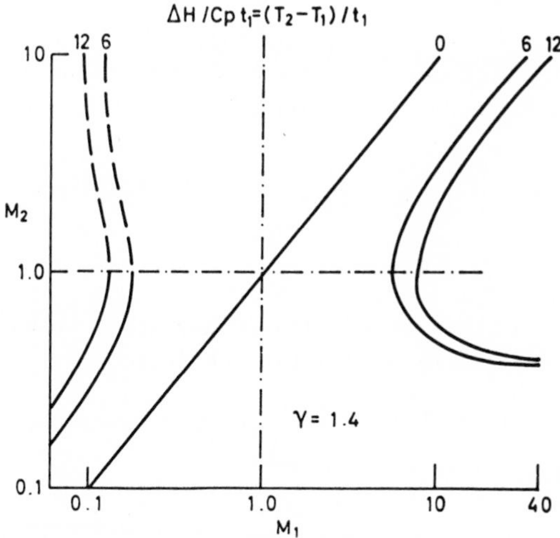

Fig. 4. Relation between Mach number before heat addition (M_1) and Mach number after heat addition (M_2).

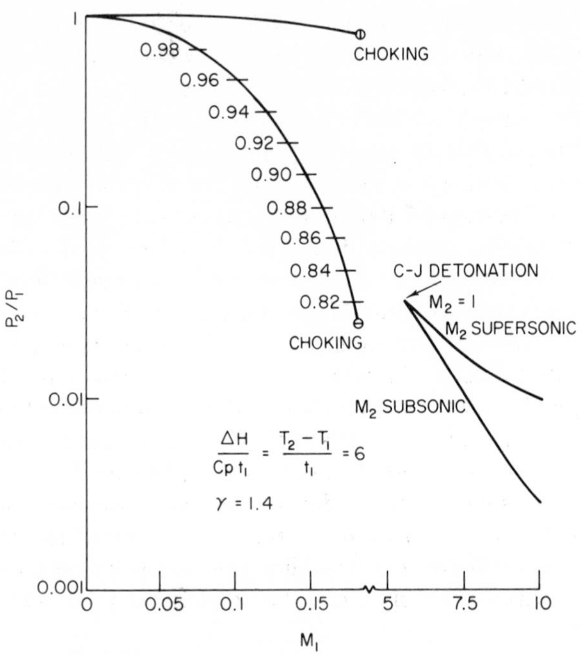

Fig. 5. Total pressure ratio across a combustion wave (P_2/P_1) versus Mach number (with respect to wave) M_1.

If the area of the burner is not constant and the fuel is introduced at some angle ψ, the momentum equation (15) is modified and

$$W_2V_2 + A_2p_2 = W_1V_1 + A_1p_1 + (W_fV_f + A_fp_f)\cos\psi + \int_1^2 p\,dA - \text{wall drag}, \tag{21}$$

where the subscript f denotes fuel. The equations can be solved readily for constant pressure combustion, since Eq. (21) becomes

$$W_2V_2 = W_1V_1 + W_fV_f\cos\psi - \text{wall drag},$$

or approximately

$$V_2 = V_1, \tag{22}$$

that is, constant velocity combustion. However, the solution becomes more complex for an arbitrary pressure–area relation.

III. HEAT ADDITION TO A REAL GAS*

As the temperature of a gas mixture such as air increases, the number of degrees of freedom f' of the molecules increases from translational to include rotational and vibrational modes, and at sufficiently high temperatures the molecules dissociate and finally ionize. These internal changes show up as a change in the ratio of specific heats,

$$\gamma = C_p/C_v = (f' + 2)/f' \tag{23}$$

(where f' is the number of degrees of freedom), and an increase in the specific heat. In the temperature range up to about 2500°K, dissociation and chemical composition changes are generally less than 1%, and it is often adequate to use an average value of γ and C_p in calculations. At the higher temperatures, gases deviate so much from ideal behavior that real gas methods must be employed. In the present section it will be assumed that

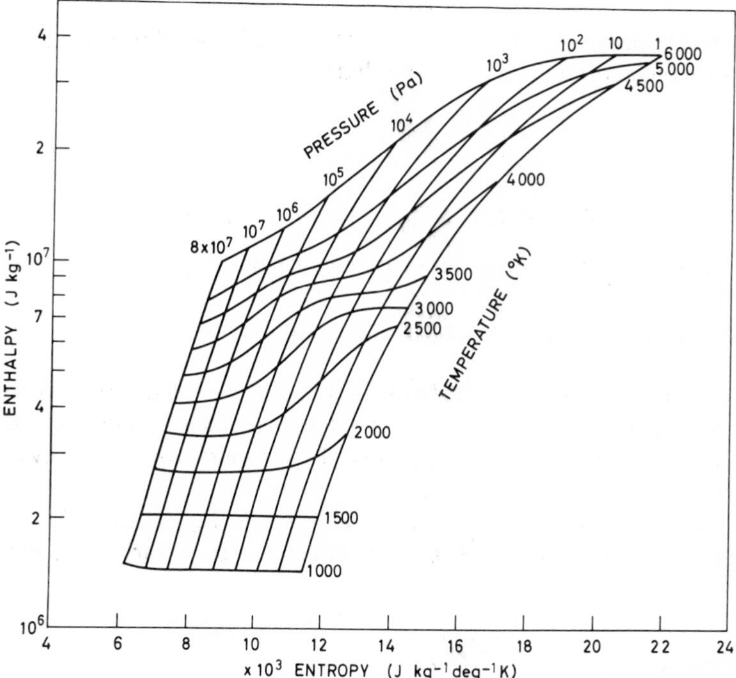

Fig. 6. Mollier diagram for air. [From Banes *et al.* (1967), with permission.]

* See Swithenbank (1968).

the chemical kinetics are sufficiently fast that the gas is in equilibrium; its state is given by a Mollier chart (Fig. 6). In practice the chart is often represented as a computer subroutine using either equilibrium constants or interpolation on stored tables.

The Mollier chart relates the variables h, p, t, S, Z (where Z is the ratio of the undissociated molecular weight to the actual molecular weight). Given any two of the variables h, p, t, S, the others may be found (Banes *et al.*, 1967).

The specific heat ratio may be obtained by numerical or graphical differentiation at constant entropy, from

$$\gamma/(\gamma - 1) = (\Delta \log p/\Delta \log tZ)_s. \tag{24}$$

This can then be used to give the speed of sound (for shifting equilibrium), where

$$c^2 = \gamma Rt, \tag{25}$$

and hence the Mach number

$$\mathrm{M} = V/c. \tag{26}$$

At low temperatures the gas is near ideal, and the following equations are convenient representations of the charts:

$$h = C_{\mathrm{p}}t + h^0, \tag{27}$$

$$p = \rho Rt, \tag{28}$$

$$S/R = S^0/R + (C_{\mathrm{p}}/R)\ln(t/t^0) - \ln(p/p^0), \tag{29}$$

where the superscript denotes reference conditions.

A. Isentropic Flow in a Variable-Area Duct

Due to the shifting equilibrium, heat may be transferred between chemical and translational modes in the flow through a variable-area duct. This is automatically incorporated when Mollier charts are used to evaluate the flow behavior.

Consider the flow through a nozzle, as shown in Fig. 7. Given A_1, p_1, t_1, V_1, and p_2, we can evaluate the exit velocity V_2, gross thrust F_2, and net

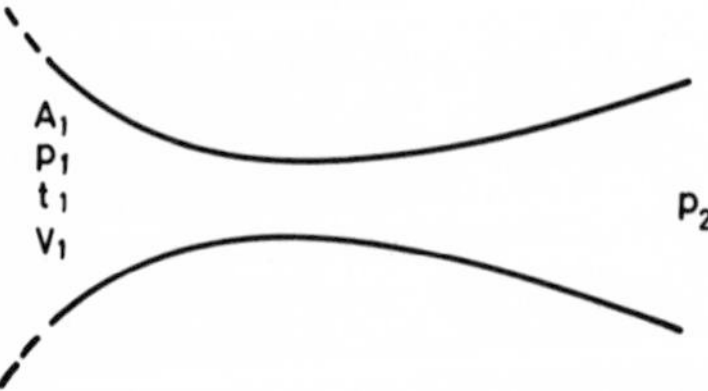

Fig. 7. Flow through a nozzle.

thrust F_n as follows:

From the Mollier chart

$$h_1 = f(p_1, t_1), \qquad S_1 = f(p_1, t_1).$$

Energy

$$H_2 = H_1 = h_1 + \tfrac{1}{2}V_1^2. \tag{30}$$

Isentropic flow

$$S_2 = S_1. \tag{31}$$

From the Mollier chart

$$t_2 = f(S_2, p_2), \qquad h_2 = f(S_2, p_2).$$

Exit velocity

$$V_2 = [2(H_2 - h_2)]^{1/2}. \tag{32}$$

To obtain thrust, the density and hence mass flow must first be evaluated. If available on the Mollier chart, $\rho_1 = f(p_1, t_1)$, otherwise molecular weight ratio $Z_1 = f(p_1, t_1)$. Then

$$\rho_1 = p_1/RZ_1t_1. \tag{33}$$

The density ρ_2 is evaluated similarly. Therefore

$$W_2 = W_1 = \rho_1 A_1 V_1$$

and

$$A_2 = W_2/\rho_2 V_2.$$

Gross thrust

$$F_2 = W_2 V_2 + A_2 p_2. \tag{34}$$

Net exit thrust

$$F_n = F_2 - p_\alpha A_2. \tag{35}$$

B. Constant-Pressure Heat Addition

When heat ΔH is added to a gas with no change in composition, then $H_2 = H_1 + \Delta H$, and the real gas calculations can be carried out with a single Mollier chart. In general, however, the heat is produced by the combustion of fuel, and separate Mollier charts for air and combustion products must be used before and after heat release, respectively. Care must be taken that both charts use the same reference enthalpy level.

For the particular case of heat addition at constant pressure, the velocity must remain constant (i.e., $V_2 = V_1$) and a simple solution may be obtained. Given t_1, ΔH, $p_1 = p_2$, and $V_1 = V_2$, we wish to find t_2.

From the Mollier chart, $h_1 = f(p_1, t_1)$. Since the velocity is constant,

$$\begin{aligned} h_2 &= h_1 + \Delta H, \\ t_2 &= f(p_2, h_2). \end{aligned} \tag{36}$$

As shown above, it is useful to evaluate the Mach number so that thermal choking can be considered. Since the mass flow ρAV is constant, ρ is inversely proportional to A and

$$A_1/A_2 = \rho_2/\rho_1 \approx t_1/t_2 \approx \mathrm{M}_2^2/\mathrm{M}_1^2. \tag{37}$$

Thus in the case of supersonic heat addition, the area increases while the Mach number decreases, and sonic exit conditions are possible. On the other hand, in subsonic constant-pressure combustion, the Mach number again decreases and thermal choking is avoided. The actual Mach numbers can be evaluated from the charts using Eqs. (24)–(26).

If allowance is made for the addition of a fuel–air ratio to provide the source of heat, then

$$V_2 = V_1/(1 + \alpha), \tag{38}$$

$$H_2 = (H_1 + \Delta H)/(1 + \alpha), \tag{39}$$

and Eqs. (36) and (37) must be modified slightly.

C. Constant-Area Heat Addition

Referring to the model shown in Fig. 3, p_1, t_1, V_1, α, and H_f are given (where H_f is the heating value of the fuel). As before

$$H_1 = h_1 + (V_1^2/2), \tag{40}$$

$$H_2 = (H_1 + \alpha H_f)/(1 + \alpha), \tag{41}$$

$$\omega_1 = \omega_2/(1 + \alpha), \tag{42}$$

$$f_2 = f_1 = p_1 + \omega_1 V_1 = p_1 + \rho_1 V_1^2, \tag{43}$$

where from the chart, $\rho_1 = f(p_1, t_1)$. The exit conditions must be evaluated by simultaneous solution of the following equations:

$$\rho_2 = p_2/RZ_2t_2, \tag{44}$$

$$p_2 = f_2 - (\omega_2^2/\rho_2) \qquad \text{(Rayleigh relation)}, \tag{45}$$

$$h_2 = H_2 - (\omega_2^2/2\rho_2^2) \qquad \text{(Fanno relation)}, \tag{46}$$

where

$$t_2 = f(p_2, h_2), \tag{47}$$

$$Z_2 = f(p_2, h_2). \tag{48}$$

The Rayleigh and Fanno curves intersect at two points, representing supersonic and subsonic combustion, respectively, as illustrated in Fig. 8 (see also Fig. 4).

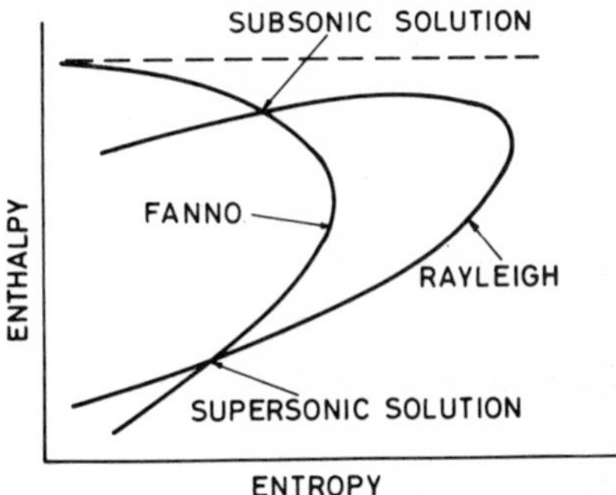

Fig. 8. Solutions of balance equations.

The supersonic case is only relevant for supersonic initial conditions, and can be found by inserting $p_2 = p_1$ to calculate ρ', and then taking $\rho_{\text{first guess}} = 1.05\rho'$. A graphical or numerical iterative procedure is then used to locate the solution.

The subsonic intersection is close to $p_2 = f_2$ and $h_2 = H_2$, so these conditions can be used to locate the solution quickly.

If the curves do not intersect, then heat has been added beyond the choking point and the inlet conditions would be modified (e.g., by shock formation).

Once the solution has been found, the exit velocity and Mach number can be evaluated from Eqs. (32) and (24)–(26).

The above procedure is most readily illustrated by an example: Calculate the supersonic solution for constant-area combustion of stoichiometric hydrogen (neglecting thrust gain or shock loss due to injection).

Given

$$p_1 = 10^5 \quad \text{N/m}^2, \qquad V_1 = 3000 \quad \text{m/sec}, \qquad t_1 = 2000°\text{K},$$

$$\alpha = 0.02928, \qquad H_f = 1.164 \times 10^8 \quad \text{J/kg}.$$

From the Mollier chart for air (Fig. 6):

$$h_1 = f(p_1, t_1) = 2.7 \times 10^6 \quad \text{J/kg};$$

from Eq. (40),

$$H_1 = 7.2 \times 10^6 \quad \text{J/kg};$$

from Eq. (41),

$$H_2 = 10.35 \times 10^6 \quad \text{J/kg};$$

At 2000°K, dissociation is negligible, therefore $Z_1 = 1$. Universal gas constant

$$R_0 = 8314.3 \quad \text{J/deg-kg-mole};$$

$$\text{molecular weight of air} = 28.965;$$

$$\text{molecular weight of stoichiometric hydrogen–air} = 28.905;$$

from Eq. (33),

$$\rho_1 = 0.174 \quad \text{kg/m}^3;$$

from Eq. (42),

$$\omega_2 = 507 \quad \text{kg/sec m}^2;$$

from Eq. (43),

$$f_2 = 1.67 \times 10^6 \quad \text{N/m}^2.$$

Rayleigh relation:

$$p_2 = 1.67 \times 10^6 - (2.57 \times 10^5)/\rho_2. \tag{49}$$

Fanno relation:

$$h_2 = 10.35 \times 10^6 - (1.285 \times 10^5)/\rho_2{}^2. \tag{50}$$

Taking $p_2 = 0$, Eq. (49) gives $\rho' = 0.164$ kg/m³. Therefore, for supersonic solution assume $\rho_{\text{first guess}} = 1.64 \times 1.05 = 0.172$ kg/m³. This is then inserted into Eqs. (49) and (50) and the resulting values of p and h are used to obtain t from the Mollier charts for stoichiometric hydrogen–air

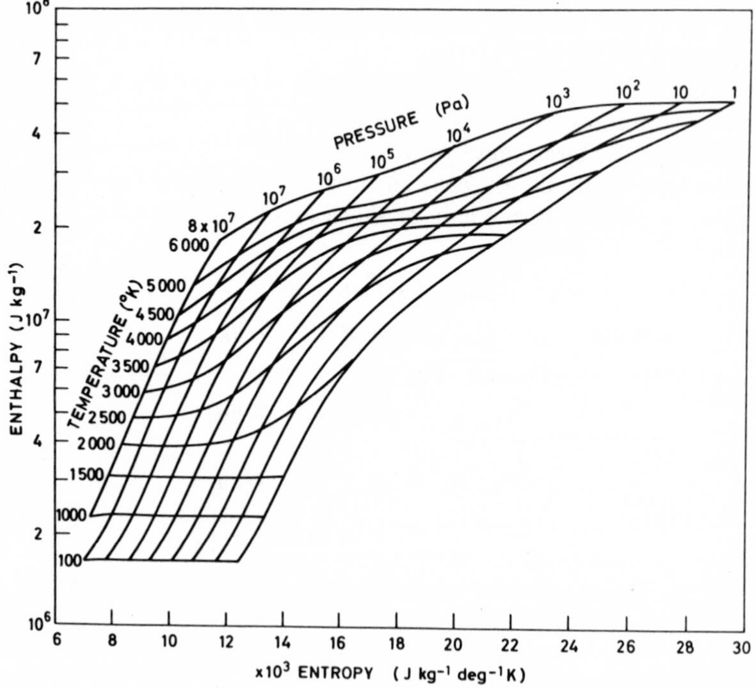

Fig. 9. Mollier diagram for stoichiometric hydrogen-air. Equivalence ratio 1. [From Banes *et al.* (1967), with permission.]

combustion products (Fig. 9). Inserting p and t in Eq. (44) then gives

$$\rho = 0.218 \quad \text{kg/m}^3.$$

The value of ρ_{guess} is then decreased until the true value (when $\rho = \rho_{\text{guess}}$) is obtained, yielding

$$\rho_2 = 0.167 \quad \text{kg/m}^3, \qquad t_2 = 2700°\text{K},$$

$$h_2 = 5.74 \times 10^6 \quad \text{J/kg}, \qquad V_2 = 3020 \quad \text{m/sec}.$$

$$p_2 = 1.3 \times 10^5 \quad \text{N/m}^2,$$

IV. TURBULENCE*

Almost all flames of technological importance are turbulent, hence the mechanism and production of turbulence and the turbulent flame merit particular attention. In this section some of the relevant features of turbulence theory are presented; further details can be found in texts such as that by Hinze (1959).

Turbulence is the irregular fluctuation of small masses of fluid (fluid particles) whose motion is superimposed on the mean flow of the fluid. Because the turbulent motion is random, it has some similarities with molecular motion; however, the analogy cannot be taken too far. For example, the turbulent quantities corresponding to molecular mass and mean free path are not constants of the medium, nor can turbulent motion be sustained independently of the mean motion of the fluid. However, as turbulent "mass" and "mean free path" are so much larger than those in molecular motion, turbulent transport is very much more effective than molecular diffusion.

From the point of view of turbulent combustion, the most important features of the turbulence are the velocity distribution—most conveniently represented by the turbulent energy spectrum (Fig. 10) and the scale of turbulence.

Turbulent motion is a consequence of eddy formation in the shear flow produced by mixing of streams of different velocity in viscous fluids. The streams may be produced by individual jets including zero velocity jets from the base of bluff bodies, or in boundary layers on bodies. In combustion systems, the free boundary layer produced in the former case is more important.

* See Hinze (1959).

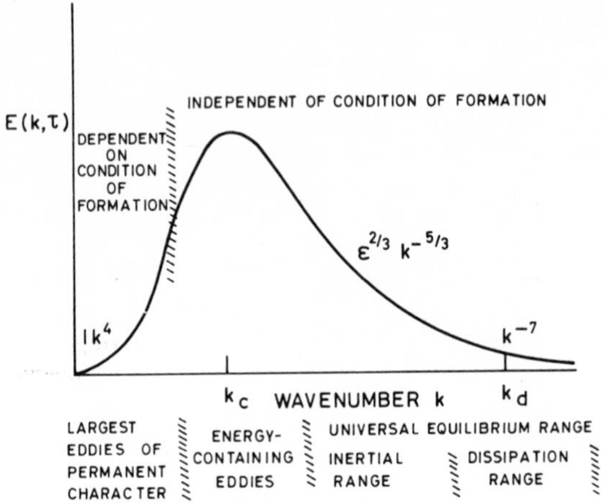

Fig. 10. Isotropic turbulent energy spectrum.

The large eddies produced when the streams first interact are highly anisotropic, but these eddies soon interact with each other, producing a complete spectrum of eddies such as that indicated in Fig. 10, and the turbulence becomes nearly isotropic (i.e., there is no preference for any specific direction). The smallest eddies, which have the highest wavenumber (k), are the ones which viscously dissipate the turbulence to heat. The loss from the larger eddies is predominantly by transfer of energy to the smaller eddies, and the region of the spectrum noted as the universal equilibrium range is independent of the condition of formation. It is important to realize that this range includes the small eddies of the dissipation region, since it is in this region that mixing takes place at the molecular level. Unless gases are mixed at the molecular level they are not truly *mixed* and combustion reactions cannot go to completion. This point is illustrated in Fig. 11, which shows that, due to unmixedness, a turbulent jet flame is typically 50% longer than the point at which the time mean concentration on the axis becomes stoichiometric.

Although the application of turbulence theory to internal flow problems is still in an early stage of development, due to mathematical complexity,

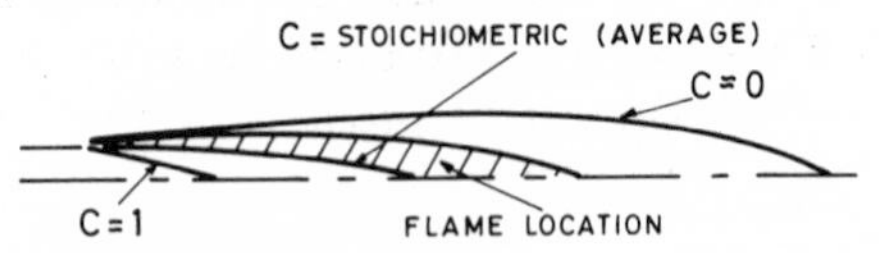

Fig. 11. Location of turbulent diffusion flame.

and in many cases only an order of magnitude solution can be obtained, it is nevertheless essential to understand the fundamental principles in order to interpret mixing and combustion phenomena correctly.

In the conventional manner (Hinze, 1959), we define gas dynamic variables in terms of steady and fluctuating components such as

$$\rho = \bar{\rho} + \tilde{\rho}, \tag{51}$$

$$U_i = \bar{U} + u_i, \tag{52}$$

and when substituted into the equations of motion, besides the terms referring to the mean motion, turbulence terms $\rho\overline{u_i u_j}$ known as Reynolds stresses are produced. The average value of the fluctuating component u_i is zero, and its root mean square value is u', where

$$u_i' = (\overline{u_i^2})^{1/2}. \tag{53}$$

The corresponding one-dimensional turbulent kinetic energy component is

$$\rho \int_0^\infty E(k_1)\, dk_1 = \tfrac{1}{2}\rho u_1'^2, \tag{54}$$

where the wavenumber k_1 is related to the eddy frequency n by

$$k_1 = 2\pi n/\bar{U}_1. \tag{55}$$

The turbulent energy spectrum function $E(k)$ represents the average amount of energy of the turbulent motion between wavenumbers k and $k + dk$. For isotropic turbulence, the rms fluctuations in the three dimensions are equal, and the total turbulent kinetic energy (KE) per unit mass is

$$\int_0^\infty E(k)\, dk = \tfrac{1}{2}(u_1'^2 + u_2'^2 + u_3'^2) = \tfrac{3}{2}u'^2, \tag{56}$$

that is,

$$\text{KE} = \tfrac{3}{2}\rho u'^2 \qquad \text{per unit volume.} \tag{57}$$

The important feature of the energy is that the total energy in the system must be conserved, and it will be shown that the maximum kinetic energy of turbulence can be obtained from the energy balance, without evaluating the local shear stresses, eddy viscosities, and so on. The validity of Eq. (56) can be justified by the fact that about 80% of the turbulence energy is contained outside the low-frequency range, hence the energy distribution is dominated by the isotropic region. This is particularly true in the region where mixing and combustion are being completed—that is, away from the initial eddy-forming region. It will also be noted that when the turbulence is isotropic, terms $\overline{u_i u_j}$ with $i \neq j$ are absent, and no average turbulent shear stress can exist. The normal stress (when $i = j$) is then given by

$$\overline{u_n^2} = \tfrac{1}{3}\overline{u_i u_i} = \tfrac{2}{3}\text{KE} \qquad \text{per unit mass.} \tag{58}$$

The three-dimensional spectrum function can be divided into several regions. In the lowest wavenumber band the energy function for isotropic turbulence increases according to k^4:

$$E(k) \sim k^4. \tag{59}$$

As pointed out above, for real systems, this region is not isotropic. However, since only a small amount of energy is involved, errors are not significant except close to the baffle, jet, or other turbulence generator. If a single large eddy is formed, as for example in a recirculation region, this can be represented as shown by the dashed line in Fig. 12, where it can be seen that the energy involved is comparatively small.

The energy spectrum function is given by the von Kármán interpolation formula as

$$E(k) = Ik_e^4(k/k_e)^4/[1 + (k/k_e)^2]^{17/6}. \tag{60}$$

This formula extends over the whole wavenumber region, excluding the highest wavenumbers where viscosity effects lead to dissipation of turbulent kinetic energy. Equation (60) reduces to

$$E(k) = k^4 \qquad \text{when} \quad k \ll 1,$$

and to

$$E(k) \sim k^{-5/3} \qquad \text{when} \quad k \gg 1.$$

The curve shown in Fig. 12 is plotted from Eq. (60) with $I = 1$ and $k_e = 1$.

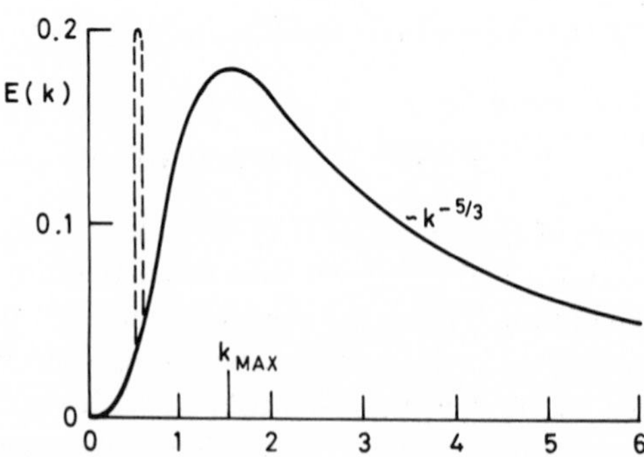

Fig. 12. Turbulent energy spectrum with recirculation vortex. Plotted from Eq. (60), $k_e = 1$.

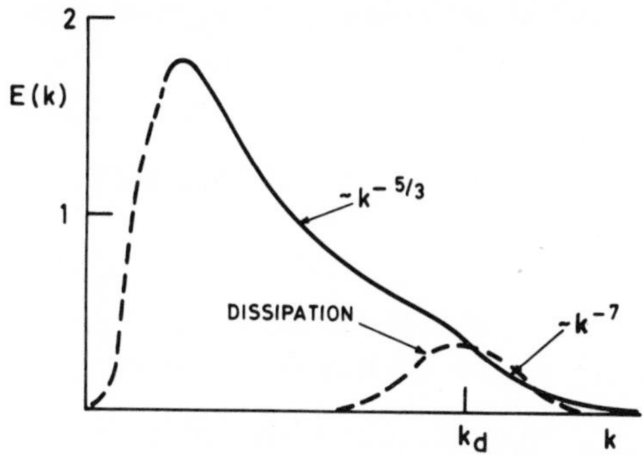

Fig. 13. Turbulent energy spectrum showing dissipation.

In this case, differentiating the equation gives the location of the maximum value $k_{max} = 1.55$.

The dissipation range of the spectrum is characterized by the wavenumber k_d. In this region, the turbulent kinetic energy is dissipated to heat, and the energy spectrum function, valid in the inertial and viscous region, is

$$E(k) = (8/9\alpha)^{2/3}[(e\nu^5)^{1/4}(k/k_d)^{-5/3}]/[1 + (8/3\alpha^2)(k/k_d)^4]^{4/3}. \quad (61)$$

Equation (61) reduces to $E(k) \sim k^{-5/3}$ in the inertial range, and to $E(k) \sim k^{-7}$ in the viscous range. The value k_d occurs at the changeover (Fig. 13). The wavenumber of the dissipative eddies can be obtained from the Reynolds number (equal to the ratio of inertial forces to viscous forces), which must be of the order unity. Following Kolmogoroff (1941), dimensional reasoning gives

$$\eta = (\nu^3/e)^{1/4} \qquad \text{(for the length scale)},$$

$$v = (\nu e)^{1/4} \qquad \text{(for the velocity scale)}. \quad (62)$$

Thus the Reynolds number $= v\eta/\nu = 1$ as required, and the wavenumber $k_d \approx \eta^{-1}$. [Townsend (1951) showed $k_d \approx 1/(5\eta)$.] Therefore

$$k_d = \tfrac{1}{5}(e/\nu^3)^{1/4}. \quad (63)$$

The rate of dissipation can be obtained from either

$$e = Au'^3/l_e, \quad (64)$$

or

$$e = 15\nu u'^2/\lambda_g^2,$$

where l_e is the average size of the energy-containing eddies and A is a numerical constant of the order unity.

From Eqs. (63) and (64) we obtain

$$k_d = \tfrac{1}{5}(Au'^3/l_e\nu^3)^{1/4}. \tag{65}$$

Example. Assuming

$$u'/\bar{U} = 0.1, \qquad \bar{U} = 30 \quad \text{m/sec},$$

$$l_e = 0.01 \quad \text{m}, \qquad \nu = 15 \times 10^6 \quad \text{m}^2/\text{sec},$$

then the wavenumber of the dissipation eddies $k_d = 6000\ \text{m}^{-1}$, which corresponds to an eddy size $\eta = 0.167$ mm. The frequency is given by

$$n_d = k_d\bar{U}/2\pi = 29 \quad \text{kHz}. \tag{66}$$

It will be noted that the size of the viscous dissipation eddies is considerably larger than the mean free path of the molecules (ca. 10^{-7} m). Similarly, for the assumed size of the energy-containing eddies:

$$k_e = 1/l_e = 100 \quad \text{m}^{-1} \tag{67}$$

and the frequency

$$n_e = 480 \quad \text{Hz}.$$

It now remains to consider the dynamic behavior of the energy spectrum. If energy is added to the flow,

$$\begin{bmatrix}\text{change in KE in}\\ \text{wavenumber range}\end{bmatrix} = \begin{bmatrix}\text{energy transferred}\\ \text{to other wavenumbers}\end{bmatrix}$$

$$- \begin{bmatrix}\text{energy dissipated}\\ \text{by viscous effects}\end{bmatrix} + \begin{bmatrix}\text{energy}\\ \text{in}\end{bmatrix},$$

$$(\partial/\partial\tau) \int_0^k E(k, \tau)\, dk = \int_0^k F(k, \tau)\, dK - 2\nu \int_0^k k^2 E(k, \tau)\, dk + H_k(k, \tau), \tag{68}$$

where $F(k, \tau)$ is the three-dimensional transfer spectrum function. The energy input term is most significant for the lower values of k in the near wake region of a baffle, and elsewhere as $k \to \infty$, the change in total KE of the turbulence is equal to the dissipation of turbulence to heat:

$$(\partial/\partial\tau) \int_0^\infty E(k, \tau)\, dk = -2\nu \int_0^\infty k^2 E(k, \tau) = e. \tag{69}$$

The existence of the k^2 factor in the dissipation term demonstrates readily that dissipation is concentrated at the higher wavenumbers. It can also be shown that when a region of turbulence decays, the shape of the energy spectrum decays more rapidly at the higher wavenumbers.

If we consider the creation and decay of turbulence behind a grid of bars or baffles (see Fig. 14), then in the near flow field the jets interact with the wakes and recirculation regions to produce a field of turbulence. As shown in the figure, the velocity peaks decay fairly rapidly (according to x^{-1} for axisymmetric jets) and the jet velocity excess (or wake velocity deficit) is down to about 30% of its initial value in about 10 mesh widths downstream. Since the energy is proportional to the square of the velocity, the jet has then lost a large part of its initial energy. Thus much of the baffle pressure loss energy is converted to turbulence at an apparent origin of the turbulence about 10 mesh widths downstream. Measurements on the three components $u_1'^2$, $u_2'^2$, and $u_3'^2$ have also shown that the residual turbulence becomes practically isotropic when $x/d_m = 10$ to 15.

We therefore have a clear picture of the turbulence energy spectrum.

(a) The total area under the curve is equal to the kinetic energy of turbulence ($\text{KE} = \frac{3}{2}\rho u'^2$), and the maximum possible value of the area can be calculated from an energy balance across the turbulence generation system.

(b) The low wave number end is very nonisotropic in the region where the initial, large turbulent eddies are being created. These can only be created where the stream has significant velocity gradients.

(c) The spectrum extends from eddies whose dimensions are comparable with the size of the baffles or jets creating the turbulence, through a peak at the energy-containing eddies, decreasing according to $k^{-5/3}$, then more rapidly in the dissipation region, at which the eddies are two or three

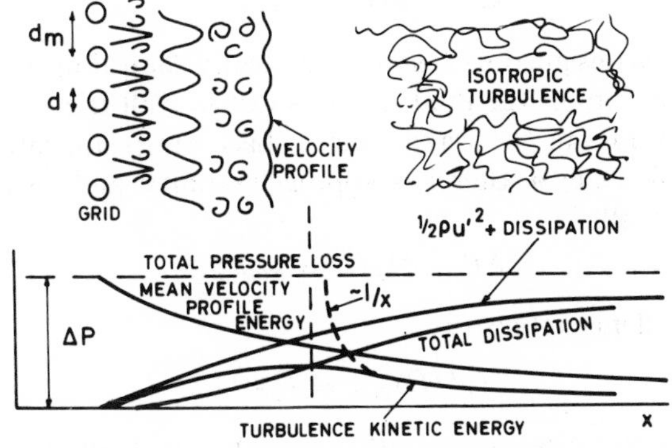

Fig. 14. Diagram illustrating grid-generated turbulence.

orders of magnitude greater than the molecular mean free path. The energy-containing eddies usually occur at a few hundred hertz, and the dissipation eddies at 10^4 to 10^5 Hz.

(d) These points are illustrated in Fig. 14 which shows the stages of the transfer of energy from the total pressure of the gas, first to velocity, then to turbulence, and finally to heat. For free boundary layers, all the energy which goes to heat must follow this path; however, shock waves and wall boundary layers can induce heating without the intermediate formation of turbulent eddies. It should also be noted that the gas temperature rise due to the decay of turbulence is extremely small, and can generally be neglected.

(e) It is interesting to consider the size of the smallest eddies which can be conceivably created. In this case the energy-containing eddy size l_e will be identical with k_d^{-1} and Eq. (67) becomes

$$\eta = k_d^{-1} = 5(\eta\nu^3/u'^3)^{1/4} = 5^{4/3}(\nu/u'). \tag{70}$$

Such eddies have a Reynolds number of order 8.5 and will only survive for a few "revolutions." If we increase this by an order of magnitude to allow a significant mixing life, we obtain

$$\eta \approx 85(\nu/u').$$

Thus at atmospheric conditions and $u' \approx 1$ m/sec, there is little point in trying to create mixing eddies whose dimensions are less than 1 mm.

A. Turbulence Generation by Bluff Bodies

The efficiency with which turbulence is produced may be defined by the factor

$$\eta_T = (\tfrac{3}{2}\rho u'^2_{max})/(\xi\tfrac{1}{2}\rho\bar{U}^2). \tag{71}$$

If a bluff body is designed with aerodynamically shaped upstream surfaces, then the momentum of the fluid which is lost in wall friction is negligible, and the energy loss in the wake (due to base drag) appears first as turbulence. The efficiency η_T therefore approaches unity, and the total pressure loss across the baffle is

$$\Delta P = \xi\tfrac{1}{2}\rho\bar{U}^2 = \tfrac{3}{2}\rho u'^2_{max}. \tag{72}$$

Therefore turbulence intensity

$$(u'/\bar{U})_{max} = (\tfrac{1}{3}\xi)^{1/2}. \tag{73}$$

We therefore wish to derive a relation whereby the baffle loss factor ξ can be calculated for a given geometry. We consider a baffle located between

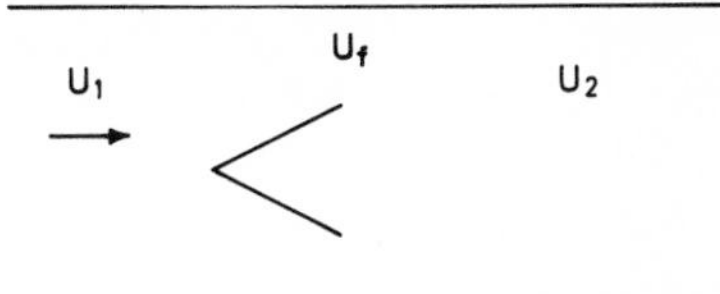

Fig. 15. Baffle in constant-area duct.

stations 1 and 2 in a constant-area duct (Fig. 15) and denote the conditions at the baffle edge by subscript f. Carrying out a momentum balance, assuming the losses to occur at the sudden expansion, gives

$$W_2U_2 + A_2p_2 = W_1U_1 + A_1p_1 - \tfrac{1}{2}C_D A_1 B_g \rho U_f^2, \tag{74}$$

where C_D is the drag coefficient, B_g is the geometrical blockage, and $A_1 = A_2$. Therefore

$$2(P_1 - P_2) = p_1 - p_2 + \tfrac{1}{2}C_D B_g \rho U_f^2.$$

But $P_1 - P_2 = \xi\tfrac{1}{2}\rho U_1^2$, $p_1 - p_2 = \tfrac{1}{2}\rho U_1^2[(U_f^2/U_1^2) - 1)]$, and $U_f^2/U_1^2 = 1/(1 - B_g)^2$. Therefore

$$2\xi = [1/(1 - B_g)^2] - 1 + [B_g C_D/(1 - B_g)^2],$$

$$\xi = \tfrac{1}{2}\{[(1 + B_g C_D)/(1 - B_g)^2] - 1\}. \tag{75}$$

Example. The drag coefficient of a hemisphere is 0.4. Therefore, assuming the geometrical blockage of the duct to be 0.25,

$$\xi = 0.48.$$

Thus about half of the upstream kinetic energy is converted to turbulence and the maximum possible turbulence intensity created would be

$$(u'/U)_{\max} = (\xi/3)^{1/2} = [\tfrac{1}{6}\{[(1 + B_g C_D)/(1 - B_g)^2] - 1\}]^{1/2} = 0.40. \tag{76}$$

For bluff bodies with streamlined forebodies and length-to-diameter ratios less than 4 (e.g., hemispheres, half-cylinders, and baffles with plunged holes), $C_D \approx 0.4$; hence this value can be used for most combustor designs. We can therefore plot the maximum turbulence intensity as a function of blockage ratio from Eq. (76), as shown in Fig. 16.

A particularly important point is attained at 60% blockage, at which the turbulence intensity $(u'/U)_{\max} \approx 100\%$. This corresponds to a well-stirred reactor (see Section VI) such as that used for gas turbine cans and other industrial high-intensity combustion systems. For this condition, the pressure loss of an optimum gas turbine can is given by $u'/U \approx 1$; therefore

$$\Delta P/(\tfrac{1}{2}\rho U^2) = \xi = 3(u'/U)^2_{\max}.$$

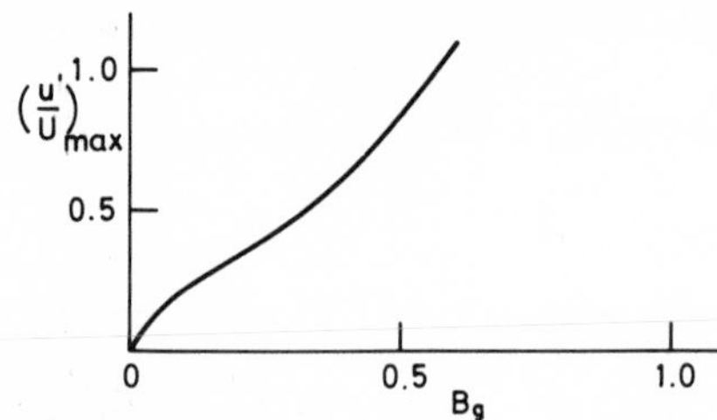

Fig. 16. Maximum turbulence intensity $(u'/U)_{max}$ versus geometric blockage ratio (B_g).

Thus a pressure loss of three dynamic heads corresponds to the minimum well-stirred reactor (this includes mixing systems as well as combustors), and it is interesting to note that gas turbine burners which have been developed empirically approach this performance. The loss can be expressed in terms of the Mach number in the chamber since

$$\tfrac{1}{2}\rho U^2 = \tfrac{1}{2}\gamma p \mathrm{M}^2.$$

Therefore

$$\Delta P/p \approx \tfrac{1}{2}(3\gamma)\mathrm{M}^2 = 2.1\mathrm{M}^2 \qquad \text{when} \quad \gamma = 1.4.$$

From this relation it follows that well-stirred combustors give high-pressure loss if the chamber Mach number is high.

V. COMBUSTOR DESIGN

To illustrate the application of these principles to combustor design, let us consider a combustor which consists of a perforated diaphragm located across a constant-area duct carrying premixed gaseous air–fuel mixture. This is shown in Fig. 17. As a result of the pressure drop across the can (or diaphragm), a multitude of jets are formed, and the pressure difference appears first as jet velocity. These jets soon degenerate to turbulence because of shear, and the turbulence then decays to heat. In the simplest two-zone model, the region in which the turbulent mixing is taking place can be considered as a stirred reactor, and the region after the turbulence

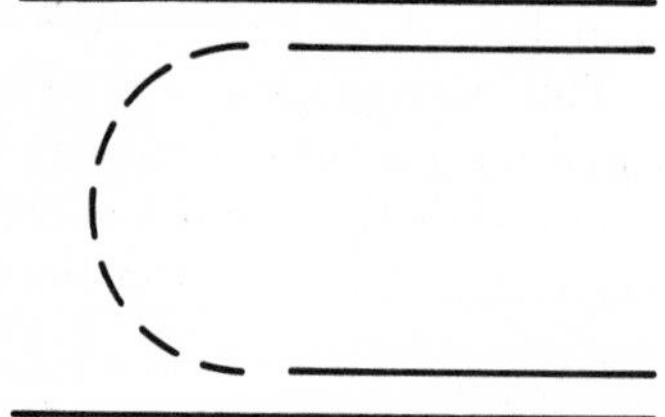

Fig. 17. Perforated colander combustor.

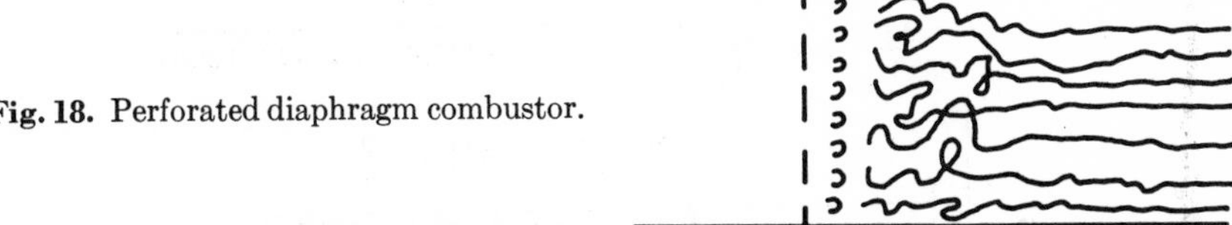

Fig. 18. Perforated diaphragm combustor.

has decayed to negligible proportions can be considered as a plug-flow reactor. These two regions are illustrated in Fig. 18.

To make the model more specific, we assume that the geometrical blockage ratio is 50% and the baffle consists of ten parallel bars distributed across the width of 10 cm (Fig. 19). From Schlichting (1962, p. 604) we find the wake of each bar becomes equal to the pitch when $x/\lambda > 4$, and when $u_1 = U_\infty - u$ the solution to the momentum equation is

$$u_1 = (U_\infty/8\pi^3)(\lambda/l_m)^2(\lambda/x)\cos(2\pi y/\lambda), \tag{77}$$

where l_m is the constant mixing length and $l_m/\lambda \approx 0.103$ when $\lambda/d = 8$ (for $\lambda/d = 2$, l_m/λ is not known at present). The velocity difference (u_1) thus decreases as x^{-1} for this geometry. (N.B.: For confined swirling flows the well-known Crya–Curtet relations could be used to derive u_1, or for complicated flow systems the distribution of u_1 could be measured experimentally.)

Since the decay of u_1 results in the creation of turbulence, we now carry out an energy balance to determine the maximum turbulence kinetic energy at any distance. It should be noted that this is the sum of the dissipation and the actual kinetic energy of turbulence at any point. Now the kinetic energy of the mean flow perturbations (KE_{av}) is

$$\begin{aligned} KE_{av} &= \frac{1}{\lambda}\int_0^\lambda \tfrac{1}{2}\rho u_1^2\,dy \\ &= \tfrac{1}{2}\rho U_\infty^2[(\lambda/l_m)^4/64\pi^6](\lambda/x)^2\tfrac{1}{2}, \end{aligned} \tag{78}$$

$$KE_{av}/q = C'(\lambda/x)^2. \tag{79}$$

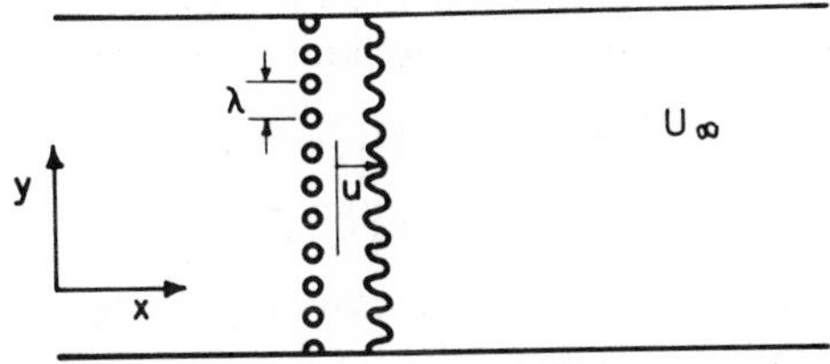

Fig. 19. Wake behind a row of bars.

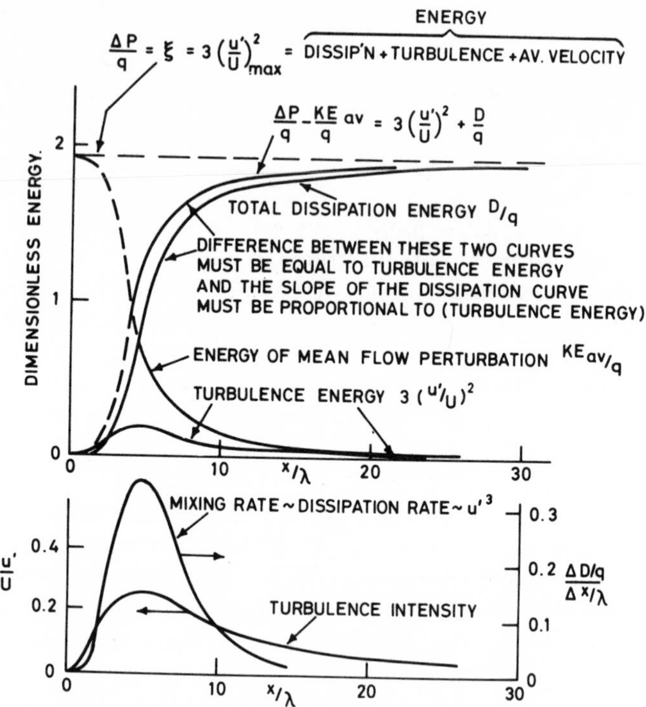

Fig. 20. Energy, turbulence, and dissipation behind a baffle.

Here the dynamic head $q = \frac{1}{2}\rho U^2$ is used to nondimensionalize the kinetic energy of the mean flow perturbation (KE_{av}). This function is plotted in Fig. 20, assuming $C' = 15.4$, and it can be seen that this energy falls very rapidly with distance. It therefore follows that the sum of turbulence and dissipation rises in a manner which can be evaluated by subtracting the mean flow perturbation energy from the overall pressure energy loss across the baffle.

Now from the overall energy balance we find that the pressure drop across the baffle is simply related to the number of dynamic heads lost across the baffle and also to the maximum kinetic energy of turbulence; thus

$$\Delta P = \xi \tfrac{1}{2}\rho U^2 = (\tfrac{3}{2}\rho u'^2)_{\max}, \tag{80}$$

$$\Delta P/q = \xi = 3(u'/U)_{\max}{}^2. \tag{81}$$

For a baffle having 50% blockage, Eq. (76) shows that $(u'/U)_{\max} = 0.8$ and hence $\xi = 1.92$. We can therefore show $\Delta P/q$ in Fig. 20, and this is the total stirring energy which is available to the system.

From the energy balance we obtain

$$(\Delta P/q) - (KE_{av}/q) = 3(u'/U)^2 + (D/q), \tag{82}$$

where D is the total dissipation energy. Now it is apparent that $D/q \to \Delta P/q$ as $x/\lambda \to \infty$ since the turbulence will eventually dissipate. It is also found experimentally that $(u'/U)^2$ decreases as x^{-1} from an apparent source at $x/\lambda \approx 10$. [This is illustrated in the report by Swithenbank (1968, Fig. 23).] The location, size, and mixedness of the stirred reactor section of the combustor must now be determined by evaluating the dissipation curve. Inspection of Fig. 20 shows that we already have the sum of the dimensionless turbulence KE and the dissipation, and subtracting the value of $3(u'/U)^2$ from the total yields the value of D/q when $x/\lambda > 10$. In the intermediate region we must obtain D/q from the well-known relation between turbulence velocity and dissipation rate. This is given by Shapiro [1954, Eq. (5-16)].

$$dD/d\tau = \rho e = \rho A(u'^3/l_e), \tag{83}$$

where l_e is the average size of the energy-containing eddies, and A is a numerical constant of the order unity. Integrating this expression gives

$$D = \rho \int (u'^3/l_e U)\, dx. \tag{84}$$

If we assume that $l_e = K\lambda$, where K is a constant of order 0.1, then the gradient of the dissipation curve is

$$\Delta(D/q)/\Delta(x/\lambda) = (2/K)(u'/U)^3, \tag{85}$$

or more conveniently,

$$\Delta(D/q)/\Delta(x/\lambda) = 20[3(u'/U)^2/3]^{3/2}, \tag{86}$$

where $3(u'/U)^2$ is the dimensionless energy form of the turbulence intensity which is plotted in the upper part of Fig. 20. It is convenient to plot the gradient $\Delta(D/q)/\Delta(x/\lambda)$ against $3(u'/U)^2$ for use in the next step, and this relationship is given in Fig. 21.

The method used to obtain the dissipation curve and the turbulence curve from the sum of the two in Fig. 20 is now straightforward. Starting at the left-hand end of the curve, the turbulence energy will initially be equal to the sum since the dissipation is zero at this stage. Moving slightly to the right (a unit increment of x/λ is convenient), the new value of $3(u'/U)^2$ is estimated. If this estimate is correct, then *both* the local gradient of the dissipation curve must agree with the estimated value of $3(u'/U)^2$ *and* the sum of the local value of total dissipation and the local value of

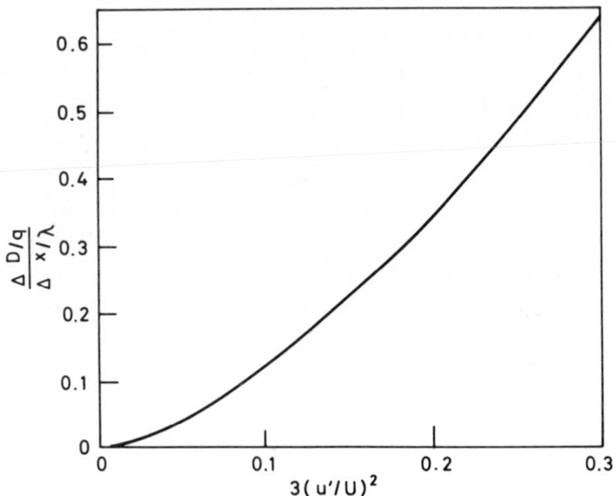

Fig. 21. Gradient of dissipation curve versus turbulence energy.

$3(u'/U)^2$ must be equal to the sum curve already derived. The result of this iterative procedure is plotted in Fig. 20.

Careful study of the curves is most instructive, and the conclusions will be discussed in turn.

(1) Since mixing is proportional to dissipation, the fact that the total dissipation curve is asymptotic to $\Delta P/q$ shows that $\Delta P/q$ is the most important parameter governing mixing—provided sufficient space is available to achieve the dissipation, and provided the energy is used in useful mixing rather than the interaction of eddies of the same composition.

(2) For this particular example the dissipation is close to its maximum value when x/λ lies between 10 and 20. Since λ is one-tenth of the duct height, the mixing requires a length only slightly greater than the duct height.

(3) The turbulence energy rises to a peak which is about an order of magnitude less than the maximum energy while the actual turbulence intensity peak is $u'/U = 0.25$ compared to $(u'/U)_{\max} \approx 0.8$ for this baffle.

(4) At some point between $10 < x/\lambda < 20$, the turbulence energy becomes greater than the energy supply available from the mean velocity perturbations. It is possible that this accounts for the changeover to the turbulence decay law where the turbulence energy varies as x^{-1}.

(5) The dissipation rate and hence mixing rate reaches a pronounced maximum at $x/\lambda \approx 5$ and is negligible when $x/\lambda > 10$. The first section of

the chamber therefore behaves as a stirred reactor—as postulated above—followed by the second section which behaves as a plug-flow reactor. The relatively sudden transition from one to the other is a result of the dependence of dissipation rate on the cube of the turbulence intensity.

(6) It should be borne in mind that the assumptions made above are slightly in error when $x/\lambda < 10$ since relations which apply to isotropic turbulence have been used, and experiments have shown that the turbulence only becomes isotropic at $x/\lambda > 10$ to 15.

(7) Because the dissipation energy curve runs so nearly parallel to the $(\Delta P/q - \mathrm{KE}_{\mathrm{av}}/q)$ curve, a first estimate of the turbulence energy distribution can be made from the slope of the latter.

(8) For the simple two-zone reactor model being used here, the stirred reaction zone can be approximated by a region extending from $x/\lambda = 0$ to 10 with a mean dissipation rate $(\Delta D/q)/(\Delta x/\lambda) = 0.17$; hence the total dissipation in this region $\Delta D/q = 1.7$.

VI. STIRRED REACTOR THEORY

As shown by Vulis (1961), the fraction of oxygen untreated (Ψ) in a stirred reactor is given by

$$\Psi = C/C_0, \tag{87}$$

where C is the actual oxygen concentration in the reactor and C_0 is the oxygen concentration of the stream entering. Then the rate of change of the dimensionless oxygen concentration in a perfectly stirred reactor, assuming a stoichiometric hydrocarbon–air mixture, is given by (Clarke *et al.*, 1962; Kretschner and Odgers, 1972):

$$\frac{d\Psi}{d\tau} = \frac{9.37 \times 10^{12}\Psi^2 \exp\{-14{,}000/[t_0 + 2100(1 - \Psi)]\}}{C_0(25.8 - \Psi)^2[t_0 + 2100(1 - \Psi)]^{1.5}}. \tag{88}$$

The mean residence time in a perfectly stirred reactor is

$$\tau_{\mathrm{PSR}} = (1 - \Psi)/(d\Psi/d\tau) \tag{89}$$

and in a plug-flow reactor it is

$$\tau_{\mathrm{PFR}} = \int (d\Psi/d\tau)^{-1}\, d\Psi. \tag{90}$$

In this latter case, note that C_0 changes for each section of the reactor so that in general numerical integration is required.

In the case of a partially stirred reactor (Vulis, 1961), the model used postulates that the reaction proceeds according to the local concentration in the mixing region surrounding oxygen-containing eddies (where oxygen is considered as the reactant). We therefore distinguish between the average oxygen concentration in the reactor C_K, the local oxygen concentration in the reaction region C, and the oxygen concentration at entry C_0. The average rate of reaction is then

$$dC/d\tau = (C_0 - C_K)/\tau_s. \tag{91}$$

But

$$dC/d\tau = [C \exp(-1/\theta)]/\tau_K, \tag{92}$$

where

$$\theta = Rt/E \tag{93}$$

and

$$\tau_K = \{C_0[5(n+1) + 2 - \Psi]^2[t_0 + 2100(1 - \Psi)]^{1.5}\}/(9.37 \times 10^{12}\Psi). \tag{94}$$

The rate of mixing is assumed to be proportional to the concentration difference $C_K - C$. Then

$$dC/d\tau = (C_K - C)/\tau_D. \tag{95}$$

Note that the temperature in the mixed region is determined by the heat release, and the reaction rate is therefore determined by the correct temperature as postulated by this model.

The completeness of combustion is obtained from Eqs. (91)–(93) as

$$1 - \Psi = \{1 + \tau_{sD}^{-1} + [\exp(1/\theta)/\tau_{sK}]\}^{-1}, \tag{96}$$

where $\tau_{sD} = \tau_s/\tau_D$ and $\tau_{sK} = \tau_s/\tau_K$ are the two dimensionless time ratios which determine the combustion characteristics. For the stoichiometric hydrocarbon mixture assumed in the present example, the appropriate characteristics are plotted in Fig. 22.

From Eq. (94) we see that the value of τ_K depends on Ψ, so that if we want complete combustion in the stirred-reactor section of the combustor, then $\Psi \to 0$ and $\tau_K \to \infty$; thus the size of the reactor would become very large to retain a reasonable value for τ_{sK}. In the particular example being considered (i.e., a duct 10 cm high and 10 cm wide), the length of the stirred-reactor region is also about 10 cm. The volume is therefore approximately 1 liter.

As the flow rate through a very well-stirred reactor is gradually increased, we find that the completeness of combustion decreases until excessive supply causes blowout. In the case of a partially stirred reactor that is mixing limited, the completeness of combustion is given by Swithenbank (1970,

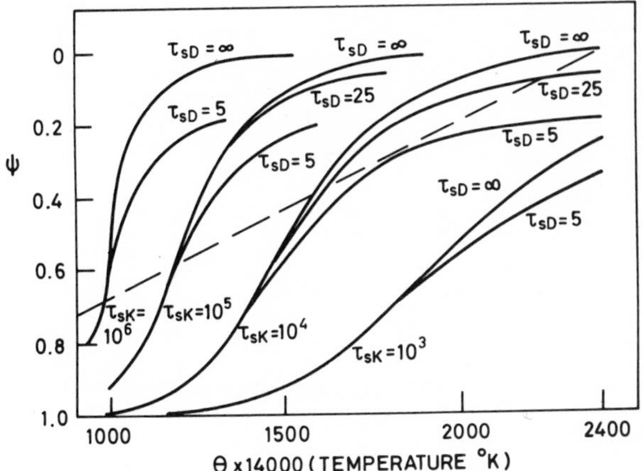

Fig. 22. Partially mixed reactors, outlet oxygen concentration versus temperature for various values of τ_{sD} and τ_{sK}. Heat removal is shown by the dashed line.

Eq. 9-36):

$$1 - \Psi = \tau_s/(\tau_s + \tau_D) = [1 + (1/\tau_{sD})]^{-1}. \tag{97}$$

The completeness of combustion in the first stage is therefore controlled by mixing and/or kinetics; however, Fig. 23 illustrates the fact that decreasing the mixedness increases the range of mixture strength over which the burner will operate although it reduces the maximum loading. In the present example, we wish to calculate how the efficiency of the first stage combustion varies with throughput, and how the mixing affects the loading curve. Only the first of these two problems is considered herein; the second is discussed in another study [see Dixon (1970)].

The stay time τ_s is given by

$$\tau_s = V'/v \quad \text{sec},$$

where V' is the volume of the stirred reactor, that is, 1 liter, and v is the volumetric flow rate at the mean reactor pressure and temperature. Hence τ_{sK} can be evaluated as a function of the throughput.

The diffusion time for a gaseous premixed system depends on the dissipation rate as indicated above:

$$[\text{dissipation rate per unit mass}] = e = (d/d\tau) \int E(k)\, dk$$

$$= (d/d\tau)(\tfrac{3}{2}u'^2)_{\max} = u'^3/l_e. \tag{98}$$

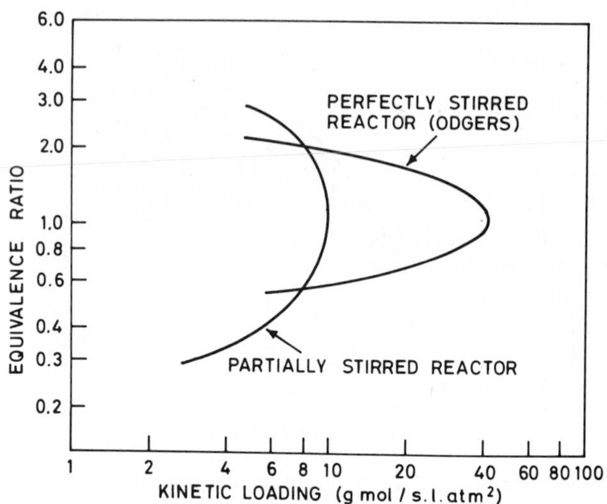

Fig. 23. Stability loops, showing equivalence ratio versus burner loading. $t_0 = 300°K$.

Therefore,

$$\int d\tau = \int (l_e/u'^3)\ d(\tfrac{3}{2}u'^2),$$

$$\tau = (l_e/(\tfrac{2}{3})^{3/2}) \int Z^{-3/2}\, dZ,$$

where

$$Z = \tfrac{3}{2}u'^2,$$

$$\tau = (2l_e Z^{-1/2})/(\tfrac{2}{3})^{3/2},$$

$$\tau = 3(l_e/u'). \tag{99}$$

Since the mixing time is taken to be proportional to the dissipation time, we can write

$$\tau_D \sim l_e/u' \tag{100}$$

and

$$\tau_{sD} \sim (X/l_e)(u'/U). \tag{101}$$

This result can be compared with that given in Eq. 9-39 of the report by Swithenbank (1970). Dimensional analysis suggests that in general,

$$\tau_{sD} \sim (X/l_e)^a (u'/U)^b, \tag{102}$$

where the exponents a and b together with the proportionality constant should be verified by experiment. Until this comparison is available values of these constants of the order unity appear to be reasonable.

At this point it may be noted that when the fuel is supplied as droplets, the characteristic diffusion time is the "sum" of the aerodynamic diffusion and the droplet evaporation times. The latter is approximately the droplet lifetime (i.e., $\tau \sim r^n$, where n varies between 1 and 2, depending on whether the droplet is heated by radiation, conduction, or convection).

In the present example $l_e \approx 0.2\lambda \approx X/50$ and $u'/U \approx 0.8$; hence $\tau_{sD} \approx 40$ at *all* values of throughput and the system behaves as an almost perfectly stirred reactor.

Reference to Eq. (97) shows that unmixedness would limit the combustion completeness to

$$1 - \Psi = [1 + (1/40)]^{-1} = 0.975. \tag{103}$$

Since comparatively little dissipation can occur in the plug-flow region of the combustor, to a first approximation the combustion efficiency would be only about 98%, limited by unmixedness. Since the residence time and hence molecular diffusion in the plug-flow part of the combustor would increase as the flow velocity decreases, a gradual increase in the overall combustion efficiency may be anticipated as the flow velocity is reduced.

Considering next the effect of throughput on the blowoff limit, we must first evaluate the ratio τ_{sK}. For a first order reaction, τ_K is independent of the conversion; however, for this particular reaction, τ_K is a function of Ψ as given by Eq. (94) (plotted in Fig. 24). It is therefore more convenient to

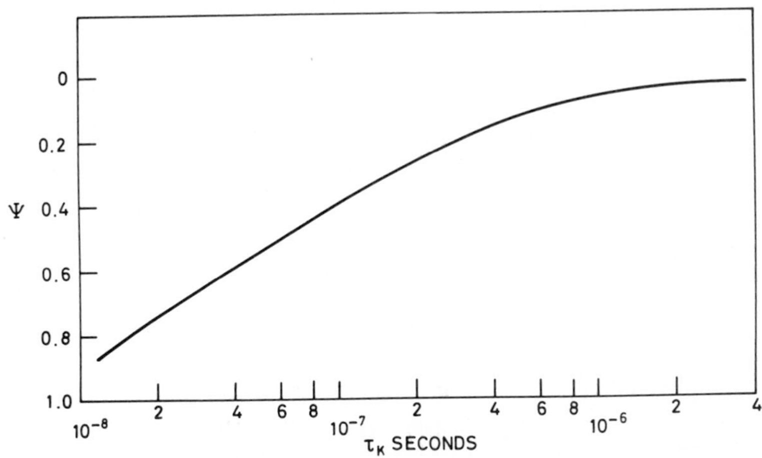

Fig. 24. Variation of kinetic time with conversion. $t_0 = 300°K$.

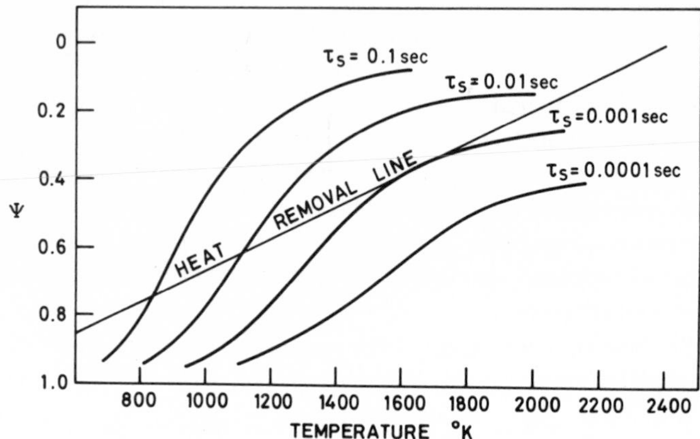

Fig. 25. Oxygen conversion, residence time, and heat removal. $t_0 = 300°K$.

dimensionalize the curves, as shown by Dixon (1970), with the result given in Fig. 25. Since τ_{sD} is so large, these curves are accurate for the case in question without including the mixedness factor. As shown in Fig. 24, the static temperature follows the equation

$$t = 300 + 2100(1 - \Psi). \tag{104}$$

From the intersection of these curves, it can be seen that the heat removal curve is tangent to the reactedness curve when $\tau_s \approx 0.001$ sec. The burner will therefore extinguish due to excessive supply when the mean velocity in the combustor

$$U \approx X/\tau_s \approx 100 \quad \text{m/sec.} \tag{105}$$

Here it should be noted that a small increase in inlet temperature (above the 300°K assumed in this example) would move the τ_s curves to the left and the straight line to the right, thus giving a large reduction in blowoff residence time. An increase in pressure above the 1 atm assumed would also have a similar effect.

The result given by Swithenbank [1970, Eq. (9-18)] is also illustrated in Fig. 25. We see that when the oxygen flow from the stirred reactor is 80% reacted, then $\Psi = 0.2$ and $\tau_{PSR} \approx 0.0013$ sec.

The relations above make it clear that the stirred reactor section of the combustor only results in partial combustion, and the plug-flow region—without quenching—is essential to complete the reaction.

For the plug-flow region, we make use of the equation (Swithenbank,

1970)

$$\tau_{\mathrm{PFR}} = \int (d\Psi/d\tau)^{-1}\, d\Psi. \tag{106}$$

Since $d\Psi/d\tau$ is a function of C_0 this integration must be carried out numerically assuming that the plug-flow reactor consists of a large number of perfectly stirred reactors in series.

To illustrate the result let us assume that the reaction is 80% complete as above; then the flow velocity is about 33 m/sec. The first stage of the plug-flow reactor may be taken as 1 cm thick. Thus $\tau_{\mathrm{PFRI}} = 0.00013$ sec in a local stirred reactor whose initial oxygen concentration C_0 is 0.2 that of one fed with air, and whose inlet temperature is almost 2000°K. Then

$$\frac{d\Psi}{d\tau} \approx \frac{\{9.37 \times 10^{12}\Psi^2 \exp\{-14{,}000/[2000 + 400(1 - \Psi)]\}\}}{\{C_{01}[5(m + 1) + 2 - \Psi]^2[2000 + 400(1 - \Psi)]^{1.5}\}}, \tag{107}$$

and using the same procedure described above, the exit value of Ψ may be calculated. The result is initially a very rapid approach to the final temperature since C_{01} is small and the temperature is large in Eq. (107), and almost complete combustion is achieved in a few stages. It follows that the minimum distance required before the combustor exit (or the addition of dilution air which may quench the reaction) is obtained from this calculation. If insufficient distance is available, then the loss in combustion efficiency is also calculable.

The overall result of this approach is therefore to relate the combustor performance to the geometrical variables via the fundamental processes occurring within the chamber rather than by empirical factors such as those used in conventional design methods.

List of Symbols*

a	Exponent defined by Eq. (102)
A	Area or numerical constant of order unity
b	Exponent defined by Eq. (102)
B_g	Geometrical blockage (baffle area/duct area)
c	Sound speed
C	Concentration
C_{D}	Drag coefficient
$C_{\mathrm{p}}, C_{\mathrm{v}}$	Specific heats
D	Total dissipation energy
e	Dissipation by turbulence per unit of mass

* S.I. units are used throughout.

E	Activation energy
$E(k, t)$	$2\pi k^2 E_{i,i}(k, t)$, three-dimensional energy spectrum function
f'	Degrees of freedom
f	Thrust per unit area
F	Thrust
h	Static enthalpy
H	Total enthalpy
H_f	Heating value of the fuel
I	Loitsianskii's integral
k	Wave number; k_e wavenumber range of energy-containing eddies; k_d wavenumber range of main dissipation
K	Constant
l_e	Turbulence length scale of energy-containing eddies
l_m	Mixing length
M	Mach number
n	Eddy frequency or exponent in droplet lifetime equation
p	Static pressure
p_α	Ambient static pressure
P	Total pressure
q	Dynamic pressure
R	Gas constant = (universal gas constant)/(undissociated molecular weight)
s	Static entropy
S	Total entropy
t	Static temperature
T	Total temperature
u_1	Turbulence velocity component
u'	Root mean square turbulence velocity $u' = (\overline{u'^2})^{1/2}$
U	Velocity; $\bar{U}$ time mean value. $U = \bar{U} + u$
v	Volumetric flow
V	Velocity
V'	Volume
ω	Mass flow per unit area
W	Mass flow
x	Axial distance
X	Total distance
y	Transverse distance
Z	(Undissociated molecular weight)/(actual molecular weight)
α	Fuel–air ratio or turbulent constant
γ	Ratio of specific heats
η_T	Turbulence generation efficiency
η	Eddy length scale
η_d	Size of dissipation eddy
θ	Dimensionless temperature
λ_g	Transverse dissipation scale of turbulence
λ	Pitch of bars
ν	Kinematic viscosity
ξ	Baffle loss factor (number of dynamic heads)
ρ	Density
$\tilde{\rho}$	Fluctuating component of density
τ_K	Kinetic time

τ Time
τ_s Stay time or residence time
τ_D Diffusion time
ψ Fuel injection angle
Ψ Fraction of oxygen untreated

References

Banes, B., McIntyre, R. W., and Sims, J. A. (1967). Properties of Air and Combustion Products with Hydrogen and Kerosene Fuels. AGARD Propulsion and Energetics Panel. Rolls Royce Ltd., Bristol Engine Div., Bristol, England.

Clarke, A. E., Odgers, J., and Ryan, P. (1962). *Int. Symp. Combust., 8th*, p. 982. Williams & Wilkins, Baltimore, Maryland.

Dixon, G. T. (1970). Burner Design. Ph.D. Thesis. Sheffield Univ., Sheffield, England.

Hinze, J. O. (1959). "Turbulence." McGraw-Hill, New York.

Kolmogoroff, A. N. (1941). *C. R. Acad. Sci. USSR* **30,** 301; **32,** 16.

Kretschmer, D., and Odgers, J. (1972). Modelling in Gas Turbine Combustors—A Convenient Reaction Rate Equation. *J. Eng. Power, Trans. ASME* **94,** 173–180.

Lewis, B., and von Elbe, G. (1961). "Combustion, Flames, and Explosions in Gases." Academic Press, New York.

Schlicting, H. (1962). "Boundary Layer Theory," 4th ed. McGraw-Hill, New York.

Shapiro, A. H. (1954). "Compressible Fluid Flow." Ronald Press, New York.

Swithenbank, J. (1968). Hypersonic Air Breathing Propulsion. *In* "Progress in Aeronautical Sciences" (D. Kuchemann, ed.), Vol. 8. Pergamon, Oxford.

Swithenbank, J. (1970). Combustion Fundamentals. Dept. of Chem. Eng. and Fuel Technol., Sheffield Univ., Sheffield, England, FTCE5/JS/2/70, Rept No. HIC 150. Defense Documentation Center, Washington AD710321 AFOSR/70/211 TR.

Townsend, A. A. (1951). *Proc. Roy. Soc., London* **208A,** 534.

Vulis, L. A. (1961). "Thermal Regimes of Combustion." McGraw-Hill, New York.

Weinberg, F. J. (1963). "Optics of Flames." Butterworths, London.

V

Combustion Noise: Problems and Potentials

ABBOTT A. PUTNAM

BATTELLE MEMORIAL INSTITUTE
COLUMBUS, OHIO

and

DENNIS J. BROWN

DEPARTMENT OF CHEMICAL ENGINEERING AND FUEL TECHNOLOGY
THE UNIVERSITY OF SHEFFIELD
SHEFFIELD, ENGLAND

I. INTRODUCTION

One of the problems that has plagued the development and utilization of combustion equipment over the years is that of combustion-driven oscillations. While problems with combustion-driven oscillations in solid- and liquid-propellant rocket engines have attained the greatest notoriety, a perusal of the literature will reveal that combustion-driven oscillations have been observed in just about every type of combustion system (Putnam, 1971). And each type of combustor seems to have its own detailed feedback mechanism; that is, the manner in which the oscillations lead to a variable heat release, properly timed to add energy to subsequent oscillations, varies from design to design. As a result, obtaining the detailed information required to identify the exact mechanism of driving can be costly and time consuming. In addition, the associated mathematical representations are complex; often nonlinear equations are indicated as necessary to the solution. But these mathematical and experimental vistas, while intriguing to the scientist, have little attraction for the design engineer or the user of combustion equipment whose main desire is to prevent or eliminate the problem as economically as possible.

To complicate the problem of the design engineer further, various private and governmental groups are now emphasizing the desirability of eliminating noise pollution. Thus, one must not only eliminate combustion-driven oscillations which may be destructive of equipment as well as producing an unbearable environment, but one must try to suppress or minimize the broad-band noise of the combustion roar produced by large combustion equipment. Table I presents some of the features distinguishing between combustion-driven oscillations and combustion roar. The items therein are discussed in more detail subsequently.

On the positive side, pulse combustion is becoming recognized as a phenomenon with considerable economic potential. In contrast to the traditional research and engineering approach of trying to prevent or reduce

TABLE I

COMBUSTION NOISE

Combustion roar	Combustion-driven oscillations
Broad noise spectrum	Discrete frequencies and harmonics
Low efficiency of noise production	High efficiency of noise production
Frequency criterion F/δ (?)	Driving criterion $\oint hp\,dt > 0$
Self-generated, stream turbulence and noise alter amplitude	Many types of feedback mechanism

combustion-driven oscillations, imaginative and promising means for utilizing rather than eliminating them are being found. Multiple-coupled combustors and mufflers afford means for handling the environmental noise problems. Well-planned and directed efforts should bring some of these positive uses to economic fruition.

Several extensive discussions of combustion-generated noise directed to the research scientist are available in the literature. Therefore, this discussion is directed to the designer or user of residential, commercial, or industrial combustion equipment. The prevention or elimination of combustion-generated noise is only one of his many problems. Again, if purposeful use of pulse combustion is to be made, it will be only one of his problems. Therefore, this discussion is aimed at readers with only the general background and knowledge of combustion and does not require a specialized knowledge of acoustics. Furthermore, mathematical considerations are minimized, and chemical aspects of the problem are not touched upon. Nevertheless, the information given here should also be stimulating to the researcher in the field of combustion-generated noise.

In this broad-brush treatment, the main objective is to give a "feel" for the noise-generating phenomena that will be of use in eliminating or making use of these phenomena. For this reason some liberties will be taken with scientific rigor. These liberties are not significant for the systems operating at the pressures, temperatures, and velocities common to industrial, commercial, or residential use.

Specifically, five areas are covered. First, the conditions under which combustion-driven oscillations that are acoustic in nature occur (the "driving criterion"), and the means by which these conditions are produced are discussed; combustion-driven oscillations that are nonacoustic in nature or that depend on a feedback mechanism through the control and supply system are not discussed (Putnam, 1964a). Second, the manner in which the patterns of oscillating pressure and velocities in a combustor may be predicted is considered. Third, means of suppressing or eliminating oscillations are outlined. Fourth, some comments are made concerning the lesser investigated phenomenon of combustion roar. The first three areas covered are then considered as background for the last area of coverage, in which the potential uses of combustion-driven oscillations, or "pulse combustion," are discussed. It is suggested that this last area is a more profitable one for applied research.

II. CONDITIONS FOR DRIVING

A. Rayleigh Criterion

A simple criterion for determining whether or not a combustion-driven oscillation will occur was suggested by Rayleigh (1945). Even though the purist can prove inadequacies in the criterion in high-velocity systems, it has been found adequate for dealing with hydrocarbon–air systems operating at atmospheric pressure, as discussed herein. Rayleigh noted that if, in an acoustic oscillation, heat is added when the pressure is above average and extracted when the pressure is below average, the amplitude of the oscillation will increase. In mathematical form, the oscillation will grow when

$$\oint hp\, dt > 0, \tag{1}$$

where h is the instantaneous rate of heat input, p is the difference between the instantaneous and average pressure, t is the time, and $\oint$ indicates integration over one cycle (Rayleigh, 1945). In the simplest picture of the driving mechanism, the periodic heat-release term is usually assumed to lag the periodic fuel supply term by a time lag τ. If there is damping, and there always is, some positive value will replace zero in the right-hand side of Eq. (1).

Rayleigh's criterion has been expanded for handling of combustion-driven oscillations in rockets by the NASA staff in the form of the response factor "N" (Heidmann and Wieber, 1966).

If there is any driving mechanism in a combustion system which will satisfy the Rayleigh criterion at any of the natural acoustic frequencies of the system, combustion-driven oscillations can occur. The wonder, then, is that they are not observed more often than they are. This is especially true if one considers the oft overlooked second part of the Rayleigh criterion concerning the frequency shift caused by that portion of the periodic heat release not in phase with the pressure. It is easily shown that the shift is always such that combustion-driven oscillations will be encouraged. As a result, for a given rig and mode, and with low damping, more than half the possible conditions will produce oscillations.

B. Feedback Mechanisms

Possible feedback mechanisms that may be involved in combustion-driven oscillations can be considered on the basis of the above-mentioned criterion. One mechanism often mentioned depends on the natural tendency for the rate of heat release in a combustible mixture to vary with pressure, with only a small time lag (Crocco and Cheng, 1956). Thus, for all frequencies below some cutoff frequency, the inequality above (1) is satisfied, and a mechanism for driving exists. Fortunately, this mechanism is rather weak compared with the usual damping effects in combustors operating near atmospheric pressure, with hydrocarbon fuels and air.

Several more pertinent mechanisms come about through the action of the acoustic component of the velocity. One common action is for the velocity of the primary premixed fuel and air to vary periodically in such a manner that subsequent burning of the mixture adds energy to the acoustic oscillation. This, in turn, maintains the periodic variation in fuel–air supply. This mechanism, common in multiple-port gas-fired residential heating equipment, is one of the best understood (Speich and Putnam, 1959). Alternative versions depend on periodic variations in supply of fuel or air, or both (de Saint-Martin *et al.*, 1963; Pariel and de Saint-Martin, 1965; Putnam *et al.*, 1967) in diffusion flames. These are commonly known as "singing flames," and deserve additional attention because of their widespread occurrence in industry.

Higgins (1802) reported the first "singing flames" from hydrogen diffusion flames from a fuel source in tubes open at both ends. Table II presents the matrix of values of supply line length S and termination conditions, frequency of oscillation f (and wavelength, λ), and time lag τ (i.e., the time between the fuel entering the combustion chamber and being burned) that explains the singing flame oscillations in the laboratory-sized equipment used by Higgins and subsequent investigators. This matrix also seems to explain many of the combustion-driven oscillations in large installations such as blast furnace stoves. However, recent comparisons between theoretical values of time lag and the time lag assumed to permit matching of the matrix prediction for industrial combustors lead one to doubt that the "singing flame" mechanism is always involved in the large industrial units. Rather, observations that in some instances the flame is flashing periodically

TABLE II

CONDITIONS FOR ACOUSTIC DRIVING

	Time lag condition[a]	
	$0 < f\tau - m < \frac{1}{2}$	$\frac{1}{2} < f\tau - m < 1$
Acoustically closed inlet to fuel supply line	$\frac{1}{2} < (2S/\lambda) - n < 1$	$0 < (2S/\lambda) - n < \frac{1}{2}$
Acoustically open inlet to fuel supply line	$0 < (2S/\lambda) - n < \frac{1}{2}$	$\frac{1}{2} < (2S/\lambda) - n < 1$

[a] $n = 0, 1, 2, \ldots; m = 0, 1, 2, \ldots.$

many meters in the large combustors from one stable flame position to another suggest another mechanism of driving. But, interestingly enough, computation of a pseudo-time lag for the flashing flame on the basis that the oscillating velocity component periodically causes blowoff and flashback leads to an agreement within the range of time lag values given by the matrix. In a sense, this is necessary since the matrix is based on the Rayleigh criterion.

There are other examples where periodic variation in the velocity may change the flame shape significantly in a periodic manner and thus change the heat-release rate of the flame; again with the proper phasing, combustion-driven oscillations occur. In an oil-fired unit, that portion of the air velocity which varies periodically relative to a fuel spray can change either the particle size of the droplets or the fuel distribution. Either of these effects can lead in turn to combustion-driven oscillations. The periodic variation in velocity can also shift periodically the local gaseous fuel–air ratios, with similar results. This mechanism appears common in gun-type oil-fired furnaces (Speich and Putnam, 1960). In combustors with natural vortex shedding, the velocity change may act as a triggering mechanism, with a subsequent timed heat release rate which leads to combustion-driven oscillations. Single-port gas-fired residential heating units behave in this manner (Speich and Putnam, 1959). A varying velocity over a heated surface, as in the Rijke tone apparatus, can also lead to oscillations (Merk, 1956–1957).

The variety of possible mechanisms of driving is such that it is hopeless to try to treat them all in detail in less than a book; therefore, the cited references and others should be consulted for specific detailed treatments

(Putnam, 1964b). However, it is evident that the specific characteristics of a particular combustor might be quite significant in determining the specific driving mechanism.

III. ACOUSTIC PRESSURE AND VELOCITY PATTERNS

Another matter of interest is the form of the acoustic oscillation in a combustion system. In the combustion systems being considered, one can assume that the acoustic mode is essentially of a standing-wave type.* In other words, at all times the amplitude of the oscillating component of the pressure throughout the system relative to some index amplitude is only a spatial function, while the absolute amplitude at any position is time dependent and periodic. The same is true of the oscillating component of the velocity.

A. Oscillations in Ducts

Figure 1a illustrates this point. With a simple organ-pipe-type oscillation in a tube of length L, closed at both ends, the gases can be visualized as sloshing back and forth from one end of the tube to the other. In the fundamental mode of oscillation, the velocity amplitude will be maximum at the center and zero at the closed ends. The shape of the curve will be sinusoidal. When the gases are at one end, the pressure will be maximum at that end and minimum at the other. A half-cycle later, the opposite will be the case. Again, the variation along the length at any time is sinusoidal. As to the timing, when the pressure is maximum at one end and minimum at the other, the velocity is zero. On the other hand, when the pressure is uniform the velocity has reached its maximum amplitude.

The frequency of oscillation f is the reciprocal of the time it takes for a pulse to travel a wavelength λ at the velocity of sound c, which in this case is the length of the tube and back. Therefore

$$f = c/\lambda = c/(2L). \tag{2}$$

If two of these units are put end to end, and the timing and amplitude in them are such that the pressure reaches the same maximum amplitude at the same time in the adjacent ends, then the end walls can be removed (as in Fig. 1b) without affecting the oscillation. At any one time, a full wave-

* In some supply lines, traveling-wave-type oscillations are predominant; traveling waves are characterized by having acoustic pressure and velocity in phase.

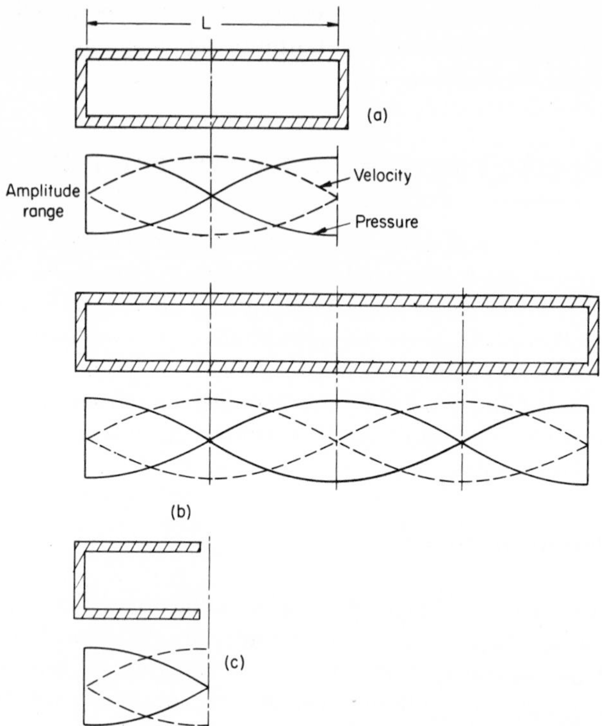

Fig. 1. Modes of standing wave oscillations in tubes. (a) Tube closed on both ends, fundamental mode. (b) Tube closed on both ends, second mode. (c) Tube closed at one end, fundamental mode.

length of pressure is in the tube. The wavelength of the standing wave is thus equal to the tube length in the second mode of oscillation. Addition of more length, with like argument, shows that for a tube of length L, closed at both ends, the possible wavelengths λ of oscillation are

$$\lambda = 2L/n, \qquad n = 1, 2, 3, \ldots. \tag{3}$$

Any or all of these modes may be present at a given time. For small pressure amplitudes, they do not interact and are merely additive.

The oscillating component of the pressure does not vary in the center of the tube of Fig. 1a. Thus the tube may be cut in two and one end removed, as in Fig. 1c, and the oscillation will continue (neglecting acoustic end losses to the surrounding atmosphere). In actuality, correction for the open end should be added. With most systems, for one end, this correction term is about 0.3 times the tube diameter (Kinsler and Frey, 1962).

TABLE III

VALUES OF CONSTANT α_{mn}

m	n = 0	1	2	3
0	0.000	1.220	2.333	3.238
1	0.586	1.697	2.714	3.726
2	0.972	2.135	3.193	4.192
3	1.337	2.551	3.611	4.643

The addition to the closed end of this tube of sections of equal frequency closed at both ends, followed by wall removal, gives

$$\lambda = 4L/(2n + 1), \qquad n = 0, 1, 2, \ldots, \tag{4}$$

for the natural mode wavelengths of a tube open at one end and closed at the other.

By a similar process, a relation like that for the tube closed at both ends is obtained for a tube open at both ends.

Since, for the closed tube, the transverse dimensions do not enter into the problem, it follows that, for a right parallelepiped, modes corresponding to the width and height may also be present. However, in addition, diagonal model may also be present; for example, the gases might slosh between two diagonally opposite corners. Because of the corner convergence, gases flowing across the center will more rapidly pressurize the corner. This decreases the characteristic time and thus the characteristic wavelength. All these modes are represented, however, by the equation

$$2/\lambda = [(n_l/L_l)^2 + (n_{\mathrm{w}}/L_{\mathrm{w}})^2 + (n_{\mathrm{h}}/L_{\mathrm{h}})^2]^{1/2}, \qquad n_i = 0, 1, 2, \ldots. \tag{5}$$

For a cylindrical system, the equation for the characteristic wavelengths is

$$2/\lambda = [(\alpha_{mn}/R)^2 + (n_{\mathrm{h}}/L_{\mathrm{h}})^2]^{1/2}, \qquad m, n, n_{\mathrm{h}} = 0, 1, 2, \ldots, \tag{6}$$

where R is the duct radius and α_{mn} is the quantity given in Table III.

B. Helmholtz-Type Oscillations

While the acoustic systems discussed above correspond in shape to many combustion systems [and to room shapes, which can be important in certain instances (Speich *et al.*, 1957)], there are cases in which the individual components of a combustion system may each be considerably smaller than the wavelength of the observed mode. The classical example of the Helmholtz resonator illustrates this case.

Figure 2 shows a simple Helmholtz resonator. The usual manner for determining the fundamental frequency of this unit is to consider the gases in the neck of the resonator to be acting as a solid mass against the gases in the body of the resonator that are behaving as a spring (Kinsler and Frey, 1962). However, the relation can be derived on the basis of considerations already discussed.

The resonator is considered to be composed of two sections of ducts. First, the volume portion is considered to be equivalent to the closed end of a quarter-wave tube. As has been noted, the pressure and velocity curves have a sinusoidal shape. Therefore,

$$v_{(x=V_c/A)} = \bar{v}_c \sin(2\pi V_c/A\lambda) \cong \bar{v}_c(2\pi V_c/A\lambda). \tag{7}$$

The approximation is possible because of the small value of x relative to the wavelength. The pressure changes little in the distance, which is short compared with the wavelength, and to the same degree of approximation it

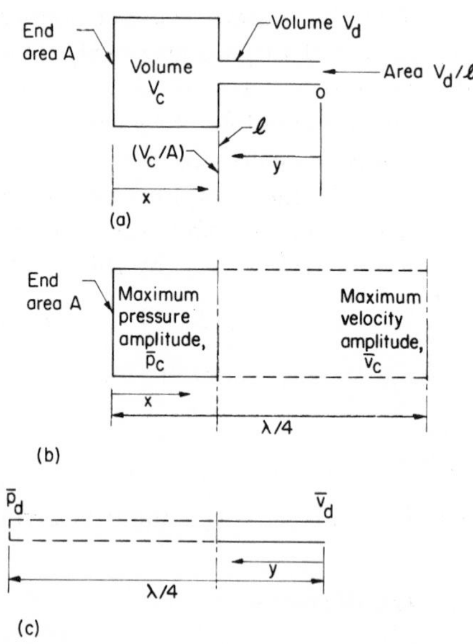

Fig. 2. Helmholtz resonator and components. (a) Helmholtz resonator. (b) Quarter-wave tube with same fundamental wavelength and large cross section. (c) Quarter-wave tube with same fundamental wavelength and small cross section.

is a constant value $\bar{p}_c$. For the neck portion of the resonator,

$$p_{(y=l)} = \bar{p}_d \sin(2\pi l/\lambda) \cong \bar{p}_d(2\pi l/\lambda), \tag{8}$$

while the velocity changes little from $\bar{v}_2$. The pressures at the interface are equal, that is,

$$p_{(y=l)} = \bar{p}_c, \tag{9}$$

and the volume displacements at the interface are equal, that is,

$$\bar{v}_d(V_d/l) = Av_{(x=V_c/A)}. \tag{10}$$

Therefore, if a universal relation is assumed between $\bar{p}$ and $\bar{v}$ of $\bar{p} = K\bar{v}$, all pressures, velocities, and K can be eliminated, and the following is obtained:

$$\lambda/2\pi l = (V_c/V_d)^{1/2}. \tag{11}$$

In a combustion problem, the temperature of the gas in the combustor itself is often different from that in the neck. From any standard acoustics reference it may be determined that K is the "characteristic acoustic resistance" of the fluid, ρc. Making use of this relation, with ρ and c differing in the two sections, it may be shown that the wavelength λ corresponds to the gas temperature in the neck.

For a Helmholtz resonator with two openings (which resembles many combustion systems in shape), the easiest manner in which to handle the probably different temperatures in the two necks is to note again that

$$\lambda f = c, \tag{12}$$

and to consider the volume portion of the resonator to be in two sections and the frequency of oscillation to be the same in both sections. It can rapidly be deduced that

$$2\pi f = [(V_{d1}/V_c)(c_1/l_1)^2 + (V_{d2}/V_c)(c_2/l_2)^2]^{1/2}. \tag{13}$$

The extension to the case of multiple necks is obvious.

C. Further Comments

Some distortion of this simple picture may be caused by a fan in the air-supply system, some possible vortex-shedding effects in the air- and fuel-supply systems, and leakage in the exhaust ducting. The first two distortions are usually not significant in considering pressure and velocity amplitudes in the class of combustion systems covered herein. Leakage in the exhaust ducting, if large, may cause significant changes in predicted frequencies and require careful study of the acoustic system.

From the inequality presented earlier, it may be noted that a flame in a region of appreciable pressure amplitude is more likely to produce oscillations. But, for the occurrence of many of the driving mechanisms that have been discussed, the acoustic component of the velocity must also be appreciable. The information about the Helmholtz resonator added to the conventional understanding of momentum effects in a flow of fluid from a nozzle reveals that both these conditions are often fulfilled at the entrance of a burner into a furnace.

Finally, the fact that energy is added to the oscillating gases in the region of the flame, while energy is dissipated at several other locations in the combustion system, leads to the conclusion that superimposed on the standing-wave system there must be a traveling-wave component to distribute the energy. But, as noted before, this component is small compared with the standing-wave component in the systems being considered.

IV. SUPPRESSION OF PULSATIONS AND OSCILLATIONS

Since few mechanisms of driving are understood well enough for the designer to design away from them, a more promising course of action for the cure of many pulsating combustors is to use one of a variety of damping schemes (Putnam, 1964c). Some of the common schemes have been use of (a) quarter-wave tubes, (b) Helmholtz resonators and multiple Helmholtz resonators, (c) sound-absorbing batting, (d) orifices, ports, or holes, and (e) baffles.

A. Quarter-Wave Tubes

One of the easiest suppression systems to understand is the use of a quarter-wave tube (Hanby, 1969). Within one-eighth wavelength of a point of maximum pressure amplitude, a tube such as that shown in Fig. 1c is inserted. A pressure pulse will then travel from the combustor to the closed end of the tube, reflecting from the end, and return at the time of minimum pressure in the combustor. Thus it will tend to cancel out the pressure variation. If the tube is too small in diameter compared with the combustion chamber or ducting to which it is attached, it will not be sufficiently effective. If the tube is too large, it will become a part of the general combustion configuration and may act only to shift the natural frequencies of the combustor. Because of the sharpness of their response, they are generally only suitable for laboratory installations where constant attention can be given to their tuning.

B. Helmholtz Resonators

The Helmholtz resonator (Fig. 2) acts much the same as the quarter-wave tube. A recent adaptation of this device, first in afterburners and then in rockets, is to use a double wall, with orifices through the inner wall (Phillips, 1968). This is essentially a multiplicity of Helmholtz resonators with connected chambers. For a variety of reasons, the response frequency broadens out as the number of orifices is increased, so this system can damp a wide range of frequencies. It has been found convenient to fill these resonators partly with a porous material such as an open-cell urethane foam; while this cuts the maximum damping effects somewhat, it improves the range of damping considerably.

C. Sound-Absorbing Batting

The multiple Helmholtz concept leads naturally to the idea of using damping materials, such as glass-wool blanket, in regions where temperatures are not excessive. In the higher frequency range, even small amounts, properly placed, can be quite effective. But in the low-frequency range, say below 200 Hz, such materials are of little use except in borderline cases.

D. Orifices, Ports, and Holes

Orifices, ports, and holes are excellent when they can be tolerated in or near a region of high pressure amplitude. While their presence can affect the amount and distribution of air supplied to the combustor, it was shown several years ago that the acoustic damping effect is a separate aspect (Sage and Schroeder, 1960). A more recent study has shown that the average flow rate, and thus the average pressure differential across these holes, is also important. There is an optimum pressure differential that can more than double the effectiveness of an orifice (Gordon and Smith, 1965). This optimum pressure differential gives a flow velocity through the orifice equal to twice the product of frequency and the sum of orifice length and diameter.

E. Baffles

Baffles are used in rockets to interfere with specific modes of oscillation, to increase "surface-to-volume ratio" losses, or to shift the minimum acoustic frequency upward. Since there is a tendency for driving effects to decrease with increasing acoustic frequency, this shift can be of great help. However, it does not appear too practical a solution in most combustor applications. Entrance and exit baffles and shrouds have also been found effective on occasion, but this is a highly erratic effect.

F. Other Comments

Sometimes a solution is found in moving the combustion region away from the region of maximum acoustic pressure amplitude. This is often possible in residential gas-fired heating units of modern design.

If a fan is used, a backward-pitch fan will ordinarily be more stabilizing than a forward-pitched fan. Increasing the fan diameter or speed and narrowing the fan, in order to move the operating point farther from the unstable peak-pressure operating point, also helps to reduce oscillations.

As noted above, the cure of a specific problem of combustion-driven oscillation or pulsation in "one-of-a-kind" or "few-of-a-kind" combustors is usually best approached by using some damping scheme. However, for a manufacturer of combustors in large quantities, such as in home heating units, there is a long-range economic advantage in determining in some detail the driving mechanism, so that the source of the pulsation can be specifically attacked.

V. COMBUSTION ROAR

Some of the problem areas associated with combustion roar are now considered. Since this is a newer area of research than that of combustion-driven oscillations, some general remarks are made before considering specific problems. But one must note first that combustion roar is produced by a moving flame front.

A. Review of Status

The first comments on combustion roar appear to have been made by Gaydon and Wolfhard (1953), followed by the first significant paper by Bragg (1963). Since then, there have been several single papers issued by various people, but since about 1966 to 1967 the group at Shell's Thornton Research Center (Thomas and Williams, 1966; Hurle *et al.*, 1968; Price *et al.*, 1969) and the group at Battelle (Putnam, 1968; Giammar and Putnam, 1970, 1971) appear to be the only continuous publishers of research in this area. The group at Shell has shown that the noise output of small turbulent premixed flames can be directly related to the rate of change of flame surface area. These investigators, and others, had also generally found that the noise output varied approximately with the square of the firing rate, indicating a monopole-like acoustic source.

Flames, principally turbulent diffusion flames, of much higher firing rates compared to previous investigations, up to nearly 0.3 MW, were studied at Battelle. The results showed several features that had not been observed in smaller flames. For instance, while the noise output seemed to vary about with the second power of the firing rate at lower firing rates, it appeared to vary as the first power of the firing rate at higher firing rates. Although the evidence is not conclusive, it appears that in thrust-controlled turbulent diffusion flames, the noise output varies with the first power of the firing rate and the efficiency of noise production is constant. This may be related to the constancy of flame size under these conditions. For buoyancy-controlled flames then, the variation with the square of the firing rate is approximated. (For small, premixed flames, a flame periodicity may exist that correlates the flame surface convolutions and leads to the monopole-like output.)

Figure 3 shows the acoustic spectrum obtained using three natural gas-firing rates from a pair of opposed fuels spuds aspirating their own air. The use of an anechoic chamber minimized the modal effects of enclosure and supporting interior structure. The similarity to these profiles of acoustic profiles obtained with natural gas flames from other burners after adjustment for other sources of noise, enclosure resonances, and flame size, is strongly suggestive that the basic profile is a function of the combustible mixture. Following this (unconfirmed) hypothesis further, the peak frequency of the noise spectrum might be closely related to the ratio of the

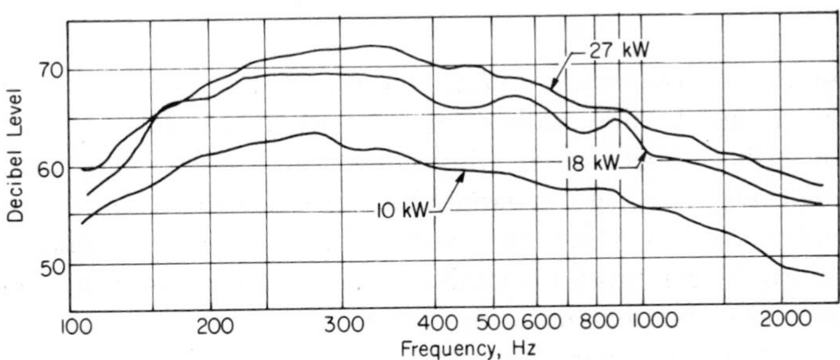

Fig. 3. Spectral curves for three firing rates of No. 52 spud with 0.127-m spud spacing, measurements made at 90° in anechoic chamber. [After Giammar and Putnam (1970).]

laminar burning velocity of the mixture to the laminar flame thickness δ. Now, whether or not these assumptions are correct, they give a starting point for analysis of the spectrum data as observed. It should be pointed out that even noise tests of premixed flames are not simple to make. The mere fact that burners are usually axially symmetric immediately adds a dimensional complication to the analysis. Furthermore, since ducts are used to supply the mixture, there is often noise generated in the ducting as well as the jet noise at the exit. At low flow rates, a flow periodicity can also show up. So the problem of obtaining data on the noise output of simple premixed flames for a range of fuel–air ratios and fuels to confirm or disprove the F/δ relation is not as straightforward as it might first appear.

Diffusion flames have certain obvious disadvantages for study, but they also have certain advantages. The edges of turbulent diffusion flames are the fastest burning mixtures and the strongest producers of sound. However, the difference of aspiration rate of large buoyancy-controlled turbulent diffusion flames as contrasted to the thrust-controlled turbulent diffusion flames leads to a difference in the rate of change of noise output with firing rate.

In the case of premix burners, recent studies of a series of sizes of similar accelerating nozzle burners (with small pilot flames), used singly and in pairs, indicated that the efficiency of conversion of chemical energy to acoustic energy varied essentially linearly with the pressure drop across the burners (Giammar and Putnam, 1971). Thus, for a single size of burner, an increase in firing rate would result in a noise power output varying about as the cube of the firing rate. Less extensive data seemed to indicate a similar result for nozzle-mix burners.

B. Noise Sources in Industrial Burners

The sources of noise from an industrial burner may now be considered. Starting in the supply lines, valves, orifices, bends, jets, obstacles, and the shear force of the velocity profile all produce noise (which may propagate to the outside and/or affect the flame) and turbulence (which decays more rapidly and which may affect the flame). With various transition effects, these pass into the combustion chamber. The turbulence intensity tends to amplify the noise, but seldom in direct proportion to the intensity. Noise in the high-frequency range (above 10 kHz) appear to reduce the low-frequency combustion roar. But even without these upstream effects, the normal combustion roar of the flame is amplified by resonance and shape effects of the combustion chamber itself. The noise and flame-generated turbulence and upstream turbulence next pass, with more transition effects, into the furnace. In the furnace, there is a repetition of these phenomena, especially when one considers that many flames are only partly enclosed within a burner. Finally, the noise passes into the interior room or factory space, and to whatever sensor system may be present. Some of these phenomena are now considered in more detail.

1. Enclosure Effects

The simplest enclosure effect to understand results from multiple paths of a signal from the source, such as a burner, to a sensor, such as a human ear or a microphone. As an example, if a burner is near a floor, then one signal path is a straight line, while a second is the reflected path from the floor. Considering frequency as varying over a range, there is variation back and forth between amplification and partial cancellation of the signal. With a source between two parallel walls, modes related to the spacing are reinforced. And finally, by following the same line of reasoning, one can see that room modes will be reinforced in the signal. The case of a combustion chamber may be visualized in another way. Assuming that the combustion process acoustically is composed of a set of monopole sources, which act without feedback from the surroundings, these can be considered then as a series of pistons acting on the volume of gases in the combustion chamber. As is well known, the various frequency inputs will be amplified an amount depending on their proximity to mode frequencies and on the damping.

Noise generated in a duct is not all propagated to the environment surrounding the end of the duct. As the frequency is lowered, a cutoff point is reached for each mode at which an increasing attenuation of the signal takes place as it exits from the duct, as the frequency decreases. It is rather interesting to note that for a monopole source in a long duct, the amplification of the source noise by the presence of the duct is exactly balanced by the reflection at a sharp-edged exit. The power of the outside signal is the same as the unamplified monopole source signal. The amplified portion either goes directly upstream or is reflected upstream from the exit. However, in most combustors, there is an upstream restriction that causes multiple reflections and further amplification.

2. Jet Noise

It would seem that jet noise effects would be predictable from the literature and not be much of a problem. However, it was discovered (1) that jet noise from small and/or oddly shaped spuds and openings that are typical of burners does not follow the usual assumption relative to variation of the Strouhal number at peak frequency with velocity and size; (2) that jet noise can occur in such a range of frequencies as to mask combustion roar; (3) that combustion amplification of jet noise can occur; (4) that the increase in flow velocity from combustion with constant mass flux can increase jet noise; and (5) that other amplifying effects are present when flow from an orifice is restricted by walls. In fact, in one case an orifice near an exit was found to combine with the exit in producing a "crow call."

Data from independent studies seemed to indicate that high-frequency noise, such as one might get from a restriction or jet in a line, would shift the noise output from the low-frequency combustion roar to an amplification of the jet noise. Thus an ultrasonic noise source might be used to suppress the noise that is normally heard, while a lower (but still high) frequency noise might make a burner sound worse by putting the energy in a range more sensitive to the human ear.

3. Noise from Blockage Plates

A series of tests were run using 50% blockage plates upstream various distances from a premixed burner. Tests were run with and without mufflers between the plates and the exit. It was found that the principal effect was from the turbulence generated by the grids. The higher the turbulence intensity at the burner, the higher the combustion roar. Turbulence scale did not seem to have any direct effect. However, the greater the number of holes, the closer the grid could be put to the exit because of the decay of intensity in a shorter distance. Since the noise spectrum of these grids was in a relatively low-frequency range where it was masked by the combustion roar, during combustion, it could not be determined within the accuracy of measurement whether acoustic suppression of the combustion or combustion amplification of the noise had occurred.

4. Effect of Flame Size

The effect of flame size was shown by a series of tests with two impinging fuel jets, at a constant firing rate, but with different orifice sizes, and spud spacings changed proportionally. The firing rate was such that the flames were buoyancy controlled rather than thrust controlled. The flames were all in a yellow–blue transition region, and were all of similar thermal volumes as near as one could tell, even though of different eccentricities, considering the flames as spheroids. The results are shown in Fig. 4. A cutoff at the high-frequency end of the spectrum penetrated to lower frequencies with increase in spud diameter, and thus spud spacing, in a manner strongly indicating cancellation of higher frequencies because of

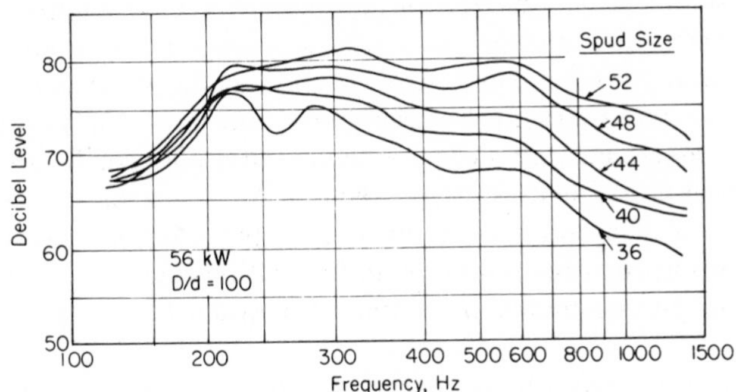

Fig. 4. Spectral curves for five spud sizes for a firing rate of 56 kW, spud spacing to spud diameter of 100, measurements made at 90°. [After Giammar and Putnam (1970).]

the large flame size. In fact, the cancellation agrees with simple theory based on the idea that the monopole noise sources in the flame are distributed on the flame surfaces toward the spud ends.

C. Summary of Combustion Roar

There are three sides to the problem of combustion roar: (1) the production of noise by the movement of the turbulent flame surface itself; (2) the production of noise by the combination of flow and configuration in the range of frequencies where either or both interaction or masking can occur; and (3) the production of the transmission, resonance, and damping effects that shape the noise spectrum that is observed.

VI. USES OF PULSE COMBUSTION

"Is there anything whereof it may be said, See, this is new? It hath been already of old time, which was before us" (Ecclesiastes 1 : 10). Happily, pulsating combustion cannot be dismissed in this way. All is not known or understood, and new ideas and devices continue to appear, although a dramatic breakthrough has yet to occur (Brown, 1971).

Pulse combustion devices have been proposed for a wide variety of applications. Perhaps the most important are first, the proposals for steam raising, water and air heating; second, those for aircraft propulsion; and last, those for constant-volume, pressure-rise combustion for gas turbines.

Other suggested applications are for gas compressors or pumps, gasifiers, sound producers, earth movers, aerosol generators, outboard motor, blowtorches or steam guns for paint removal, production of carbon monoxide synthesis gases and synthetic hydrocarbons, and even the production of electricity by squeezing a piezoelectric device with pulse combustion movement (Griffiths *et al.*, 1963; Putnam, 1964c). With a rising interest in the beneficial effects of pulsations on chemical engineering processes, such as extraction, absorption, distillation, and fluidization, there may be another application for pulse combustion in pulse generation (Milburn and Baird, 1970).

In the European area, the attraction of pulse combustion appears to lie in the promise of very high combustion intensities, the improved heat transfer in the presence of pulsations, and the direct conversion of thermal energy into kinetic energy or thrust. Elsewhere, there appears in addition an interest in elimination of fan and fan power, and in the case of aerodynamically valved units, a drastic reduction in total number of moving parts.

A. Basic Principles of Pulse Combustion

There are two basic types of pulse combustors for purposeful application, the mechanically valved units and the aerodynamically valved units. An array of flapper valves of some sort is used in the entrance region of a mechanically valved unit to prevent backflow through the inlet. In the aerodynamically valved units, an inlet is used which is so shaped that the pressure drops in the two flow directions through the inlet are greatly different at the same flow rate. In both types of pulse combustors, the mechanism of operation is basically the same.

Figure 5 shows schematically the operation of an idealized pulse combustor without mechanical valving. The four phases are shown: combustion, expansion, purge and recharge, and recharge and precompression. In sketch 1–2, it is assumed that a combustible mixture is present in the system. When there is an explosion within the combustion chamber, the gaseous products of combustion move outward in both directions, toward the entrance and the exit of the combustion chamber. However, because of the particular configuration, far more gas moves through the exit than through the entrance.

As the gases expand and move outward, the pressure drops in the combustion chamber. Eventually the pressure in the combustion chamber falls below atmospheric pressure and gas flow reverses. Although some of the combustion gases come back from the exit duct, the majority of the inward

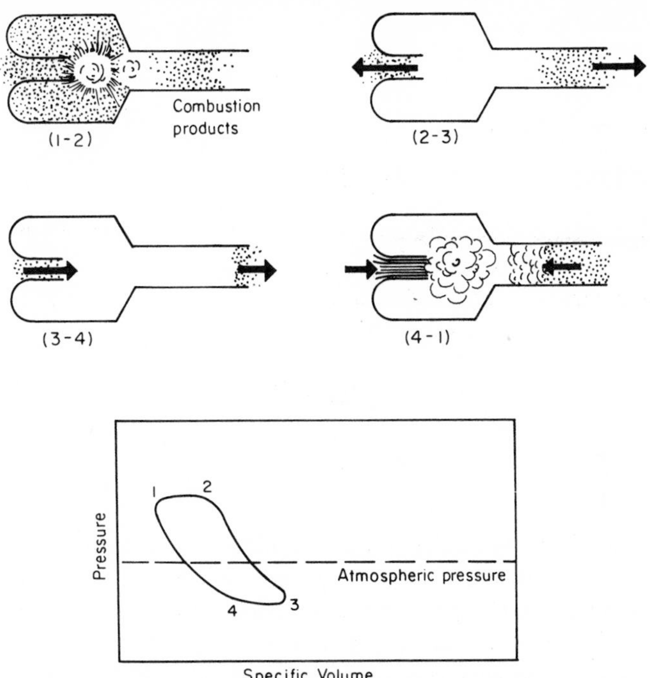

Fig. 5. Operating phases of idealized pulse combustor. (1–2) Combustion, (2–3) expansion, (3–4) purge and recharge, (4–1) recharge and precompress.

flow is through the entrance; most of the gas entering the combustion chamber is fresh air.

Gaseous fuel can be (1) premixed with the entering air, (2) fed continuously to the combustion chamber, (3) supplied from a plenum through a separate aerodynamic or mechanical valve, or (4) supplied from a tuned chamber. Liquid fuel can be (1) carbureted into the entering stream of air, (2) picked up by the entering airstream as it passes over a pool of liquid fuel, or (3) injected into the combustion chamber continuously throughout the cycle. Pulverized fuel can be (1) supplied continuously, or (2) carried in an airstream handled in one of the methods above. In any case, as the gases move back in and build up the pressure in the combustion chamber, a critical concentration of hot combustion products, fuel, and fresh air is reached at some point. The mixture ignites, and the cycle is repeated.

Inlet valve design is very important and much work has been done on both flapper valves and aerodynamic valves. A typical flapper valve is shown in Fig. 6. Air flows in the direction shown by the arrows through a

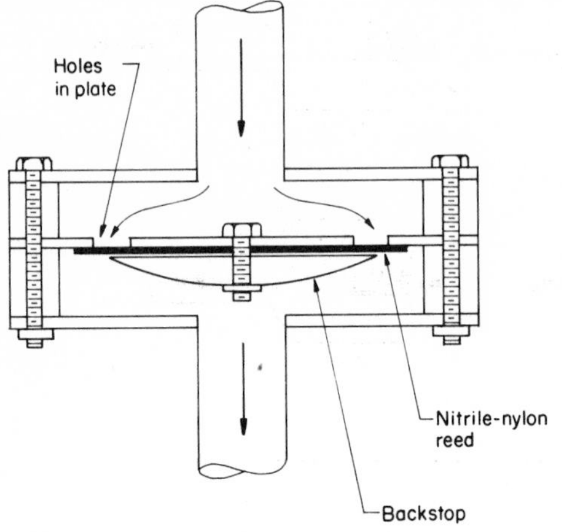

Fig. 6. Typical flapper valve.

series of holes in a steel disk and into the combustion chamber by displacing a nitrile–nylon reed, while chamber pressure is atmospheric or below. When combustion occurs, the pressure rises and the reed is forced back against the disk and airflow through the holes is either prevented or much reduced. Flapper valves are now reliable, long-lived, and permit use of air inlet silencing (Hanby and Brown, 1968; Katsnel'son *et al.*, 1969; Alebon *et al.*, 1963). However, more interest has been shown in aerodynamic valves; since they have no moving parts, they are suitable for operation at high temperature. The arrow indicates the direction of airflow into the combustor for the cup-shaped aerodynamic valve shown in Fig. 7. With this valve, the ratio of forward-to-reverse flow was about 5 : 1 for the same pressure drop (Bertin *et al.*, 1953). In a recent proposal for a spiral tube pulse combustor, the air inlet is at the center where the tube has the greatest curvature and smallest diameter (Severyanin and Lyskov, 1969a). Both these features help to reduce return flow and permit the use of simple valves or none at all. No valves are specified in another proposal for a ring- or torus-shaped combustor (Severyanin and Lyskov, 1968), similar to a much earlier proposal (Bodine, 1957a,b). For a two-chamber pulsating combustor, a freely oscillating damper diverts air first to one chamber and then to the other without stopping the airflow (Severyanin *et al.*, 1970); flow diversion by fluidic valves is also being studied.

Fig. 7. Example of aerodynamic valve. [From Bertin *et al.* (1953).]

B. Steam Raising, Water and Air Heating

1. Steam Raising

The achievement of high combustion intensities offers the possibility of economies in size and capital cost of steam-raising devices. Thus, pulverized coal, oil, and gas have all been tried in pulse combustors developed for this purpose.

The 16.7-MW coal combustor studied by Sommers (1961) is shown in Fig. 8. Either coal dust or pulverized coal was fed from a hopper and pneumatically metered into the downward-pointing conical combustion chamber, where it was mixed with 200–400°C primary air from a fan. Secondary air could be added at the lower end of the resonance tube. It was necessary to heat up the combustion chamber, resonance tube, and slag chamber with gaseous fuel before coal-dust combustion. The total length of combustion chamber and resonance tube was 4.5 m; the pulsation frequency of 30–40 Hz corresponded to quarter-wave organ-pipe operation. The resonance tube and slag chamber were cooled by the tubes of a Lamont system and this proved to be sufficient protection for the walls. A grate in the form of three vertical rows of pipes was provided at the end of the slag chamber. The old boiler used was a tilted tube nest with two steam superheaters arranged between. The 12% ash content coal dust used had a calorific value of 28,800 kJ/kg, with particle size 38% greater than 90 μm. The combustion intensity was 21 MW/m^3 in the resonance tube and 2.7

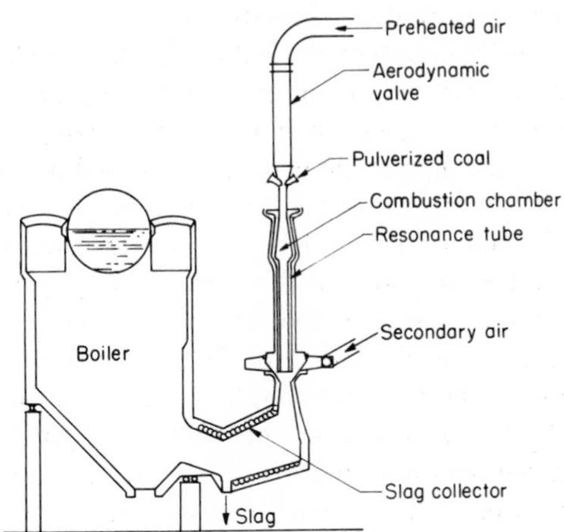

Fig. 8. Pulse combustor–boiler system tested with coal dust and with pulverized coal by Sommers (1961).

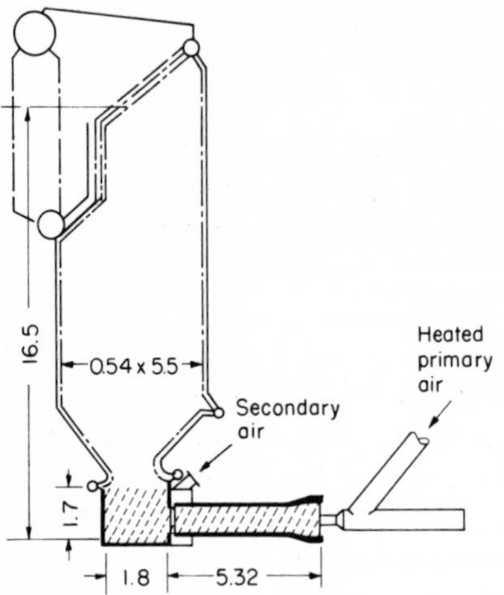

Fig. 9. Dual-pulse combustor-boiler system tested with fuel oil. [From Yu. L. Babkin, *Thermal Engineering* **12**(9), 31–36 (1965), by kind permission of the Department of Education and Sciences. *Thermal Engineering* is a cover-to-cover translation of the Russian periodical *Teploenergetika* and is available through Pergamon Press, Oxford.]

MW/m^3 in the resonance tube plus slag collector. About 20% of the total heat transfer was in the resonance tube with a heat flux of 454 kW/m^2, and 32% in the resonance tube plus slag collector, with a heat flux of 174 kW/m^2. Slag flowed as a colorless liquid from the edge of the tube, through the slag chamber into the sump. [Little trouble with slag has been experienced in either this or other smaller scale work (Hanby and Brown, 1968; Severyanin, 1969).] The gas oscillation produced in the tube was damped relatively quickly in the flue flow path but, despite this, the crown of the boiler and other components vibrated considerably.

The 49-MW oil-fired system studied by Babkin (1965) is shown in Fig. 9 (dimensions are in meters and the cross-hatching shows the zone of visible flame). In the figure, two coupled combustors working under stable counterphase conditions are shown installed under a medium pressure boiler. The discharge from the combustors was directed into the inoperative slag hopper and then into the boiler furnace. The combustors were charged with primary air at 150°C and a pressure of 3 kN/m^2. The slightly fuel-rich mixture necessary for good pulsation stability was found to burn cleanly, but secondary air had to be admitted radially around the exit of each resonance tube to permit combustion to be completed. Acoustic damping was achieved by means of blind branch pieces adjusted to be slightly above the working frequency. Refractory lining was used only in the ignition chambers. The resonance tubes were fully screened and the screen tubes connected to the power station's hot water supply system. Despite a large number of defects in design and installation, the design output was achieved in test runs. With a fuel consumption of 4200 kg/hr of high-sulfur fuel oil, the combustion intensity was 7.2 MW/m^3, and the heat flux in the resonance tube was 300 kW/m^2, similar to that quoted by Sommers for coal. The total amplitude of the pressure oscillations was 15 kN/m^2; the noise in the boilerhouse was similar to that from the fan.

For efficient mixing after entry and before combustion, the importance of matching the momentum of fuel and airstreams has been recognized (Hanby and Brown, 1968), as has the usefulness of tangential entry of fuel and air into the combustion chamber (Hanby and Brown, 1968; Severyanin, 1969; Katsnel'son *et al.*, 1969). Both an expanding cone shape of chamber and a stabilizer bar in a cylindrical chamber have been found to improve recirculation, ignition, and combustion (Hanby and Brown, 1968; Francis *et al.*, 1963). Preheating chambers outside but coaxial with the combustion chamber have improved volatile release and ignition and combustion performance (Severyanin, 1969; Severyanin and Lyskov, 1969b).

Very little attention has been paid to boiler design. Coupling together and silencing of combustors are discussed later.

2. Water Heating

A successful application of pulse combustion is the Lucas–Rotax Pulsamatic 29.3-kW gas-fired furnace and hot water boiler for space heating (Alebon *et al.*, 1963; Greensteel Hydronics Limited, 1963) shown in Fig. 10.

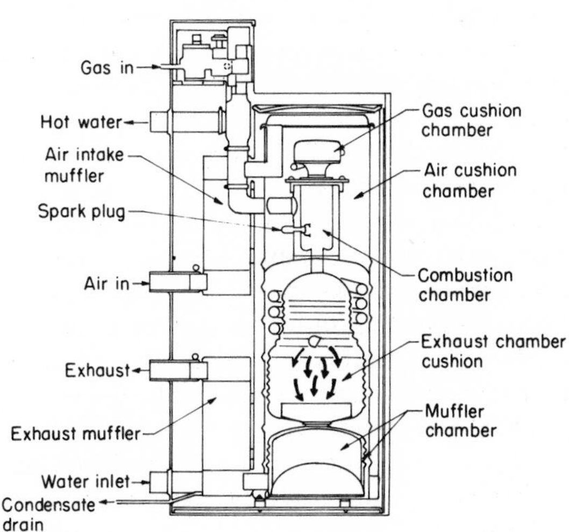

Fig. 10. Lucas–Rotax Pulsamatic residential hot water boiler. [From Greensteel Hydronics, Ltd. (1963).]

The American Gas Association has carried out extensive work on modifications of the design, not only for residential use in hot-water and hot-air heating, but also for small commercial use (Griffiths and Weber, 1969). This heating unit is much smaller than a comparable furnace with standard gas firing and requires neither special foundations nor chimney. The burned gases are exhausted by means of a 5-cm plastic pipe through the most suitable outside wall, and fresh air is supplied in a similar manner. Since the exhaust gases may be cooled below their dew-point temperature, a condensate drain pipe to the sewer is necessary.

Automatic start-up is initiated by a thermostat on heat demand. This signal starts simultaneous operation of a scavenging fan, a spark plug, and a seven-stage motor-operated relay switch. After 11 sec of fresh-air purging, this switch admits a single metered fuel charge to the combustion chamber by means of a solenoid-operated gas valve. The gas–air mixture is ignited by the spark and thus initiates the pulse combustion cycle. Within seconds, the initial pulsations reach resonant frequency and the combustion chamber pressure reaches its optimum level. At this point, the pressure switch of the control system cuts off the power supply to the scavenging fan and to the spark plug and to the motor of the relay switch, while maintaining the gas valve in an open position; combustion then becomes self-supporting. On reaching the desired temperature in the heated space, the thermostat interrupts the power supply to the gas valve and this shuts off the furnace. The furnace is also automatically shut down when fuel or power supplies fail or when boiler water temperature rises above a preset limit.

The gas and air cushion chambers are above the minimum volume required to ensure stable operation and are made rigid and sufficiently dense to reduce noise transmission to a minimum. The annular ring-impregnated Dacron flapper valves oscillate at 70 Hz with an amplitude of 0.056 cm on the gas and 0.142 cm on the air side.

The bigger and thicker valve, the one on the air side, weighs only 1 gm. The fuel–air mixing head ensures complete mixing of the air and gas streams immediately downstream from the inlet valves. The flame trap embodies a solid core specially designed to direct the flow alongside the combustion chamber walls. With the combustion chamber and adjoining jet pipe resonating as a quarter-wave organ pipe at 70 Hz, the combustion chamber temperature is 1280°C and the average gas temperature is 530°C. There are no components with the sole function of heat transfer from the hot gases but combustion chamber, jet pipe, exhaust cushion chamber, and high-pass muffler pot are all water cooled and made of copper or copper alloy to obtain good heat-transfer rates. By contrast, components with the principal or only purpose of suppressing noise include the air intake, exhaust mufflers, primary air cushion chamber, exhaust cushion chamber, and high-pass muffler pot. Noise suppression was extremely effective, the final overall sound pressure level being 62 dB.

3. Air Heating

After many years of development, the pulse combustion air heater shown in Fig. 11 is now used for buses and other service vehicles. It has also been suggested for use as an immersion or flow-through heater, mixing tube, and turbo heater (Huber, 1964, 1965, 1969; Eberspacher, 1963).

The mode of operation may be explained by reference to the numbers in the figure. For starting, the preheater plug 13 and the spark plug 3 are actuated at the same time as air is supplied through valves 27 and 28 into the mixing chamber 10 and carburetor 31 from a diaphragm air pump. The carburetor is connected by air tube 23 to the sealed gasoline tank 22 to permit its pressurization. The fuel is pushed through tube 25, fuel filter 24, and pipe 32 to the regulator valve 34. As soon as knob 8 is unscrewed, regulator valve 34 is opened and gasoline flows to atomizer nozzle 30, as does air. This mixture and more air coming through valve 27 flow through into the mixture chamber. Warmed up by preheater plug 13, the mixture flows toward spark plug 3. Ignition takes place there, causing an explosion in the combustion chamber 2. The pressure rises also in mixing chamber 10 and pulsation or resonance tube. Air valves 27 and 28 and diaphragm valve 12 all close.

A small amount of gas flows through throttle 29 and atomizer nozzle 30 to the carburetor. In the atomizer nozzle 30, gasoline is sucked in and carried on to carburetor 31. Most of the gas is pushed out through resonance tube 1. The mass of gas flowing in the tube then reverses, after the pulse, in

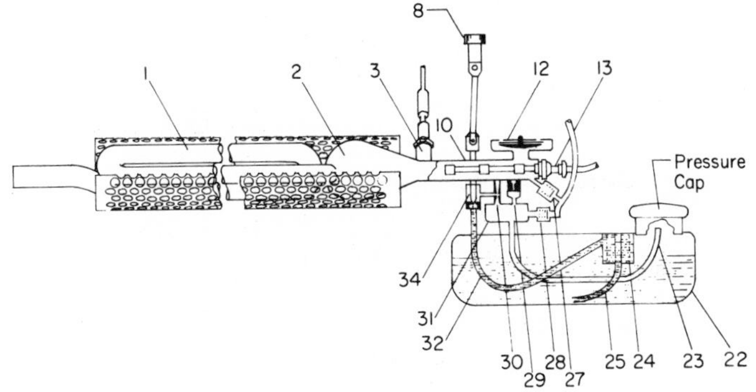

Fig. 11. Huber air heater. [From Huber (1964).]

combustion chamber 2 and mixing chamber 10, and diaphragm valve 12 opens again. The air flowing through the diaphragm valve toward the mixing chamber 10 mixes with gasoline sucked from atomizer 30 by gas flowing from carburetor 31. Since throttle 29 prevents backflow of gas, there is an excess pressure in carburetor 31 and gasoline tank 22 which maintains the gasoline flow.

In practice, this pulse combustor is used either to heat fresh air flowing over the resonance tube portion, or the resonance tube portion is immersed in a liquid to be heated. In addition, the exhaust gas can be used for ejecting air from an enclosure, and thus providing the through-flow energy for moving warmed air through an enclosure. For drying operations, the jet action of the exhaust gases may be used to aspirate cooler air into the jet. At some point, depending on the use, a silencer must also be provided.

4. Air Heating for Drying and Conveying Applications

Details have been reported of an 88-kW combustor designed to dry and convey maize (corn) (Muller, 1967) and on combustors for lignite drying and conveying (Ellman *et al.*, 1966, 1969). The arrangement of the 205-kW propane-fired combustor incorporated into an entrained lignite drying system is shown in Fig. 12. Drying gas temperatures ranging from 370 to 790°C were obtained by regulating the fuel rate and the amount of secondary air. The drying column was a 20-cm-diameter, 15.25-m-long, hairpin-shaped tube which ended at a 60-cm-diameter product separation cyclone. Although lignite particles of 2-cm size were easily entrained, drying of finer material

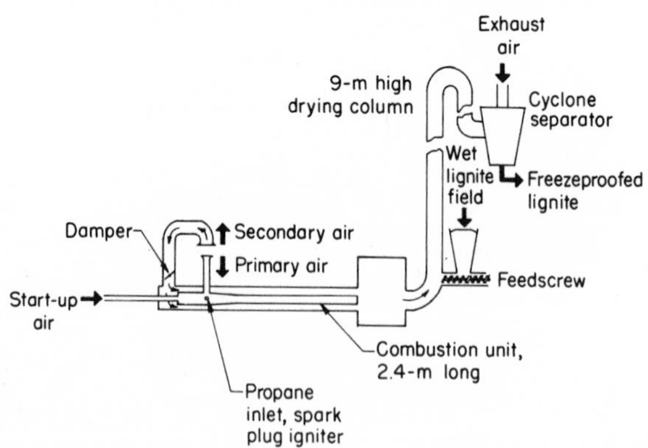

Fig. 12. Experimental drying system for freezeproofing lignite. [From Ellman *et al.* (1966).]

was more effective. The moisture content of the dried lignite ranged from 10% at a feed rate of 320 kg/hr to 30% at a feed rate of 3200 kg/hr.

A larger, 7.3-MW pulse jet has been built and operated on propane, residual oil, and on a mixture of lignite with 3% by weight of residual oil. The combustion intensity attained was 5.2 MW/m^3 but no figures have been given for drying tests. The apparatus had an overall length of 10.7 m, the main tube being 9.15-m-long. Powdered lignite, a waste product of the lignite mining operation, was introduced by dropping it down a stainless steel standpipe projecting to the center of the refractory-lined combustion chamber. The high-amplitude pulsations were found to facilitate ash discharge; the molten ash formed a pool in the bottom of the combustion chamber and was ejected as it flowed into the tailpipe section. The noise, mainly at low frequency, was found to be less of a problem than expected.

C. Aircraft Propulsion Applications

The SNECMA "Escopette" (Blunderbuss) pulse jet (Bertin *et al.*, 1953, 1957; Marchal and Servanty, 1963) for the direct propulsion of target aircraft is an improvement, with aerodynamic valves, of the Schmidt–Argus mechanically valved engines used to power the V1 "buzz-bomb" in World War II (Schmidt, 1957; Gosslau, 1957). A later development, the "Ecrevisse" (Crayfish) is shown in Fig. 13. The inlet is bent to face the same way as the exhaust in order to use not only the main exhaust thrust but also that developed by reverse flow through the aerodynamic valve. The use of a separate short tube, designed as a thrust augmenter, was found to improve aspiration. The fuel consumption of the Escopette was 1.8 kg/hr per kilogram of thrust produced. Pulse jets have also been suggested as propulsive units for helicopter blade tips (Anon., 1953) but there has been no commercial development on these lines.

Lockwood (1964) used two augmenters on the Escopette pulse jet in developing a vertical lift device; this would be particularly valuable where helicopters and similar craft would throw up dirt which might then move into the engine interior. But this direction of research eventually led to large drying units composed of multiples of the Escopette–augmenter sets. The pulse action gave faster drying of particulate material dropped through the

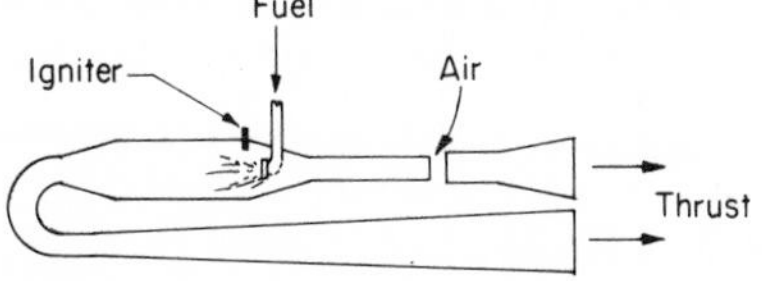

Fig. 13. Ecrevisse pulse jet.

hot exhaust gases. Mufflers were used on the inlet and exhaust of entire systems.

D. Constant-Volume Pressure-Rise Combustion for Gas Turbines

A tireless advocate of the use of pulsating combustion for gas turbines was Reynst (1961). He was attracted by this way of burning fuel with previously compressed air to give a gas pressure rise before expansion into the turbine, instead of the pressure drop and loss of efficiency which occurs with the conventional constant-pressure combustion system. The idea was not new, of course, having been studied in and before the year of Reynst's birth, 1909. Karavodine (Roberson, 1951) used four chambers pulsating at a frequency of 38 Hz, equally spaced around the turbine; the pressure fluctuated between 34.5 kN/m^2 above and 13.8 kN/m^2 below atmospheric but the turbine output was only 746 W and the overall thermal efficiency only 2.8%. The Holzwarth explosion turbine was much more complicated, depending on hydraulically operated valves and external compression of air and fuel to give a working top pressure of as high as 960 kN/m^2 but was developed over a period of 23 years into a useful and efficient machine. A six-chamber oil-fired machine installed for the Prussian State Railways in 1920 had an output of 350 kW at 3000 rpm and a two-chamber design was put into operation in a Hamborn steelworks, using waste blast furnace gas as fuel. These and the oil-fired machine installed by Jendrassik in Budapest in 1937 had a thermal efficiency between 20 and 30%. Reynst believed that the chief problem in the construction of a pressure producer was to ensure that the expansion energy of the hot gases was transmitted to the cooling air, which was to be compressed with least loss. He suggested the use of Pescara's free-piston device in which all the air is compressed from atmospheric pressure to turbine inlet pressure by means of a piston pump driven by a two-stroke diesel engine (Anon., 1951; Meltzner, 1957). The combustion air reacts stoichiometrically with the fuel in the combustor, the extra air being mixed with the exhaust gas at turbine inlet pressure during the exhaust phase of the diesel engine.

In recent work, the emphasis has been on the development of pressure-gain combustors with aerodynamic valves, that is, no moving parts, and of multiple-combustor assemblies with operation phased so as to smooth out damaging pressure fluctuations at the turbine inlet and so preserve the high efficiency of modern turbines and compressors (Porter, 1958; Marchal and Servanty, 1963, 1966; Marchal, 1968; Muller, 1969). The pressure rise reported by the SNECMA workers (Marchal, 1968) is only 1% but represents an increase in efficiency of 6–8% when compared to the 5–7%

pressure loss in conventional gas turbines. There is thus some incentive for further work. Progress has been slow in work on multiple-combustor assemblies. Four engines with a common outlet were connected and gave regular operation with about 90° phase difference between them (Porter, 1958). Bench and flight tests with groups of 2, 3, 6, 9, and 30 "Escopette" engines showed that the engines worked in phase opposition, with successive ignition and never two simultaneous explosions (Marchal and Servanty, 1966). The overall performance of particularly the large groups was, however, poor (Marchal and Servanty, 1963).

E. Gasification

Only small-scale experimental work has been reported on gasification (Traenckner, 1953). The device used was 1.6 m long and had a diameter of 35 mm except for a pear-shaped chamber at one end. A minimum chamber temperature of 1250°C was needed and best results were obtained with preheated fuel and air; reduction of heat losses enabled a larger excess of fuel to be used. Gasoline was used for some test work but more interest was shown in lignite gasification with an efficiency of 63% and a carbon conversion of 93%. With high-volatile bituminous coal, the once-through gasification efficiency was only 36% and the carbon conversion 62%. In both cases, only low-quality gas with a heating value of 3700 kJ/m^3 was produced. Similar results were obtained by the U.S. Bureau of Mines in steam–oxygen gasification of coal at Morgantown, West Virginia.

More recent work, but no large-scale development, has been reported from the U.S.S.R.

F. Earth Moving

A recent development in earth moving is interesting because, although not strictly a pulse combustor, it may lead to the application of pulse combustors in developing REDSOD (Repetitive Explosive Device for Soil Displacement) (Anon., 1969; Clark *et al.*, 1969), which consists of one or more combustion chambers, exhaust valves, and exhaust outlets. Mounted on the front of a tractor, the exhaust outlets are pushed into the soil. The combustion chambers are charged with a mixture of compressed air and gasoline, the exhaust valves being held closed by the pressure force resulting from the difference in area between the valve poppet and valve piston. The fuel–air mixture is then ignited and burned, causing the pressure and temperature of the gas in the chamber to rise (with a charge pressure of

690 kN/m^2 above atmospheric, the combustion pressure reaches 6200 kN/m^2, and the gas temperature 2500°C). After combustion is complete, the exhaust valves are opened very rapidly and the expanding exhaust gases rupture the soil and blow it from the crater.

G. The Noise Problem

In all applications of pulse combustion, the problem of unwanted noise arises. The pulsating flow leaving the exit pipe radiates as a monopole source, of strength controlled by the change in volume flow, for which the sound power at the fundamental frequency is given by (Harris, 1957; Lassiter, 1952; Powell, 1953)

$$W = (\pi\rho' c'/2c_0^2) V^2 f^2 S^2 \quad \mathrm{W}, \tag{14}$$

where p', c' are the assumed average values for density and sound velocity at the tube exit, and ρ_0, c_0 are the density and sound velocity in the surrounding atmosphere. The velocity

$$V \approx \tfrac{1}{2}(V_{\max} - V_{\min}),$$

where $V_{\max}$ is the peak velocity during the exhaust half-cycle, and $V_{\min}$ is the peak velocity during the intake half-cycle. The firing frequency is f and the exit pipe area $S = \pi d^2/4$. Approximating,

$$\log_{10} W = 2 \log_{10} V + 4 \log_{10} d + 2 \log_{10} f - 10.5,$$

with V in meters per second, and d in centimeters; $V = 100$ m/sec, $d = 10$ cm, $f = 100$ Hz, $\log_{10} W = 1.5$, and $W = 32$ W.

Assuming uniform hemispherical radiation over a nonabsorbing surface, the measured sound pressure level in decibels relative to 2×10^{-5} N/m^2 is given by

$$\mathrm{SPL}/10 = 12 + \log_{10} W - \log_{10} 2\pi r^2$$

with r in meters. A sound power of 32 W corresponds, therefore, to a sound pressure level at 1 m of 135 dB. For spherical radiation, such as that from an unbaffled tube exit, the decibel level will be three lower.

It is of interest to relate the sound power to the power available in the combustor and to derive the acoustic efficiency. This efficiency has been calculated and is approximately 10^{-4} for several of the pulse combustors discussed previously (Francis *et al.*, 1963; Reay, 1969; Alebon *et al.*, 1963; Muller, 1967; Ellman *et al.*, 1966, 1969) with a typical exit velocity of 100 m/sec. Aerodynamic noise must be considered also. Whereas the pulse-jet exhaust noise increases as $V_{\max}^2\, d^4$, the aerodynamic noise increases as

$V^6 d^2$ so that the aerodynamic noise should increase relatively as V is increased and d decreased. This has been found to occur with single pulsating combustors and with two coupled combustors, operating with a 180° phase lag. Operation of two 17-cm-diameter combustors coupled together gave an overall noise reduction of 8 dB and a 20-dB reduction at the fundamental frequency compared with the two combustors working separately (Harris, 1957), and similar results have been obtained elsewhere. The configuration preferable for gas turbine application thus holds promise for general use also. The noise problem is not an insoluble one, as has been shown by the successful application of conventional silencing methods (Alebon *et al.*, 1963; Huber, 1964, 1965; Eberspächer, 1963) and phase lag operation (Babkin, 1965) to pulse combustors.

References

Alebon, J., Lee, G. K., and Geller, L. B. (1963). A Pulsating Combustion System for Space Heating. *Proc. Boyar Conf.*, p. 61. Univ. of Montreal, Quebec, Canada.

Anonymous (1951). *Motortechn. Z.* **18,** 108, 195.

Anonymous (1953). Pulse-Jet Para-Copter. *Aeroplane* **85,** 481.

Anonymous (1969). REDSOD. S.W. Res. Inst., San Antonio, Texas.

Babkin, Yu. L. (1965). Pulsating-Combustion Chambers as Furnaces for Steam Boilers. *Thermal Eng.* **12,** 31.

Bertin, J., Paris, F., and LeFoll, J. (1953). The SNECMA Escopette Pulse-Jets. *Interavia* **8**(6), 343.

Bertin, J., Paris, F., and LeFoll, J. (1957). U.S. Patent 2,812,635.

Bodine, A. G., Jr. (1957a). Sonic Burner Heat Engine with Acoustic Reflector for Augmentation of the Second Harmonic. U.S. Patent 2,796,734.

Bodine, A. G., Jr. (1957b). Acoustic Jet Engine with Flame Deflection Fluid Pumping Characteristics. U.S. Patent 2,796,735.

Bragg, S. L. (1963). Combustion Noise. *J. Inst. Fuel* **36,** 12.

Brown, D. J. (ed.) (1971). *Proc. Int. Symp. Pulsating Combust., 1st, September 1971.* Univ. of Sheffield, Sheffield, England.

Clark, J. M., Jr., Hemion, R. H., and Brown, R. W. (1969). U.S. Patent 3,461,577.

Crocco, L., and Cheng, S. I. (1956). Theory of Combustion Instability in Liquid-Propellant Rocket Motors, AGARDograph 8. Butterworths, London.

de Saint-Martin, L., Pariel, J. M., and Himber, F. (1963). Combustion Pulsatoire aux Cowpers. *Rev. Met. (Paris)* **60,** 631.

Eberspächer, J. (1963). Eberspächer Pulsating Combustor. Information Sheet R-1862, P.O. Box 289, Esslingen/N, West Germany.

Ellman, R. C., Belter, J. W., and Dockter, L. (1966). Adapting a Pulse-Jet Combustion System to Entrained Drying of Lignite. Paper 66, *Proc. Int. Coal Preparation Congr., 5th,* p. 463. Pittsburgh, Pennsylvania.

Ellman, R. C., Belter, J. W., and Dockter, L. (1969). Operating Experience with Lignite-Fueled Pulse-Jet Engines. *Amer. Soc. Mech. Eng.* Paper 69-WA-FU-4.

Francis, W. E., Hoggarth, M. L., and Reay, D. (1963). A Study of Gas-Fired Pulsating Combustors for Industrial Applications. *J. Inst. Gas Eng.* **3,** 301.

Gaydon, A. G., and Wolfhard, H. G. (1953). "Flames," 1st ed., p. 158. Chapman & Hall, London.

Giammar, R. D., and Putnam, A. A. (1970). Combustion Roar of Turbulent Diffusion Flames. *J. Eng. Power* **92,** 157.

Giammar, R. D., and Putnam, A. A. (1971). Summary Report—Noise Generation by Turbulent Flames. Catalog No. M00080, Amer. Gas Ass., Arlington, Virginia.

Gordon, C., and Smith, P. W., Jr. (1965). Acoustic Losses of a Resonator with Steady Gas Flow. *J. Acoust. Soc. Amer.* **37** (2), 257.

Gosslau, F. (1957). Development of V-1 Pulse Jet. AGARDograph No. 20: History of German Guided Missiles Development (Th. Benecke and A. W. Quick, ed.), p. 400. Appelhaus, Salzhitter-Lebenstedt, West Germany.

Greensteel Hydronics Limited (1963). Brochure: Lucas-Rotax Pulsamatic Boiler. Winnipeg, Manitoba, Canada.

Griffiths, J. C., and Weber, E. J. (1969). The Design of Pulse Combustion Burners. *Amer. Gas Ass.* Lab. Res. Bull. No. 107.

Griffiths, J. C., Thompson, C. W., and Weber, E. J. (1963). New or Unusual Burners and Combustion Processes. *Amer. Gas Ass.* Res. Bull. No. 96.

Hanby, V. I. (1969). Convective Heat Transfer in a Gas-Fired Pulsating Combustor. *J. Eng. Power* **91,** 48.

Hanby, V. I., and Brown, D. J. (1968). A 50 lb/h Pulsating Combustor for Pulverized Coal. *J. Inst. Fuel* **41,** 423.

Harris, C. M. (1957). "Handbook of Noise Control." McGraw-Hill, New York.

Heidmann, M. F., and Wieber, P. R. (1966). An Analysis of the Frequency Response Characteristics of Propellant Vaporization. NASA TN D-3749.

Higgins, B. (1802). On the Sound Produced by a Current of Hydrogen Gas Passing through a Tube, by Editor, Mr. Nicholson, with a letter from Dr. Higgins respecting the time of its discovery. *J. Nat. Philos. Chem. The Arts* **1,** 129.

Huber, L. (1964). Pulsator-Fired Heater for Service Vehicles. *Automobiltechn. Z.* **66** (2), 31.

Huber, L. (1965). New Developments With the Eberspächer-Pulsating Combustor. *Automobiltechn. Z.* **67** (9), 296.

Huber, L. (1969). Gas-Fired Pulsating Combustor. *Heiz-Luft-Haustechn.* **20,** 436.

Hurle, I. R., Price, R. B., Sugden, T. M., and Thomas, A. (1968). Sound Emission from Open Turbulent Premixed Flames. *Proc. Roy. Soc.* (*London*) *Ser A* **303,** 409.

Katsnel'son, B. D., Marone, I. Ya., and Tarakanovsky, A. A. (1969). An Experimental Study of Pulsating Combustion. *Thermal Eng.* **16** (1), 4.

Kinsler, L. E., and Frey, A. R. (1962). "Fundamentals of Acoustics," 2nd ed., Chapter 8. Wiley, New York.

Lassiter, L. W. (1952). Noise from Intermittent Jet Engines and Steady-Flow Jet Engines with Rough Burning. NACA Tech. Note 2756.

Lockwood, R. M. (1964). Pulse-Reactor Low Cost Life-Propulsion Engines. AIAA Paper No. 64-172.

Marchal, R. (1968). Gas Turbines with Pulsating Combustion Chambers. *Entropie* (22), 15.

Marchal, R., and Servanty, P. (1963). Note on the Development of Valveless Pulse-Jets. Session of the Association Technique Maritime et Aeronautique, pp. 1–20. ATMA, Paris.

Marchal, R., and Servanty, P. (1966). Gas Turbines with Pulsating Combustion Chambers. *Entropie* (11), 37.

Meltzner, H. (1957). The Free-Piston Turbine, a New Type of Power Plant for Ships and Industry. *Motortechn. Z.* **18,** 108.

Merk, H. J. (1956–1957). Analysis of Heat-Driven Oscillations of Gas Flow—II. On the Mechanisms of the Rijke-Tube Phenomenon, *Appl. Sci. Res.* (*Netherlands*) *Sect. A* **6,** 402.

Milburn, C. R., and Baird, M. H. I. (1970). Glow Pulsation Generator for Pilot-Scale Studies. *Ind. Eng. Chem. Process Des. Dev.*, **9,** 629.

Muller, J. L. (1967). The Development of a Resonant Combustion Heater for Drying Applications. *S. Afr. Mech. Eng.* **16** (7), 137.

Muller, J. L. (1969). Theoretical and Practical Aspects of the Application of Resonant Combustion Chambers in the Gas Turbine. Council for Sci. and Ind. Res. (CSIR) Rep. MEG 831, Pretoria, South Africa.

Pariel, J. M., and de Saint-Martin, L. (1965). Elimination of Combustion Instabilities in Blast Furnace Stoves. *Rev. Mét.* (*Paris*) **62,** 537.

Phillips, B. (1968). Effects of High-Wave Amplitude and Mean Flow on a Helmholtz Resonator. NASA TMS-1582.

Porter, C. D. (1958). Valveless-Gas-Turbine Combustors with Pressure Gain. ASME Paper No. 58-GTP-11.

Powell, A. (1953). The Noise of a Pulse Jet. *J. Helicopter Ass. Great Brit.* **7,** 32.

Price, R. B., Hurle, I. R., and Sugden, T. M. (1969). Optical Studies of the Generation of Noise in Turbulent Flames. *Int. Symp. Combust., 12th,* p. 1093. Combust. Inst., Pittsburgh, Pennsylvania.

Putnam, A. A. (1964a). General Considerations of Autonomous Combustion Oscillations. *In* "Non-Steady Flame Propagation, AGARDograph 75" (G. H. Markstein, ed.), Chapter F. Pergamon, Oxford.

Putnam, A. A. (1964b). Experimental and Theoretical Studies of Combustion Oscillations. *In* "Non-Steady Flame Propagation, AGARDograph 75" (G. H. Markstein, ed.), Chapter G. Pergamon, Oxford.

Putnam, A. A. (1964c). Practical Considerations of Combustion Oscillations. *In* "Non-Steady Flame Propagation, AGARDograph 75" (G. H. Markstein, ed.), Chapter H. Pergamon, Oxford.

Putnam, A. A. (1968). Flame Noise From the Combustion Zone Formed by Two Axially Impinging Fuel Gas Jets. *Fuel Soc. J. Univ. Sheffield* **19,** 8.

Putnam, A. A. (1971). "Combustion-Driven Oscillations in Industry." American Elsevier, New York.

Putnam, A. A., Hyatt, R. S., and Rodman, C. W. (1967). Elimination of Combustion-Driven Oscillations in a Large Air Heater. ASME Paper 67-WA-FU-2.

Rayleigh, Lord (1945). "The Theory of Sound," Vol. 2, p. 226. Dover, New York.

Reay, D. (1969). The Thermal Efficiency, Silencing, and Practicability of Gas-Fired Industrial Pulsating Combustors. *J. Inst. Fuel* **42,** 135.

Reynst, F. H. (1961). "Pulsating Combustion" (M. W. Thring, ed.). Pergamon, Oxford.

Roberson, E. G. (1951). "The Industrial Gas Turbine." Temple Press, London.

Sage, R. W., and Schroeder, J. F. (1960). Noise Suppression in Oil Burners. *ASHRAE J.* **2,** 53.

Schmidt, P. (1957). On the History of the Development of the Schmidt Tube, "AGARDograph No. 20. History of German Guided Missiles Development" (Th. Benecke and A. W. Quick, eds.), p. 375. Appelhaus, Salzhitter-Lebenstedt, West Germany.

Severyanin, V. S. (1969). The Combustion of Solid Fuel in a Pulsating Flow. *Thermal Eng.* **16** (1), 9.

Severyanin, V. S., and Lyskov, V. Ya. (1968). U.S.S.R. Patent No. 222580.

Severyanin, V. S., and Lyskov, V. Ya. (1969a). U.S.S.R. Patent No. 235893.

Severyanin, V. S., and Lyskov, V. Ya. (1969b). U.S.S.R. Patent No. 237324.

Severyanin, V. S., Lyskov, V. Ya., and Khidiyatov, A. M. (1970). U.S.S.R. Patent No. 251742.

Sommers, H. (1961). Experiences with Pulsating Tube Firing in an Experimental Installation. *In* "Pulsating Combustion, The Collected Works of F. H. Reynst" (M. W. Thring, ed.), pp. 262–274. Pergamon, Oxford.

Speich, C. F., and Putnam, A. A. (1959). Oscillations in Gas-Fired Heating Equipment and Their Suppression. Amer. Gas Ass. P.A.R. Rep. 127-DR.

Speich, C. F., Dennis, W. R., and Putnam, A. A. (1957). Acoustic Coupling of Residential Furnaces with Their Surroundings. *Heating, Piping, Air Conditioning* **29,** 251.

Speich, C. F., and Putnam, A. A. (1960). Acoustical Systems Determine Oil Burner Pulsations and Their Amplitudes, *ASHRAE J.* **2,** 63.

Thomas, A., and Williams, G. T. (1966). Flame Noise: Sound Emission from Spark Ignited Bubbles of Combustible Gas. *Proc. Roy. Soc.* (*London*) *Ser. A* **294,** 449.

Traenckner, K. (1953). Pulverised-Coal Gasification Ruhrgas Processes. *Trans. ASME* **75,** 1095.

VI

Heat Transfer from Nonluminous Flames in Furnaces

H. C. HOTTEL, A. F. SAROFIM, AND
I. H. FARAG

DEPARTMENT OF CHEMICAL ENGINEERING
MASSACHUSETTS INSTITUTE OF TECHNOLOGY
CAMBRIDGE, MASSACHUSETTS

The most important parameters required for calculation of heat transfer from nonluminous flames in furnaces are the radiative properties of CO_2 and H_2O, the two principal contributors to the nonluminous radiative flux. The importance of the radiative transfer from CO_2 and H_2O was first recognized by Schack (1924) and Hottel (1927) who provided charts on gas emissivities based on the work of Paschen, Rubens, Hertz, *et al.* The early charts were superseded by direct measurement of emissivities and absorptivities in the U.S. (Hottel and Mangelsdorf, 1935; Hottel and Egbert, 1942) and Germany (Schmidt, 1932; Eckert, 1937). A critical evaluation and compilation of the data by Hottel led to total emissivity charts (Hottel, 1954) which are standards to the present. More recently, interest in the emission from rocket exhaust plumes has stimulated extensive studies of spectral emissivities of CO_2 and H_2O which have provided data which can be utilized to extend the range of utility of the original charts to path lengths, temperatures, and pressures outside the range of the earlier measurements on total emissivities. It is the purpose in this chapter to review the data on both total and spectral emissivities, and to indicate their range of applicability. First, it is necessary to review a few basic definitions and derivations related to the spectroscopic models.

I. THEORETICAL BACKGROUND

A. Bouguer–Lambert Law

Consider the passage of a collimated beam of monochromatic radiation through an absorbing medium. The attenuation $-dI$ across a differential element dl is proportional to the intensity I of the beam and to the path length traversed

$$-dI = KI\,dl, \tag{1}$$

and the proportionality constant K is known as the attenuation or absorption coefficient. Integration of Eq. (1) for the case of a homogeneous medium yields the familiar exponential attenuation law

$$I = I_0 e^{-Kl}, \tag{2}$$

where I_0 is the initial intensity of the beam. This relation, first derived by Bouguer and later independently by Lambert, is valid if the wavenumber interval under consideration is small enough that any variation of K with wavenumber can be neglected. For CO_2 and H_2O at ambient temperatures the spectral absorption coefficient may vary significantly over wavenumber intervals of 0.1 cm^{-1} and very high resolution spectrometers would be required to satisfy the constraints on the applicability of the Bouguer–Lambert law. It is more common to use theory to determine average transmissivities in terms of basic spectroscopic parameters and to determine the spectroscopic parameters utilizing low-resolution spectrometers. (One recorded exception involves the line-by-line calculation of the transmissivity of hot water vapor in which 2.3 million lines were considered at a computational cost of 40 hr of IBM 7094 time!)

B. Beer's Law

Beer postulated that the absorption coefficient is proportional to the concentration of the absorbing species, that is,

$$K = k_c c = k_p p, \tag{3}$$

where k_c is used in combination with molecular concentration and k_p with partial pressure of the absorbing species. In this chapter, the value of k geared to partial pressure is preferred. For those substances obeying Beer's law, the emissivity or absorptivity of a column of a mixture containing the absorbing species is a unique function of the concentration length or pL product for any combination of concentration or p and L. For H_2O and, to a lesser degree CO_2, Beer's law is not strictly applicable and a factor in addition to pL must be introduced to allow for the effect on emissivity of total pressure and gas composition.

C. Gas Emissivity and Absorptivity

The standard gas emissivity is defined as the ratio of intensity of emission from an isothermal gas element of fixed length to the intensity of a blackbody at the gas temperature. It thus equals the ratio of incident radiation onto a surface element dA from a narrow pencil of rays of length L to incident radiation, in the same small divergence angle, from a blackbody at gas temperature. Since the ratio is independent of the angle of incidence, the standard emissivity may also be thought of as the radiation from an isothermal gas hemisphere to a point on the center of its base, expressed as a ratio to hemispherical blackbody radiation. Similarly, the standard gas absorptivity is the fractional absorption, by an isothermal gas, of blackbody radiation emitted from a bounding surface element dA through a path of fixed length L.

From Eqs. (2) and (3),

$$\alpha_\lambda = (I_0 - I)/I_0 = 1 - \exp(-k_\lambda pL), \tag{4}$$

where k_λ is the absorption coefficient at the wavelength of interest. The spectral emissivity ϵ_λ is equal to the absorptivity α_λ provided that the population of the energy levels within the gas molecules corresponds to thermal equilibrium—valid for most practical cases. The total emissivity or absorptivity of the gases is obtained in a manner similar to that for the corresponding surface properties by integrating over all wavelengths the spectral absorptivity or emissivity weighted by the Planck function evaluated at the surface temperature T_s for absorptivity and at the gas temperature T_g for emissivity

$$\epsilon_g = \int_0^\infty \epsilon_\lambda E_{\lambda,g}\, d\lambda \bigg/ \int_0^\infty E_{\lambda,g}\, d\lambda = \int_0^1 \epsilon_\lambda\, df_{\lambda T_g}, \tag{5}$$

$$\alpha_{g,s} = \int_0^\infty \alpha_\lambda E_{\lambda,s}\, d\lambda \bigg/ \int_0^\infty E_{\lambda,s}\, d\lambda = \int_0^1 \alpha_\lambda\, df_{\lambda T_s}, \tag{6}$$

where f, from Table I at the appropriate λT_g or λT_s, is the fraction of blackbody spectral energy lying below λ. The value of k_λ or the equivalent value of k_ω when spectral position is designated in wavenumber ω ($\omega \equiv \lambda^{-1}$) is determined by the contributions of the lines in the vibration–rotation bands of CO_2 and H_2O. The next section presents the characteristics of an individual line in these bands, temporarily neglecting the effect of overlap between lines.

D. Single-Line Emission

The absorption coefficient k_ω for a single line is an approximately bell-shaped function of the wave number ω with a peak at the wavenumber ω_0 corresponding to the change $h\nu$ in the quantized energy level of the absorbing or emitting molecules. The area under the $k_\omega - \omega$ curve is defined as the integrated line intensity and is designated by S. The displacement $|\omega - \omega_0|$ from line center required to cause k_ω to fall to half of its maximum value is defined as the line half-width b. The broadening of a spectral line is a consequence of the unavoidable uncertainty in energy levels (natural line broadening); the thermal motion of the emitting and absorbing species along the line of emission or absorption (Doppler broadening); and the perturbation of the energy levels by collision (collision broadening). Collision broadening is the dominant broadening mechanism for most furnaces and combustion chambers but Doppler broadening becomes important at low pressures such as those encountered in engine exhaust plumes at high altitudes. The absorption coefficient for a collision-broadened line may be approximated by the Lorentz dispersion formula

$$k_\omega = (S/\pi)\{b/[(\omega - \omega_0)^2 + b^2]\}. \tag{7}$$

The shape of the line given by Eq. (7) is compared in Fig. 1 with those of lines having top-hat, triangular, and Doppler-broadened profiles, and having the same integrated line intensity S and half-width b.

The contribution by a single line to the flux density at the center of the

TABLE I[a]

FRACTION f OF BLACKBODY RADIATION EMITTED BELOW λ AS A FUNCTION OF λT

$$\left(f = \int_0^\lambda E_{B\lambda}\, d\lambda \Big/ \int_0^\infty E_{B\lambda}\, d\lambda \right)$$

λT (μ °K)	f	λT (μ °K)	f	λT (μ °K)	f	λT (μ °K)	f	λT (μ °K)	f
1000	0.0003	3050	0.2845	4100	0.4987	5300	0.6693	7800	0.8480
1100	0.0009	3100	0.2957	4150	0.5074	5400	0.6803	8000	0.8562
1200	0.0021	3150	0.3069	4200	0.5160	5500	0.6909	8200	0.8640
1300	0.0043	3200	0.3181	4250	0.5244	5600	0.7010	8400	0.8711
1400	0.0078	3250	0.3291	4300	0.5326	5700	0.7107	8600	0.8778
1500	0.0128	3300	0.3401	4350	0.5408	5800	0.7201	8800	0.8841
1600	0.0197	3350	0.3509	4400	0.5487	5900	0.7291	9000	0.8900
1700	0.0285	3400	0.3617	4450	0.5566	6000	0.7378	9500	0.9030
1800	0.0393	3450	0.3723	4500	0.5643	6100	0.7461	10000	0.9141
1900	0.0521	3500	0.3829	4550	0.5718	6200	0.7541	10500	0.9237
2000	0.0667	3550	0.3933	4600	0.5792	6300	0.7618	11000	0.9318
2100	0.0830	3600	0.4036	4650	0.5865	6400	0.7692	12000	0.9450
2200	0.1009	3650	0.4137	4700	0.5936	6500	0.7763	13000	0.9551
2300	0.1200	3700	0.4237	4750	0.6006	6600	0.7832	14000	0.9629
2400	0.1402	3750	0.4336	4800	0.6075	6700	0.7897	16000	0.9738
2500	0.1613	3800	0.4433	4850	0.6141	6800	0.7961	18000	0.9808
2600	0.1831	3850	0.4529	4900	0.6209	6900	0.8022	20000	0.9856
2700	0.2053	3900	0.4624	4950	0.6273	7000	0.8081	25000	0.9922
2800	0.2279	3950	0.4717	5000	0.6337	7200	0.8192	30000	0.9953
2900	0.2505	4000	0.4808	5100	0.6460	7400	0.8295	40000	0.9979
3000	0.2732	4050	0.4898	5200	0.6579	7600	0.8391	50000	0.9989

[a] Abstracted from a much more complete tabulation to six significant figures by Stevenson (1963), based on a value for c_2 of 1.43886 cm °K.

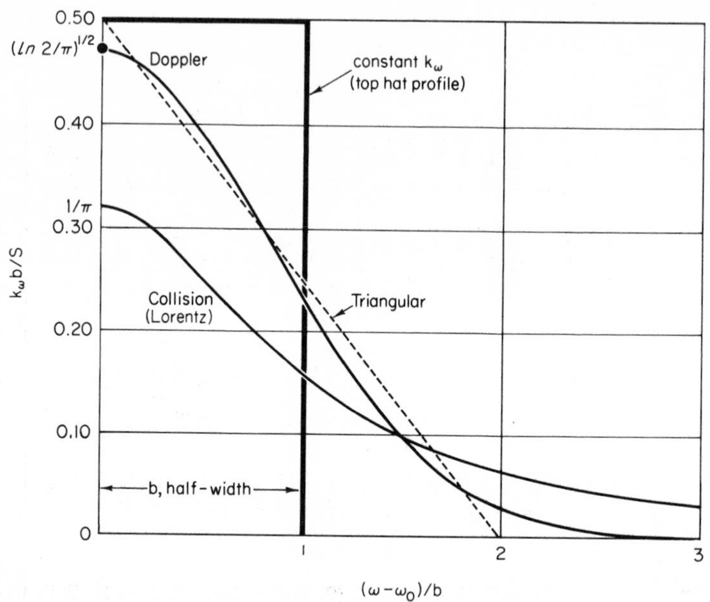

Fig. 1. Contours of spectral lines having the same integrated intensity S and half-width b: (a) Doppler broadened, (b) collision (Lorentz) broadened, (c) triangular, (d) constant k_ω (top-hat profile). [From "Radiative Transfer" by H. C. Hottel and A. F. Sarofim. Copyright 1967, McGraw-Hill, New York. Used with permission of the McGraw-Hill Book Company.]

base of a hemisphere of isothermal gas is given by

$$q = \int_0^\infty \epsilon_\omega E_\omega \, d\omega = \int_0^\infty E_\omega (1 - e^{-k_\omega pL}) \, d\omega, \tag{8}$$

where E_ω is the blackbody monochromatic emissive power at the gas temperature, given, on a wavenumber scale, by

$$E_\omega = c_1 \omega^3 / [\exp (c_2 \omega / T) - 1], \tag{9}$$

where c_1 and c_2 are the Planck first and second constants, with values 3.7418443×10^{-5} erg cm^2 sec^{-1} and 1.438833 cm °K. The major emission by a line occurs in a wavenumber range which extends a few half-widths about the line center. Since the line half-width is typically in the range of 0.1 to 1.0 cm^{-1}, the blackbody emissive power differs negligibly from the value $E_{\omega 0}$ evaluated at the line center and it can therefore be removed from within the integral in Eq. (8). (The small variation in $E_{\omega 0}$ over a line may be appreciated by noting that a width of 1 cm^{-1} corresponds to, on a wavelength scale, widths varying from 10^{-4} μ at 1 μ to 10^{-2} μ at 10 μ, or fractional changes in wavelength or wavenumber of 10^{-4} at 1 μ to 10^{-3} at 10 μ.) The flux density q per unit black emissive power is defined as the equivalent line width and denoted by A_L since it clearly represents the spectral width of blackbody radiation at the gas temperature which produces the same emission as the actual line. It is given by

$$A_L = q/E_{\omega 0} = \int_0^\infty \{1 - \exp[-(S/\pi)\{b/((\omega - \omega_0)^2 + b^2)\}pL]\}\, d\omega. \quad (10)$$

Equation (10) was first integrated by Ladenberg and Reiche (1918) to yield

$$A_L = 2\pi b f(SpL/2\pi b), \quad (11)$$

where $f(x) = x[\exp(-x)][I_0(x) + I_1(x)]$, and $I_0(\ )$ and $I_1(\ )$ are the modified Bessel functions of the zeroth and first orders. To a good approxi-

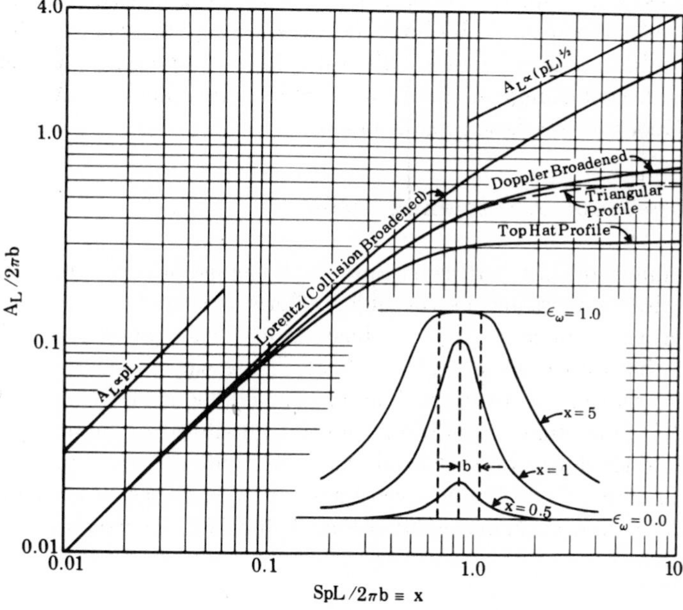

Fig. 2. Equivalent black width (A_L) of a single line for the following profiles: (a) Lorentz collision broadened, (b) Doppler broadened, (c) triangular, (d) constant k (top-hat). Inset plot shows the spectral emissivity of a collision-broadened line at path lengths corresponding to linear ($x < 0.1$), square root ($x > 5$), and intermediate region. [From "Radiative Transfer" by H. C. Hottel and A. F. Sarofim. Copyright 1967, McGraw-Hill, New York. Used with permission of the McGraw-Hill Book Company.]

mation $f(x)$ can be represented by

$$f(x) \equiv x/[1 + (\pi/2)x]^{1/2}, \tag{12}$$

which yields the correct limiting forms of $f(x)$:

$$f(x) \to x \qquad \text{as} \quad x \to 0 \tag{13}$$

and

$$f(x) \to (2x/\pi)^{1/2} \qquad \text{as} \quad x \to \infty, \tag{14}$$

and departs from the exact values by a maximum of 7.4% at an x of 0.8. Figure 2 shows the equivalent line width calculated from Eq. (10) compared with those computed using top-hat, triangular, and Doppler-broadened profiles. A characteristic of collision-broadened lines is that the wings of the lines contribute very significantly to the emission at large pL's (see inset plot in Fig. 2), a contribution which is responsible for the proportionality of A_L to the square root of pL at large pL's. By contrast, at large pL's the values of A_L for top-hat and triangular profiles approach asymptotes of $2b$ and $4b$, respectively. As pL is increased and the contribution of the wings of the lines increases a point is reached where the wings of adjacent lines overlap. Allowance for such overlap between lines for CO_2 and H_2O at high temperatures is most conveniently treated by the Mayer–Goody or statistical band model.

E. Mayer–Goody or Statistical Model

Tne number of lines that contribute to the emission in a band increases rapidly with temperature as the higher vibrational and rotational energy levels become populated. As one illustration, Plass and Stull (1960) determined that for the 4.3-μ band of CO_2, the lines with intensities exceeding 10^{-7} times that of the strongest line numbered 1400 at 300°K and 890,000 at 2400°K. Since many vibrational levels contribute to the emission the positions of the lines can be considered to be randomly distributed. A significant amount of overlap between lines can be expected at moderate and large pL's, as shown schematically in Fig. 3. The lack of correlation between

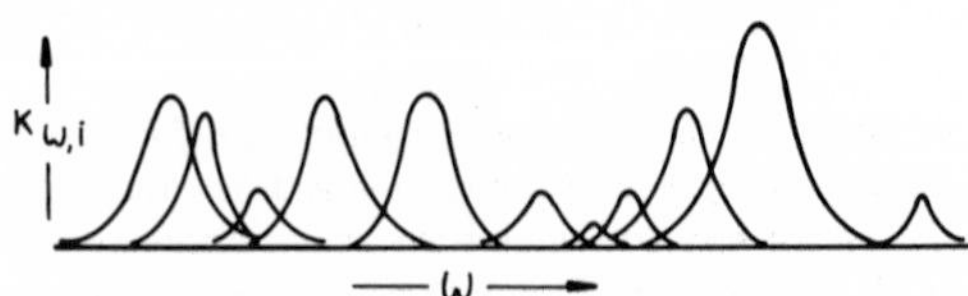

Fig. 3. Schematic representation of a random line array.

the positions of the lines contributing to k_ω leads to a great simplification in the derivation of total band emissivity. Consider a wavenumber interval nd, intensity of n lines with a mean spacing between lines of d. Since a spectral line absorbs an amount of radiation equivalent to that contained in a wavenumber interval A_L, the transmittance over the wavenumber interval nd with allowance for absorption of radiation by one line is then $1 - (A_L/nd)$. Since the lines are randomly distributed the average transmittance over the interval allowing for absorption by n lines is then given by

$$\tau_{av} = \prod_{i=1}^{n} [1 - (A_{L,i}/nd)] = [1 - (\bar{A}_L/nd)]^n, \tag{15}$$

where $A_{L,i}$ is the equivalent width of the ith line, and $\bar{A}_L$ is the equivalent width when all lines are assumed to have the same mean line intensity $\bar{S}$. If a large number of lines are present in the interval under consideration Eq. (15) becomes in the limit of large n

$$\tau_{av} = \exp(-\bar{A}_L/d). \tag{16}$$

The results obtained, making the assumption that the lines all have the same intensity, have been shown by Goody (1954, 1964) and Plass (1958) to differ insignificantly from those obtained using more complicated line intensity distributions. If the line intensities follow an exponential distribution function given by

$$P(S) = \bar{S}^{-1} \exp(-S/\bar{S}), \tag{17}$$

it can be shown that the combination of Eqs. (16) and (12) provides an exact solution. Equation (16) is the analog of the Bouguer–Lambert law. In Eq. (16) the term in the exponent represents the blocking power nA_L of n absorbing lines divided by the spectral interval nd containing them. In the Bouguer–Lambert law the term in the exponential is equal to the sum of the cross sections for absorption of the molecules along a path L per unit cross-sectional area. Examination of Eq. (16) and the definition of A_L in Eq. (11) shows that the transmissivity is a function of the two groups $2\pi b/d$ and $\bar{S}pL/2\pi b$ or, alternatively, $2\pi b/d$ and $\bar{S}pL/d$. The dependence of τ_{av} on $2\pi b/d$ and $\bar{S}pL/d$ is depicted by the solid lines in Fig. 4. When b/d is much greater than one, the overlap between lines is complete and the attenuation is given by the Bouguer–Lambert decay law with k equal to $\bar{S}/d$. At the other extreme, when b/d is much less than one, the lines act independently and the mean transmittance is given simply by $1 - (\bar{A}_L/d)$ until the pL is reached at which the wings of the line begin to overlap.

One of the merits of models such as that given by Eq. (16) is that it provides a means of reconstructing the entire $\tau_{av} - pL$ curve from limited

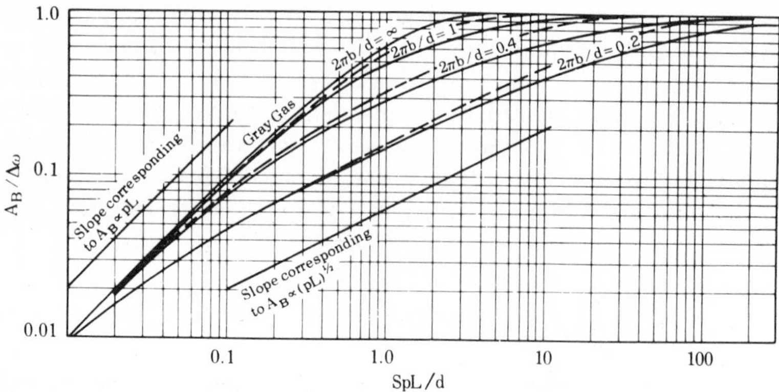

Fig. 4. Equivalent black-band width (A_B) as a function of SpL/d and $2\pi b/d$ for Elsasser (dashed curves) model and Mayer–Goody (statistical; solid curves) model. Elsasser model with $2\pi b/d = 1$ has substantially reached limiting case of a gray gas ($2\pi b/d = \infty$; $k = S/d$). [From "Radiative Transfer" by H. C. Hottel and A. F. Sarofim. Copyright 1967, McGraw-Hill, New York. Used with permission of the McGraw-Hill Book Company.]

measurements of τ_{av}. (In principle, measurements at only two pL's are required to obtain $2\pi b/d$ and $\bar{S}/d$.) The values of $\bar{S}$ and d are functions of temperature only. The value of the collision half-width b, however, is a function of the total pressure and gas composition in addition to temperature. If b were proportional to collision frequency, it could be related to absolute temperature T and total pressure P by

$$b = b_0(P/P_0)(T_0/T)^{1/2}. \tag{18}$$

Empirically determined broadening functions are, however, found to be more complicated, with greater weight being given to collisions of molecules with their own species (self-broadening). Hottel and Egbert (1942) and Howard *et al.* (1955) correlated their data on water vapor by giving twice as much weight to the water vapor content as that of the other molecules, equivalent to substituting $P + p_w$, where p_w is the partial pressure of water vapor, for P in Eq. (18). Edwards (1960) in his studies on CO_2 found that the self-broadening effect increased with increasing CO_2 concentrations and suggested using $P + p_c + (\frac{1}{2})p_c^2$, where p_c is the CO_2 partial pressure, as the pressure-broadening function for CO_2. The collision-broadening function is of greater importance for the case of H_2O than CO_2. Refinements of the functional form of the correlation of half-line width include that by Benedict and Kaplan (1959, 1964), who included the effect of resonant-dipole–dipole collisions between water vapor molecules, which has a stronger temperature dependence than conventional collision broadening. The modified function they propose has the form

$$b = b^* p_w(T_0/T) + \sum_{i=1}^{m} b_0{}^i p_i (T_0/T)^{1/2}, \tag{19}$$

where b^* is the line half-width at the reference temperature and pressure for the resonant-dipole–dipole collisions, and $b_0{}^i$ is the conventional collision broadening of water vapor molecules with species i including water vapor itself. The summation is taken over all species m. The values for the constants, selected from literature values by Ludwig *et al.* (1968) in their comprehensive study on water vapor, are $b^* = 0.44\ \mathrm{cm}^{-1}$, and, for $b_0{}^i$ at 273°K and 1 atm in units of reciprocal centimeters, 0.09 for N_2, 0.04 for O_2, 0.12 for CO_2. In the absence of data, they estimated values for $b_0{}^i$ of 0.09 for H_2O, 0.05 for H_2, and 0.10 for CO. For an H_2O–N_2 mixture, Eq. (19) gives the H_2O molecules nearly six times the weighting of N_2 at 273°K but a little less than three times the weighting at 2000°K.

In the compilation of data obtained at a given $p_w L$ by different investigations on gas mixtures of varying composition it is conventional to reduce the results to the fictitious common base of $p_w = 0$ and $p_{N_2} = 1$ atm, using either empirical correction factors or those derived from theoretical models in which a relation such as Eq. (19) is used to allow for the broadening effect of different species.

II. TOTAL EMISSIVITY AND ABSORPTIVITY DATA

The variables required to define gas emissivity are partial pressure of the emitting gas p, path length L, gas temperature T_g, total pressure P, and, to the extent that broadening parameters are functions of composition, the composition c. For gas absorptivity the surface temperature T_s is needed in addition.

The total emissivity charts in common engineering use are based on the measurements of Schmidt (1932), Hottel and Mangelsdorf (1935), Hottel and Smith (1935), Eckert (1937), and Hottel and Egbert (1942). The major experimental problem facing these investigators was that of measuring radiation from well-defined homogeneous isothermal gas masses without interference from confining windows. Schmidt (1932) used steam jets 1 to 6 cm in diameter and extended his path length by mirrored doubling and trebling. Hottel and Smith (1935) obtained CO_2 emissivities at high temperatures (1680–2350°K) from measurement on the products of combustion of premixed flames of CO–O_2 and CO–air with burner diameters ranging from 20 to 40 cm. Eckert (1937) sighted along path lengths up to 300 cm through openings in the bottom of a furnace in which the emitting gas mixture was confined. All these methods were subject to errors introduced by a net absorption of radiation from the hot core by cooler CO_2 or H_2O at the interface between the emitting gas and the ambient surroundings. The problem of absorption by cooler gases in the transition zone between hot emitting gases and relatively cold air was resolved by Hottel and Mangelsdorf (1935), who designed a furnace in which the concentration transition from emitting gas to air occurred within the isothermal furnace and the temperature transition from furnace temperature to ambient temperature took place entirely within a buffer air zone. A schematic diagram of their system in which a sharp concentration transition of the emitting species was maintained by opposed equal flows of the emitting gas and air is shown, together with a sample concentration and temperature profile, in Fig. 5.

The total emissivity of water vapor obtained by different investigators is shown in Fig. 6. It will be remembered that Schmidt's experiments were on columns of pure water vapor of different lengths (extrapolated beyond the 18-cm maximum length in Fig. 6), and that those of the other investigators were on furnaces of fixed lengths confining mixtures of different partial pressure of water vapor. The data are strictly comparable at a given temperature only when both $p_w L$ and p_w are matched, a condition satisfied by the extreme points of the curves for the furnaces of different length (these correspond to $p_w = 1.0$ atm and $p_w L = L$) and Schmidt's data at the same pL's. The difference between the terminus of a line for a fixed

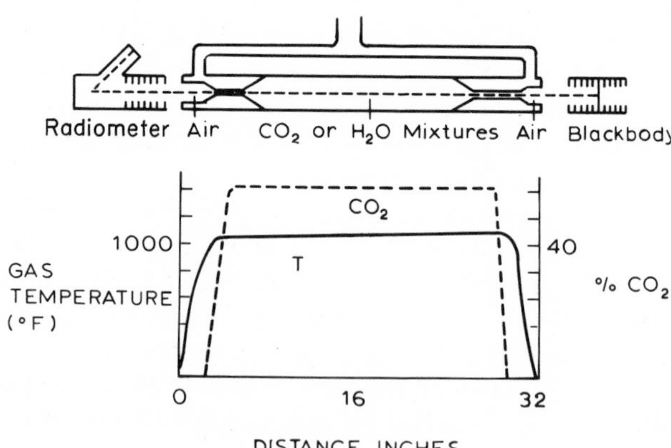

Fig. 5. Schematic representation of Hottel and Mangelsdorf furnace and representative temperature and concentration profiles.

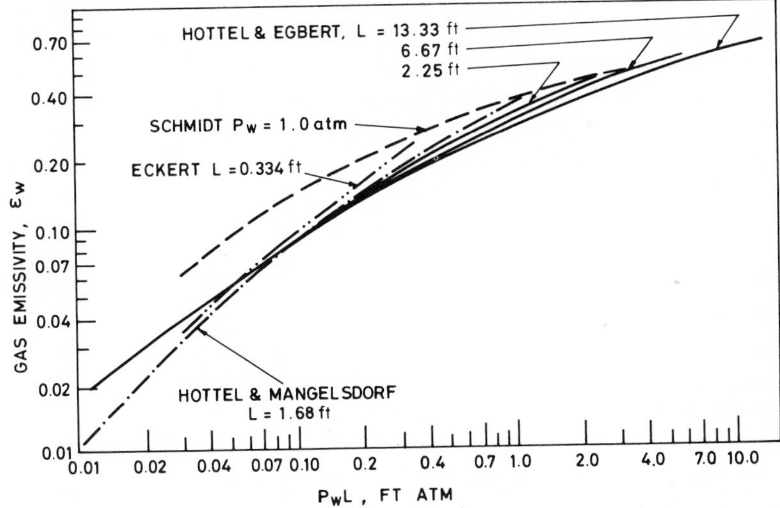

Fig. 6. Emissivity of water vapor at 700°F at various p_w's.

L furnace and Schmidt's curve at the same pL gives a measure of the uncertainty in the results introduced by instrumental error and edge cooling. The discrepancies among the results of the different investigators, summarized in Table II, show a maximum deviation of 8% and an average deviation of 4%. It should be noted that Schmidt's results were extrapolated to path lengths greater than 18 cm but that since Schmidt's study

TABLE II

COMPARISON OF EMISSIVITIES FOR $p_w = 1.0$ ATM OBTAINED BY SCHMIDT WITH THOSE OBTAINED BY ECKERT, HOTTEL AND MANGELSDORF, AND HOTTEL AND EGBERT[a]

T_g (°F)	L = 10 cm		L = 51 cm		L = 78 cm	
	Schmidt	Eckert	Schmidt	H. + M.	Schmidt	H. + E.
290	0.3	0.285	0.48	0.50	0.51	0.48
480	0.28	0.26	0.44	0.47	0.47	0.46
700	0.25	0.23	0.43	0.47	0.47	0.47
900	—	—	0.41	0.41	0.46	0.45
1100	—	—	0.38	0.40	0.44	0.42
1300	—	—	0.36	0.36	—	—

[a] Data from Schmidt (1932), Eckert (1937), Hottel and Mangelsdorf (1935), and Hottel and Egbert (1942).

came first the extrapolation was performed without benefit of the later data. The agreement between the data obtained using furnaces of different design in widely separated laboratories lends confidence to the accuracy of the results. Comparison of the data at other p_wL's provides a measure of the increase in emissivity at a fixed p_wL with increase in p_w (decrease in furnace length for a fixed p_wL). Hottel and co-workers derived a correc-

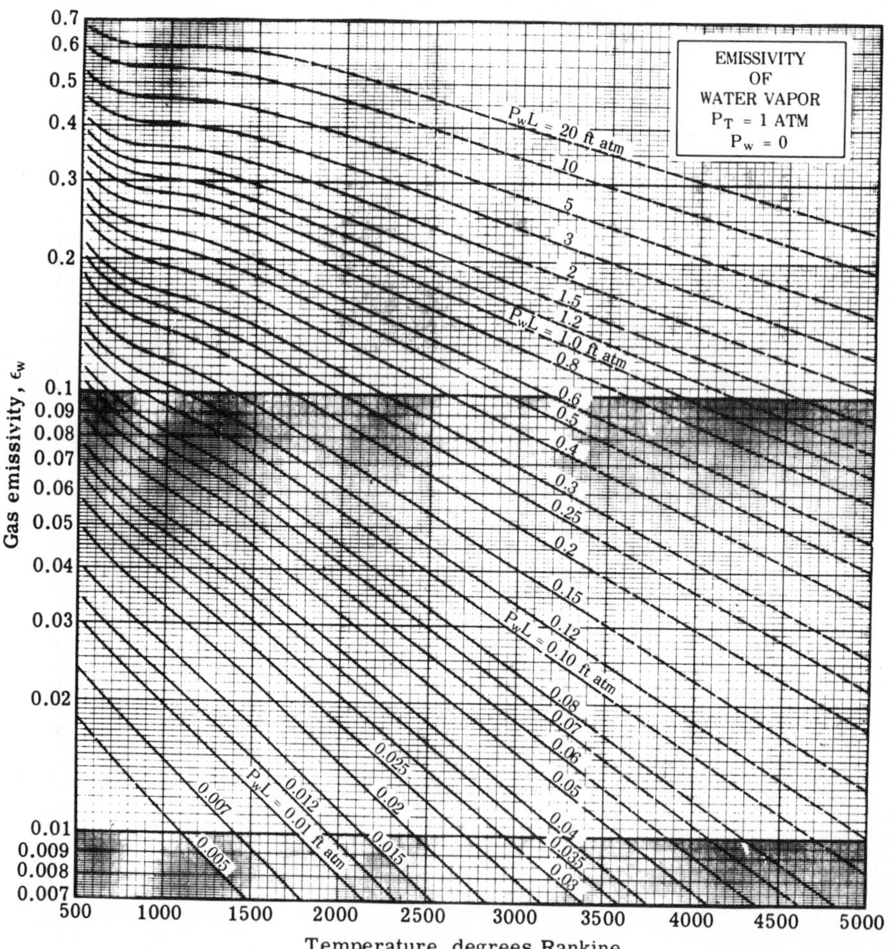

Fig. 7. Emissivity of water vapor at 1 atm total pressure (P_T) reduced to $p_w \to 0$. [From "Radiative Transfer" by H. C. Hottel and A. F. Sarofim. Copyright 1967, McGraw-Hill, New York. Used with permission of the McGraw-Hill Book Company.]

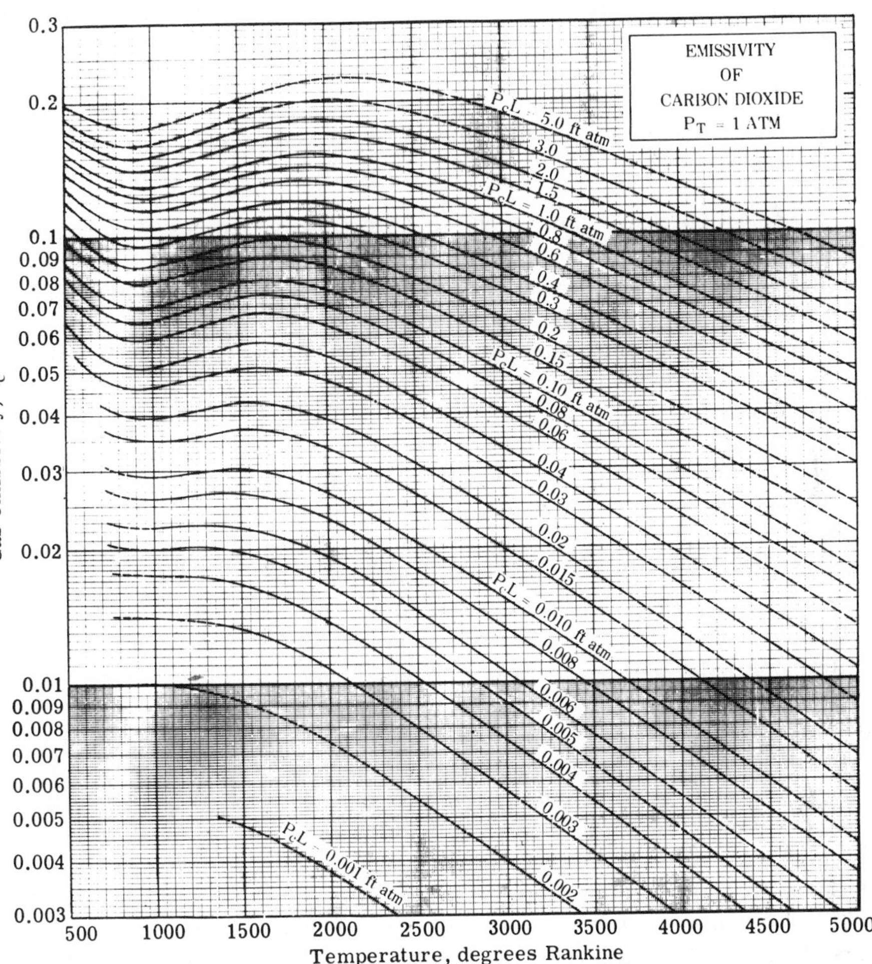

Fig. 8. Emissivity of carbon dioxide at 1 atm total pressure (p_T) (p_c corresponds to that in the experiments producing the data on which the chart is based). [From "Radiative Transfer" by H. C. Hottel and A. F. Sarofim. Copyright 1967, McGraw-Hill, New York. Used with permission of the McGraw-Hill Book Company.]

tion factor for self-broadening from these data, reduced all the data to a common p_w of zero, and developed the standard water-vapor emissivity chart (Fig. 7) on which solid lines are used to denote those portions of the curves based on measurement and the dashed portion, extrapolation. The self-broadening effect for CO_2 is relatively small and the standard CO_2 emissivity chart (Fig. 8) is based on the data mostly of Hottel and

Mangelsdorf (1935), Hottel and Smith (1935), and Eckert (1937) without any correction for self-broadening.

III. SPECTRAL EMISSIVITY MEASUREMENTS

The most extensive data on the spectral emissivity of water vapor are those by workers at General Dynamics (Ludwig *et al.*, 1968), who have reported values of $2\pi b/d$ and $\bar{S}/d$ averaged over 371 intervals of 25 cm^{-1} from 50 cm^{-1} (200 μ) to 9300 cm^{-1} (ca. 1.1 μ) determined from transmissivities of the hot products of combustion of premixed flames of H_2 and O_2 above a burner up to 20 ft long. The temperature was varied over the range 1000 to 2200°K by varying the H_2/O_2 ratio for fuel-lean conditions. The spectral data have several applications: (i) the calculation of total emissivities and absorptivities in the range of extrapolation of the Hottel chart to higher temperatures and higher pL's; (ii) the determination of the effect of total pressure on emissivity; (iii) the calculation of the correction for overlap of CO_2 and H_2O; and (iv) for calculating the emission from nonisothermal gas layers with greater accuracy than is possible with total emissivity data. The results of a comparison of the General Dynamics integrated band data with those of Schmidt and Hottel and Egbert are shown in Fig. 9. Agreement is within 10% except at low $p_w L$ and p_w, where the discrepancy is 30% at a gas temperature of 1160°R. This latter may be a consequence of either attenuation by atmospheric water vapor of the measured total emissivity data or, most probably, the error in applying Eq. (19) with the approximate half-width correction factors b_i to allow for self-broadening, or lack thereof, in generating total emissivity from the General Dynamics data. Additional reinforcement for the General Dynamics data is provided by the good agreement obtained by Simmons *et al.* (1969) between spectral emission from isothermal and nonisothermal H_2O layers and values computed from the values of $2\pi b/d$ and $\bar{S}/d$ reported by General Dynamics. It is recommended that the Hottel charts be used wherever they are based on data (solid lines in Figs. 7 and 8) and that the General Dynamics correlation be used outside that range. Total emissivities based on the high-temperature spectral data are reported by Boynton and Ludwig (1971).

General Dynamics has also published spectral tables of $2\pi b/d$ and $\bar{S}/d$ for CO_2 based in part on theoretical calculations (Ludwig *et al.*, 1966). In addition, the integrated band contributions for CO_2 have been reported by Edwards (1960), based on measurements made on a furnace of the Hottel–Mangelsdorf design in a pressure vessel operated at pressures up to 10 atm.

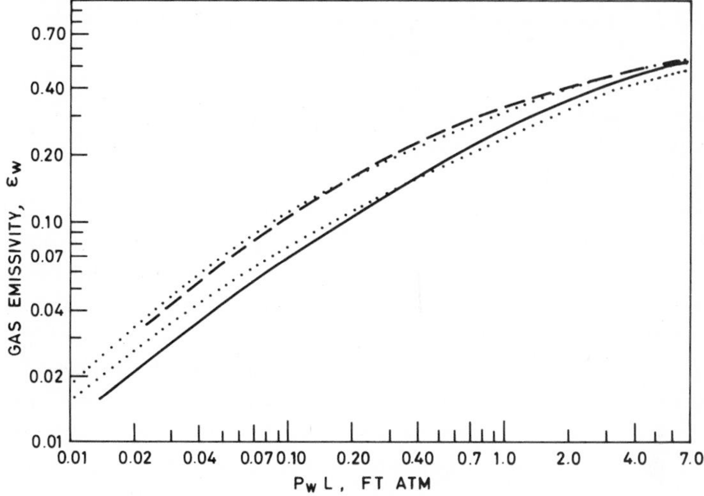

Fig. 9. Comparison of emissivities calculated from General Dynamics data with measurements by Schmidt on pure steam and by Hottel and Egbert with a 6.67-ft-long furnace. $T_g = 1560°R$. - - - Schmidt, $p_w = 1.0$ atm; —— Hottel and Egbert, $L = 6.67$ ft;···· General Dynamics, $p_w = 1.0$ atm (upper), $L = 6.67$ ft (lower).

Integration of the Edwards data provides total emissivities in good agreement with the values from the Hottel charts, confirming their validity over the range covered by the data. There is some question, however, regarding the validity of the extrapolated portions of the chart and it is recommended that in these regions values calculated from the spectral values be used (Leckner, 1971).

IV. COLLISION BROADENING

The effect of pressure or self-broadening on absorptivity depends on the spacing between lines. When $2\pi b/d$ is large, the lines overlap to an extent that the attenuation coefficient becomes a smooth function of wave number and assumes a value $\bar{S}/d$. Further increases in $2\pi b/d$ due to increase in b have no effect in this limit, as shown clearly by the top curves of Fig. 4. At the other extreme when $2\pi b/d$ is small the lines act independently and the increase in absorptance will be proportional to the increase in A_L, given by Eq. (11). At low SpL/d, A_L is independent of b and at higher values it is proportional to the square root of b. As SpL/d is increased further lines will overlap and the effect of $2\pi b/d$ will be diminished as shown by the right-hand side of Fig. 4. It should be noted here that the lines of CO_2 are spaced closer than those of H_2O and that the spacing decreases with temperature rise as the total number of lines contributing to emission increase. It is thus expected that pressure broadening will be more important for H_2O than for CO_2 and that its effect will decrease with increasing temperature.

Hottel and Egbert (1942) developed a correction factor C_w, shown in Fig. 10, by which the emissivity read from Figs. 4–7 was to be multiplied in order to allow for self-broadening. Their correction chart does not cover high-pressure or high-temperature operation. The statistical model provides a relation between spectral transmittance and line half-width which can be used together with Eq. (19) and the General Dynamics values for the spectroscopic parameters to evaluate the pressure and partial pressure dependency of total emissivity or absorptivity. An example of its application is shown in Table III, which provides emissivities calculated from the

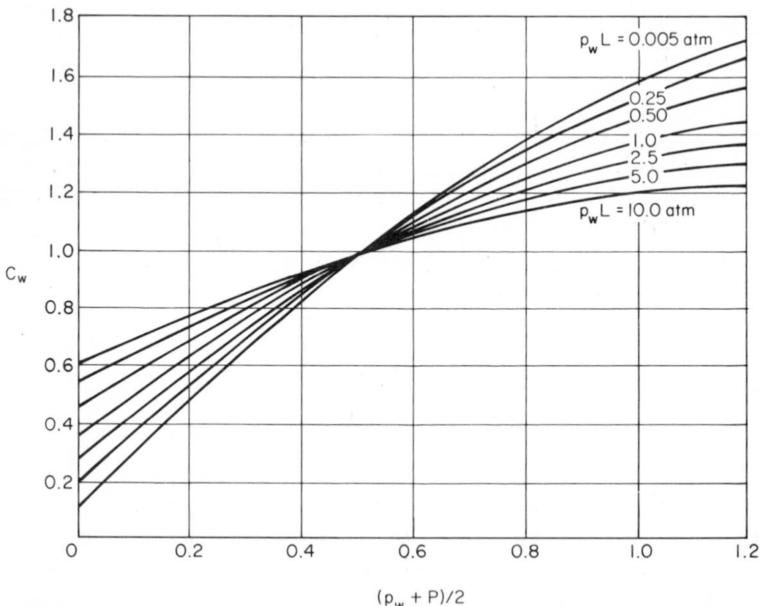

Fig. 10. Correction factor C_w for converting H_2O emissivities to values of p_w and P other than 0 and 1, respectively. [From "Radiative Transfer" by H. C. Hottel and A. F. Sarofim. Copyright 1967, McGraw-Hill, New York. Used with permission of the McGraw-Hill Book Company.]

TABLE III

PRESSURE CORRECTION FACTORS REPRESENTATIVE OF GAS TURBINE OPERATION

T (°R)	P (atm)	p_wL (ft atm)	p_w (atm)	ϵ^a	$\epsilon(p_w = 0)^b$	$C_w{}^b$	$C_w \times \epsilon_w{}^b$
2500	20	0.24	0.48	0.135	0.076	?	?
2500	1	0.24	0.48	0.111	0.076	1.28	0.097
2500	20	0.66	1.32	0.259	0.14	?	?
3500	20	0.24	0.48	0.074	0.046	?	?
3500	1	0.24	0.48	0.069	0.046	1.28	0.059
3500	20	0.66	1.32	0.160	0.092	?	?

[a] General Dynamics data.
[b] HCH data.

General Dynamics data for conditions corresponding roughly to those in a gas-turbine combustor: T = 2500 and 3500°R, p_wL = 0.24 and 0.66 ft atm, L = 0.5 ft., and P = 20 atm. The pressure-broadened half-width was calculated from Eq. (19), assuming that CO_2 and H_2O were present in equimolal quantities and that the balance of the gas was nitrogen. The results are compared in Table III with values computed at 1 atm at the same p_w and p_wL, using both the General Dynamics data and the Hottel (HCH) charts. At 1.0 atm, the emissivity at a low p_w and a p_wL of 0.24 ft atm is 0.076 at 2500°R and 0.046 at 3500°R. Increase to a partial pressure of 0.48 atm at a fixed p_wL increases the emissivity by 28%, as read from Fig. 10. (A discrepancy of 15% is noted in the emissivities from General Dynamics data and the HCH charts, but the source of the discrepancy cannot be ascertained without additional data.) As the partial pressure of the nitrogen, and to a smaller degree CO_2, is increased to bring the total pressure up to 20 atm, the emissivity is further increased by about 20% at 2500°R and 7% at 3500°R. The results are consistent with expectations that self-broadening receives more weighting than broadening by other molecules, that the effect of broadening is more important at lower temperatures, and that as b/d is increased the effect of additional increases in b/d is diminished.

The effect of total and partial pressure on emissivity of CO_2 as derived from Edwards' results is shown in Fig. 11. Because of the closer spacing of lines in CO_2, the effect is small and becomes insignificant at high temperatures. Additional calculations of the partial pressure correction factor for a total pressure of 1 atm are reported by Boynton and Ludwig (1971).

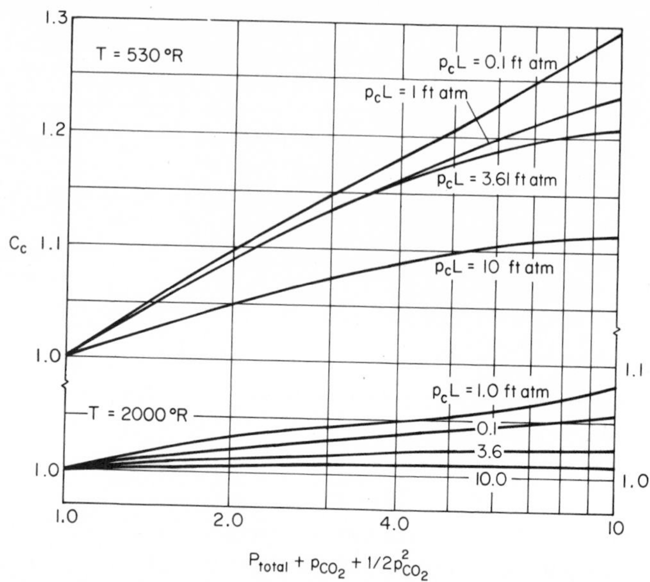

Fig. 11. Correction factor C_c for conversion of CO_2 emissivity at 1 atm total pressure and zero partial pressure to values at other total and partial pressures. (For use of plot, pressure must be expressed in atmospheres.) [From "Radiative Transfer" by H. C. Hottel and A. F. Sarofim. Copyright 1967, McGraw-Hill, New York. Used with permission of the McGraw-Hill Book Company.]

V. RADIATION ALONG A NONISOTHERMAL PATH

Consider the problem of evaluating the emission in a specific direction from a nonisothermal gas mixture with a length L along the line of sight. Let the monochromatic intensity at a distance x from the reference surface of a beam directed at the receiver be denoted by $I_\omega(x)$. On passage toward the reference surface through a differential element $-dx$ (with the minus sign introduced because the beam is traveling in a direction of decreasing x) a fraction $K_\omega(-dx)$ of the beam is absorbed and an amount $K_\omega I_{\mathrm{Bw}}(x)$ $(-dx)$ is added to its intensity in consequence of emission within $(-dx)$, where $I_{\mathrm{B}\omega}(x)$ is the monochromatic intensity emitted by a blackbody at the local gas temperature. Therefore

$$-dI_\omega(x)/dx = K_\omega I_{\mathrm{B}\omega}(x) - K_\omega I_\omega(x). \tag{20}$$

The boundary condition associated with Eq. (20) is the specification of the intensity, denoted here by $I_\omega(L)$, incident on the bounding surface at L and traveling in the direction of interest. The solution to this first-order linear differential equation is obtained by multiplying through by the integrating factor $\exp(-\int_0^x K_\omega\,dx)$. Integration then yields, for the intensity of the beam emerging from the surface,

$$I_\omega(0) = I_\omega(L)\exp\left(-\int_0^L K_\omega\,dx\right) + \int_0^L K_\omega I_{B\omega}(x)\exp\left(-\int_0^L K_\omega\,dx\right)dx. \tag{21}$$

Introduction of the equalities

$$\tau_\omega(x) = \exp\left(-\int_0^x K_\omega\,dx\right) \tag{22}$$

and

$$d\tau_\omega(x)/dx = -K_\omega\exp\left(-\int_0^x K_\omega\,dx\right) \tag{23}$$

then yields

$$I_\omega(0) = I_\omega(L)\tau_\omega(L) + \int_{\tau_\omega(L)}^{1.0} I_{B\omega}(x)[d\tau_\omega(x)/dx]\,dx. \tag{24}$$

For H_2O and to a lesser degree CO_2, $\tau_\omega(x)$ varies rapidly over narrow intervals of ω due to the line structure in the bands. It is desirable to apply Eq. (24) over wavenumber intervals $\Delta\omega$ which are reasonably attainable with conventional spectrometers, or for which the spectroscopic parameters $2\pi b/d$ and $\bar{S}/d$ are available. If these spectroscopic parameters were not functions of temperature, the integration over wavenumber would be given by τ_{av} from Eq. (16) and its derivative with respect to x. Allowance for variation of the properties with temperature may be made by the use of the following space-averaged functions in the evaluation of the τ_{av} along the nonisothermal path:

$$\tau_{av} = \exp\{-(\bar{S}\bar{u}/d)/[1 + (\bar{S}\bar{u}/4b)]^{1/2}\}, \tag{25}$$

which is identical in structure with Eq. (16) when A_L is given by the combination of Eqs. (11) and (12) and with u defined by

$$u = \int_0^x p\,dx. \tag{26}$$

The averaged properties in Eq. (25) are given by

$$\bar{S}\bar{u}/d = \int_0^u (S/d)\, du' \tag{27}$$

and

$$\bar{S}\bar{u}/4b = \left[\int_0^u (S/d)\, du'\right]^2 \Big/ \left[4\int_0^u (S/d)(b/d)\, du'\right]. \tag{28}$$

This averaging procedure yields the exact solution for the intensity emitted by a nonisothermal gas in the limiting cases of optically thin and thick gas layers and has been shown to yield good approximations to the exact solution for intermediate optical thicknesses. It is known as the Curtis–Godson approximation (Plass, 1963).

The evaluation of fluxes by use of the Curtis–Godson approximation, although straightforward, is time-consuming. For each spectral interval (371 if the General Dynamics data for H_2O are employed) the integral given by Eq. (24) can be evaluated numerically preferably dividing the path lengths into intervals each of which contributes approximately the same amount to the surface flux (the subdivision will therefore differ between spectral intervals). In order to evaluate the wavenumber averaged (over 25 cm^{-1} for the General Dynamics data) values of τ_ω to each spatial interval, the mean values of Su/d and $Su/4b$ are then obtained from Eqs. (27) and (28). The averaged values of τ_ω so obtained are used to derive the $\Delta\tau_\omega$ required for the finite difference approximation to Eq. (24). The integral is then evaluated and repeated first for different wavenumber intervals and then direction if interest is in hemispherical flux. The evaluation of intensity in a specific direction employing the General Dynamics spectroscopic parameters and ten spatial intervals for the integrations requires 11,130 separate integrations! For this reason, the Curtis–Godson approximation must be considered to provide a reference against which approximate calculations may be checked. One such approximation is obtained by integration of Eq. (24) over the entire spectrum to yield

$$I = I(L)\tau(L) + \int_{\tau(L)}^{1.0} I_B(x)[d\tau(x)/dx]\, dx. \tag{29}$$

The values of $\tau(L)$ can be equated to the complement of the emissivity evaluated at some mean temperature or from representing the emissivity by a weighted sum of gray gases and the transmissivity by

$$\tau(x) = \sum a_n \exp(-K_n x) \qquad (\textstyle\sum a_n = 1; \quad \text{one } K = 0). \tag{30}$$

Because the entire temperature dependence of emissivity is forced on the weighting factors a_n which are conventionally evaluated at the source temperature, Eq. (30) eliminates the need to allow for the effect of temperature on properties along a path, obviously at a cost of accuracy in the computed results. Unpublished comparisons of the application of Eqs. (29) and (30) and of the Curtis–Godson approximation showed differences no greater than 5% for gas paths of temperature and length in the industrially important range.

List of Symbols

a_n	Weighting factor for gray-plus-clear gas model [Eq. (30)]
A	Area
A_B (A_L)	Equivalent black-band width or band absorptance (equivalent black-line width or line absorptance)
b (b_c, b_D)	Line half-width (for collision broadening, Doppler broadening)
b^*	Line half-width at a reference temperature and pressure for the resonant-dipole–dipole collisions [Eq. (19)]
c	Particle concentration, number per unit volume
c_1, c_2	First and second Planck constants
C_c (C_w)	Correction factor for pressure broadening of radiation from carbon dioxide (water vapor)
d	Mean center-to-center line spacing
D	Diameter, characteristic dimension
E	Hemispherical blackbody flux density at a surface, or hemispherical emissive power of a blackbody. Units, energy/area × time
E_λ (E_ω)	Monochromatic emissive power of a blackbody at wavelength λ (wavenumber ω). Units, energy/area × time × wavelength or wavenumber
$E_{\lambda,g}$($E_{\lambda,s}$)	E_λ at gas (surface) temperature
f	Generally, a function
$f_{\lambda T}$	Energy fraction of blackbody spectrum lying below λ
h	Planck's constant. Units, momentum × length
I	Intensity, radiant energy flux density per unit solid angle of divergence
I_B	Intensity of blackbody radiation
I_0	Intensity on entry into system of interest
I_n( )	Modified Bessel function of order n
k	Absorption coefficient [l^{-1} × (atmospheres or concentration)$^{-1}$]
k_c (k_p)	Absorption coefficient, concentration basis (pressure base)
k_λ (k_ω)	Monochromatic absorption coefficient
K	Absorption coefficient (l^{-1})
$\bar{K}$	Mean absorption coefficient (l^{-1})
l	Length; mean free path
L	Path length; distance; characterizing dimension
m	Number of species [Eq. (19)]
n	An integer, also number of spectral lines in an interval [Eq. (15)]
p	Partial pressure of radiating gas component
p_c(p_w)	Partial pressure of carbon dioxide (water vapor)
P	Total pressure

$P(S)$	Probability function of line with integrated intensity S [Eq. (17)]
q	Flux density, many identifying subscripts (energy/area × time)
S	Integrated line intensity (spectral width per path length and pressure)
$\bar{S}$	Average integrated line intensity
T (T_g, T_s)	Temperature (of a gas, of a surface)
T	Transmittance of a system
u	Absorbing matter (l × atm) defined in Eq. (26)
V	Volume
x	$SpL/2\pi b$ [Eq. (11)]
α	Absorptivity, absorptance
$\alpha_{g,s}$	Absorptivity of a gas for radiation from a surface
ϵ (ϵ_g, ϵ_s)	Emissivity or emittance (of a gas, of a surface)
ϵ_w (ϵ_c)	Emissivity of water vapor (carbon dioxide)
λ	Wavelength. As subscript: monochromatic value at wavelength λ
ν	Frequency. As subscript: monochromatic value at frequency ν
$\prod_{i=0}^{n} a_i$	Product of terms $a_0 a_1 a_2 \cdots a_n$
σ	Stefan–Boltzmann constant
τ	Transmittance, single traversal
ω	Wavenumber λ^{-1}. As subscript: monochromatic value at wavenumber ω
ω_0	Wavenumber of line center

References

Benedict, W. S., and Kaplan, L. D. (1959). *J. Chem. Phys.* **30,** 388.

Benedict, W. S., and Kaplan, L. D. (1964). Calculation of Line Width in H_2O–H_2O and H_2O Collisions. *J. Quant. Spectrosc. Radiat. Transfer* **4,** 453.

Boynton, F. P., and Ludwig, C. B. (1971). Total Emissivity of Hot Water Vapor—II, Semi-Empirical Charts Deduced from Long-path Spectral Data. *Int. J. Heat Mass Transfer* **14,** 963.

Eckert, E. R. G. (1937). *Forschungsheft* **387,** 1.

Edwards, D. K. (1960). *J. Opt. Soc. Amer.* **50,** 617.

Goody, R. M. (1954). "The Physics of the Stratosphere," pp. 161–163. Cambridge Univ. Press, London and New York.

Goody, R. M. (1964). "Atmospheric Radiation," Vol. 1, Theoretical Basis. Oxford Univ. Press, London and New York.

Hottel, H. C. (1927). *Trans. Amer. Inst. Chem. Eng.* **19,** 173.

Hottel, H. C. (1954). *In* "Heat Transmission" (W. H. McAdams, ed.), 3rd ed., Chapter 4. McGraw-Hill, New York.

Hottel, H. C., and Egbert, R. B. (1942). Radiant Heat Transmission from Water Vapor. *Trans. Amer. Inst. Chem. Eng.* **38,** 531.

Hottel, H. C., and Mangelsdorf, H. G. (1935). *Trans. Amer. Inst. Chem. Engrs.* **31,** 517.

Hottel, H. C., and Sarofim, A. F. (1967). "Radiative Transfer." McGraw-Hill, New York.

Hottel, H. C., and Smith, V. C. (1935). *Trans. ASME* **57,** 463.

Howard, J. N., Burch, D. L., and Williams, D. (1955). Geophys. Res. Paper 40, Air Force, Cambridge Res. Center, Bedford, Massachusetts.

Ladenberg, R., and Reiche, F. (1918). *Ann. Phys.* **42,** 181.

Leckner, B. (1971). The Spectral and Total Emissivity of CO_2. *Combust. Flame* **17,** 37.

Ludwig, C. B., Thompson, J. A. L., Malkmus, W., Streiff, M. L., Abeyta, C. N., Janda, R., Actor, L., Suttie, D., and Durson, A. (1966). Study on Exhaust Plume Radiation Prediction. General Dynamics/Convair, Reps. DBE 66-001, 001a, and 017.

Ludwig, C. B., Thompson, J. A. L., Malkmus, W., Streiff, M. L., Abeyta, C. N., Janda, R., and Suttie, D. (1968). Study on Exhaust Plume Prediction. General Dynamics Corp., NASA, CR-61233.

Paschen, F. (1894). *Ann. Phys. Chem.* **53,** 334.

Plass, G. N. (1958). *J. Opt. Soc. Amer.* **48,** 690.

Plass, G. N. (1963). Spectral Band Absorptance for Atmospheric Slant Paths. *Appl. Opt.* **2,** 515.

Plass, G. N., and Stull, V. R. (1960). Theoretical Study of High Temperature Emissivities and Atmospheric Transmission. Air Force Cambridge Res. Center, Bedford, Massachusetts, TR 60-221.

Schack, A. (1924). *Z. Tech. Physik* **5,** 226.

Schmidt, E. (1932). Messung der Gesamtstrahlung des Wasserdampfes bei Temperaturen bis 1000°C. *Forsch. Gebiete Ingenieurw.* **3,** 57.

Simmons, F. S., Yamoda, H. Y., and Arnold, C. B. (1969). Measurement of Hot Temperature Profiles in Hot Gases by Emission-Adsorption Spectroscopy. NASA CR-72491, WRL 8962-18F, Infrared and Opt. Lab., Willow Run Labs., Univ. of Michigan, Ann Arbor.

Stevenson, G. T. (1963). Blackbody Radiation Functions. NAVWEPS Rep. 7261, NOTS TP2623, U.S. Naval Ordinance Test Station, China Lake, California.

VII

Radiative Exchange in Combustion Chambers

H. C. HOTTEL, A. F. SAROFIM, AND
I. H. FARAG

DEPARTMENT OF CHEMICAL ENGINEERING
MASSACHUSETTS INSTITUTE OF TECHNOLOGY
CAMBRIDGE, MASSACHUSETTS

Chapter VI illustrated the complexity of a rigorous formulation of the flux from a homogeneous but nonisothermal column of CO_2 and H_2O when the temperature and concentration distributions are known. In furnaces and combustion chambers the problem is further complicated by the need to estimate the temperature and concentration distribution of the emitting species with allowance for the effects of the flow field (when known), diffusion, chemical reaction, absorption and reradiation at any refractory surfaces, and multiple reflection at the boundaries. Rigorous treatment is out of the question. The best approach is the development of simple models followed by a test of their adequacy against either experimental measurements or the predictions of more sophisticated analytical models. Experience has shown that the predictions of models which appear to introduce gross oversimplification are often surprisingly close to the correct solutions. Integral formulations in general are tolerant to casual treatment of detail since local imperfections are averaged out in the integration process. A few simple models are presented in this chapter in sequence of increasing complexity.

I. MEAN BEAM LENGTHS

Consider first the problem of evaluating the radiative exchange between an isothermal gas and its confining walls, assumed to be black and isothermal. The flux density to a differential element on the boundary can be built from the contributions of beams of divergence $d\Omega$, length r_b, and making an angle θ with the normal (Fig. 1). The intensity of the beams, from the definition of standard emissivity, is equal to the product of the emissivity $\epsilon_g(r_b)$ evaluated for the path length r_b of the beam in question and the blackbody intensity I_B evaluated at the gas temperature. Then

$$q = \int_{2\pi} I \, d\Omega \cos\theta = I_B \int_{2\pi} \epsilon_g(r_b) \, d\Omega \cos\theta. \tag{1}$$

The total flux to the surface from the gas is obtained from integration of Eq. (1) over the enclosure surface:

$$\dot{Q}_{g \to s} = E_g \int_A \int_{2\pi} \epsilon_g(r_b) \cos\theta \, dA \, d\Omega/\pi, \tag{2}$$

where I_B has been replaced by E_g/π. The flux per unit emissive power is defined as the gas–surface exchange area and denoted by $\overline{gs}$,

$$\overline{gs} = \dot{Q}_{g \to s}/E_g. \tag{3}$$

The value of $\overline{gs}$ per unit area is a dimensionless quantity which, from Eq. (2), can be identified as an emissivity evaluated at some length denoted

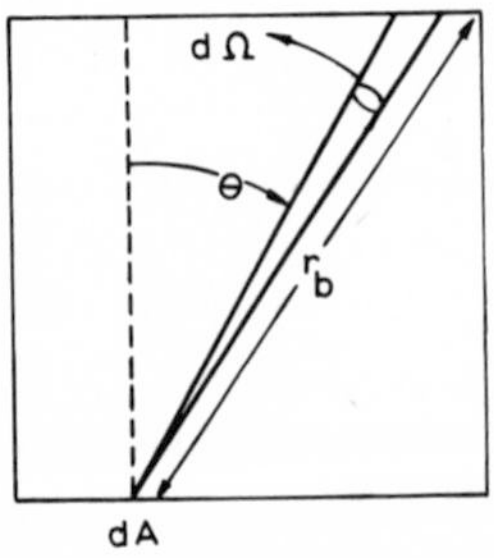

Fig. 1. Formulation of a gas-surface exchange. [From "Radiative Transfer" by H. C. Hottel and A. F. Sarofim. Copyright 1967, McGraw-Hill, New York. Used with permission of the McGraw-Hill Book Company.]

here by L_e which is a weighted mean of the lengths of all the beams contributing to the surface flux. L_e, named mean beam length, is defined as the length that validates the following equality:

$$\overline{gs}/A\,(\equiv \dot{Q}_{g\to s}/AE_g) = \epsilon_g(L_e). \tag{4}$$

From (2)

$$\dot{Q}_{g\to s}/AE_g \equiv \epsilon_g(L_e) = A^{-1}\int_A\int_{2\pi} \epsilon_g(r_b)\cos\theta\, dA\, d\Omega/\pi. \tag{5}$$

When the partial pressure of the absorbing gas tends to zero the emissivity becomes proportional to length, $\epsilon_g(L_e) = KL_e$ and $\epsilon_g(r_b) = Kr_b$, where $K = (\partial\epsilon_g/\partial r)_{p\to 0}$, can be substituted into Eq. (5), and the resulting equation integrated to yield

$$L_0 = 4V/A, \tag{6}$$

where L_0 is the limiting value of L_e as p tends to zero, V is the enclosure volume, and A is the total surface area. For higher values of p, Eq. (5) can be integrated analytically or numerically to obtain the mean beam length L_e for an enclosure of any shape and for any emissivity function. Results of the application of Eq. (5) or its equivalent to many systems support the generalization that the flux to the total envelope area may be evaluated to a good approximation, generally within 5%, by selecting an average mean beam length L_m equal to 0.88 times the value of L_0 given by Eq. (6). Better calculations of average mean beam lengths for a few selected configurations yield 0.63 diameters for a sphere, 0.94 diameters for an infinite circular cylinder, 0.60 times the side of a cube, and 1.05 times the shortest edge for a 1 : 2 : 6 rectangular parallelepiped. With the average mean beam length evaluated, the emissivity or absorptivity may be read for the particular composition and temperature from standard emissivity charts. The radiative exchange is then given by

$$\dot{Q}_{g\rightleftharpoons s} = A_s[\epsilon_g(L_m)E_g - \alpha_{gs}(L_m)E_s]. \tag{7}$$

The value of L_0 given by Eq. (6) is valid only when interest is in the flux to the entire surface boundary. When interest is in the flux to only part of the enclosure wall, recourse must be made to Eq. (3).

A. Illustration 1

The emission from a chemiluminescent reaction carried out in a cylindrical chamber is monitored by a photodiode placed at the center of the base of the cylinder. It is desired to maximize the signal recorded by the photodiode without changing the volume (i.e., residence time) of the reactor. (i) What is the height-to-diameter ratio of an upright circular cylinder that will maximize the signal from the receiver at the center of its base? (ii) How does the signal from the cylinder of optimum shape compare with those from a detector placed on the surface of a sphere or the center of the base of a hemisphere, both having the same volume as the cylinder? The emitting gas may be assumed to be optically thin ($KL \to 0$), the walls of the reactor black, and the photodiode a flat black receiver with a 2π-sr view.

Denote the height of the cylinder by h, the radius by a, and the volume by V. From the definition of L_0, Eqs. (4) and (5), and the limiting relation

$$\epsilon_g(r_b) \to Kr_b \qquad \text{as} \quad Kr_b \to 0,$$

$$\epsilon_g(L_0) = KL_0 = K\pi^{-1} \int_{2\pi} r_b \cos\theta \, d\Omega, \tag{8}$$

and, with the nomenclature shown in Fig. 2,

$$d\Omega = 2\pi \sin\theta \, d\theta, \qquad \theta_1 = \tan^{-1}(a/h),$$

$$r_b = \begin{cases} h/\cos\theta, & 0 < \theta < \theta_1, \\ a/\sin\theta, & \theta_1 < \theta < \frac{1}{2}\pi. \end{cases}$$

Substituting these values into Eq. (8) and integrating the resultant expression yields

$$L_0 = 2[(h + a) - (h^2 + a^2)^{1/2}]. \tag{9}$$

In the limit, as $a/h \to \infty$, $L_0 \to 2h$. This limit agrees with the value obtained from Eq. (6) for an infinite slab of thickness h. It is interesting to note that the value of $2h$ for a slab represents the average for beams that

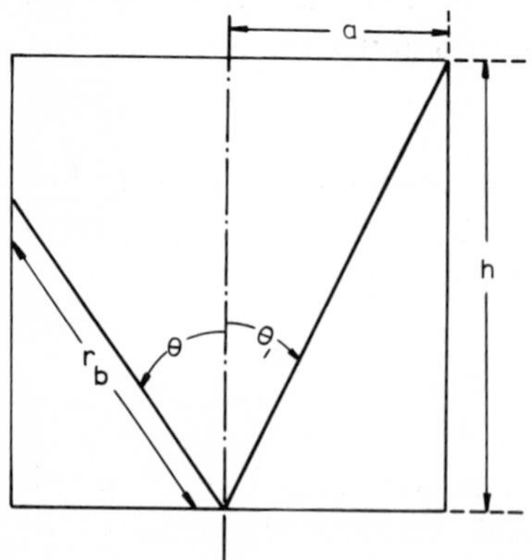

Fig. 2. Nomenclature for Illustration 1.

vary in length from h along the normal to infinity at glancing incidence. In the limit as $a/h \to 0$, the case of a semi-infinite cylinder of diameter $2a$ radiating to the center of its base, $L_0 \to 2a$. In order to determine the maximum value of L_0 for a fixed volume V $(=\pi a^2 h)$ let $a/h = z$. Then

$$h = (V/\pi)^{1/3} z^{-2/3}, \qquad a = (V/\pi)^{1/3} z^{1/3},$$

and

$$L_0/(V/\pi)^{1/3} = 2z^{-2/3}[1 + z - (z^2 + 1)^{1/2}]. \tag{10}$$

The optimum z is sought by setting the derivative of Eq. (10) to zero. The resultant equation,

$$4z^2 - 9z + 4 = 0,$$

has two roots, 1.64 and its reciprocal 0.609, the second of which can be readily shown to be the optimum value. At the optimum,

$$L_0 = 1.22(V/\pi)^{1/3}.$$

For comparison, the mean beam length for a hemisphere radiating to the center of its base is the radius or $(3V/2\pi)^{1/3}$, or $1.145(V/\pi)^{1/3}$. For a sphere the mean beam length to all parts of the bounding surface is constant and is therefore given by Eq. (6): $L_0 = 4V/A = \frac{2}{3}(\text{diameter}) = \frac{2}{3}\ (6)^{1/3}$ $(V/\pi)^{1/3} = 1.21(V/\pi)$. The relative signals from an optimized-cylinder, sphere, and hemisphere are then 1.07, 1.06, and 1.00, respectively. Considerations other than signal strength will obviously determine the design of the reactor.

II. APPROXIMATE ALLOWANCE FOR GRAY WALLS

If the walls of the enclosure are gray, with an emissivity ϵ_s, the flux calculated for black surfaces from Eq. (7) will be reduced by a factor ϵ_s' having a value between ϵ_s (when $\epsilon_g = 1.0$) and 1.0 (when $\epsilon_g \to 0$). When the emissivity is high ($\epsilon_s > 0.8$) the use of a factor $\frac{1}{2}(\epsilon_s + 1)$ as a multiplier in Eq. (7) cannot lead to much error. A better approximation for $\epsilon_g\epsilon_s'$, however, is $1/[(1/\epsilon_g) + (1/\epsilon_s) - 1]$.

III. THE LONG CHAMBER

When the gas temperature transverse to the flow direction is reasonably uniform and the chamber is long compared with its mean hydraulic radius, the opposed upstream and downstream fluxes through the flow cross section will substantially cancel. Under these conditions, the radiative contribution to the local flux may be formulated in terms of local temperature and of view factors or mean beam length evaluated as for a two-dimensional system infinite in the flow direction. The local flux density at the sink A_1 is then

$$q(T_g, T_s) = \tfrac{1}{2}(\epsilon_s + 1)(\epsilon_g E_g - \alpha_{gs} E_s) + h(T_g - T_s), \tag{11}$$

where q is written to indicate that it is a function of T_g and T_s. If $\dot{m}C_p$ is the hourly heat capacity of the gas stream, then

$$q(T_g, T_s)\, dA_1 = -\dot{m}C_p\, dT_g. \tag{12}$$

From this it is clear that the area under a curve of $\dot{m}C_p/q$ versus T_g over the interval inlet to a postulated outlet temperature $T_{g,e}$ provides the area required to transfer the energy corresponding to the differences in enthalpy between inlet and outlet. In a design calculation the outlet temperature is specified by the desired efficiency and the area under the curve may be used to calculate the length of furnace required to carry out the specified heat transfer. More often the length of the furnace is known and a trial-and-error procedure must be employed to find the outlet temperature which provides the correct area under the curve.

IV. THE WELL-STIRRED FURNACE

The simple model to be described in this section is found to make substantially correct predictions of the relation among the dominant variables for a wide range of furnace types. It is assumed that the gas mass and flame transfer heat as though at a mean temperature T_g; the gas is gray; the bounding walls of total area A_T are divisible into a heat-sink area A_1, gray and at a single temperature T_1, and an adiabatic refractory area A_r, convection to which and external loss from which are negligible. The net radiative flux from gas to sink, with allowance for multiple reflections at all surfaces and absorption and reradiation at the refractory, must be proportional to the difference in their black emissive powers. The proportionality constant is called the *total interchange area*, designated by $\overline{GS_1}$. An expression for $\overline{GS_1}$ will be derived by considering first the fluxes incident and leaving a gray surface (Fig. 3). The incident flux density is denoted by H, the reflected flux density by R, and the total leaving flux density by W. It is assumed that the surfaces reflect radiation diffusely. It is clear from Fig. 3 that

$$W = \epsilon E + R = \epsilon E + \rho H. \tag{13}$$

The net radiant flux density, defined as the net loss of energy per unit

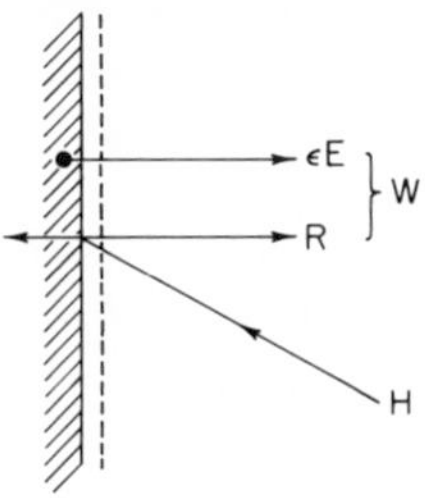

Fig. 3. Radiant flux densities at a surface. [From "Radiative Transfer" by H. C. Hottel and A. F. Sarofim. Copyright 1967, McGraw-Hill, New York. Used with permission of the McGraw-Hill Book Company.]

time and per unit area, may be formulated in terms of W and E as follows:

$$q_{\text{net}} = W - H = W - (W - \epsilon E)/\rho = (\epsilon/\rho)(E - W). \tag{14}$$

The total net flux at a surface of area A_1 is then proportional to the product of $A_1\epsilon_1/\rho_1$, which may be considered to be the analog of a conductance, by $E_1 - W_1$, which is the analog of a potential or driving force. Similarly, the flux between gas and surface sink A_1 is equal to the product of conductance $\overline{gs_1}$, defined by Eq. (3), and the difference $E_g - W_1$. The flux between two surfaces A_1 and A_r is the product of the difference in total leaving flux densities W_1 and W_r and the exchange area $\overline{s_1 s_r}$, which is the triple product $A_1 F_{1r}\tau_{1r}$, where F_{1r} is the view factor of A_r from A_1 and τ_{1r} is the mean transmittance between 1 and r. The electrical analog shown is depicted in Fig. 4 with the gas as a source of potential E_g connected to the sink and refractory by conductances $\overline{gs_1}$ and $\overline{gs_r}$ and the refractory to the sink by a conductance $\overline{s_1 s_r}$. The resistance to transfer at the sink due to departure from blackness is shown by a resistor of conductance $A_1\epsilon_1/\rho_1$ connecting the potentials E_1 and W_1. The refractory potential is represented by a floating node since there is no net transfer at the refractory surface. Summation of the resistances in series and parallel shows that

$$\begin{aligned} \overline{GS_1} &= \dot{Q}_{g\rightleftharpoons 1}/(E_g - E_1) \\ &= \{(\rho_1/A_1\epsilon_1) + [\overline{gs_1} + [(1/\overline{gs_r}) + (1/\overline{s_r s_1})]^{-1}]^{-1}\}^{-1}. \end{aligned} \tag{15}$$

The values of $\overline{gs_1}$ and $\overline{gs_r}$, from the discussion under mean beam lengths, are equal to $A_1\epsilon_g(L_{m1})$ and $A_r\epsilon_g(L_{mr})$, where the second subscript on L_m is a reminder that the mean beam lengths are slight functions of position on a surface. The term $\overline{s_r s_1}$ is the direct exchange area between the sink and refractory, including allowance for attenuation between surfaces; it is equal to the product of the direct exchange area $A_r F_{r1}$ and the average transmittance τ_{r1} between A_1 and A_r. If the small differences in the mean beam lengths associated with $\overline{gs_1}$, $\overline{gs_r}$, and $\overline{s_1 s_r}$ are neglected, $\overline{gs_1} = A_1\epsilon_g$, $\overline{gs_r} = A_r\epsilon_g$, and $\overline{s_r s_1} = A_r F_{r1}(1 - \epsilon_g)$ where ϵ_g is the emissivity evaluated at the mean beam length for the entire enclosure. Substitution in

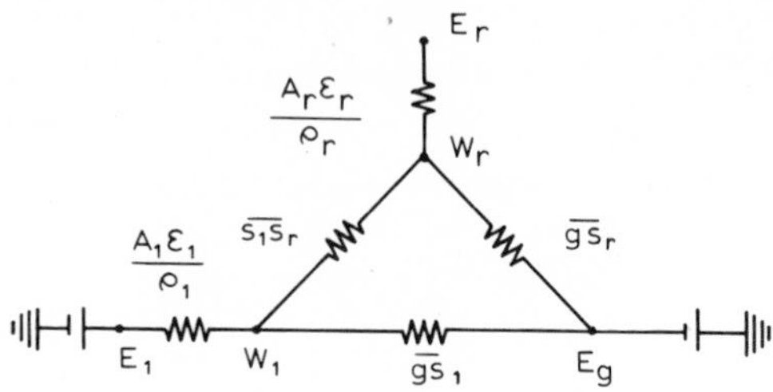

Fig. 4. Network analog for derivation of gas–surface total exchange area.

Eq. (15) yields

$$\overline{GS}_1 = A_1\{\epsilon_1^{-1} - 1 + [\epsilon_g(1 + \{(A_r/A_1)/[1 + \epsilon_g/(1 - \epsilon_g)F_{r1}]\})]^{-1}\}^{-1}. \tag{16}$$

The use of capital letters designates that, in contrast to the direct-exchange area $\overline{gs}$, $\overline{GS}_1$ is a total-exchange area, making allowance for grayness of surfaces and for the aid given by the refractory in producing net exchange between gas and surface 1. Either of two further alternative simplifications is possible. The first is to assume that the sink and refractory are interdispersed over the surface of the enclosure in such a manner that the view factor to A_1 (or A_r) from any point in the enclosure equals the fraction of the total surface area covered by A_1 (or A_r). Let C denote the cold surface fraction $A_1/(A_1 + A_r)$. Then $F_{r1} = C$, $A_r/A_1 = (1 - C)/C$, $A_1 = CA_T$, and, from substitution in Eq. (16),

$$\overline{GS}_1 = A_T[(C\epsilon_1)^{-1} + \epsilon_g^{-1} - 1]^{-1}. \tag{17}$$

The second alternative is to assume that A_1 is concentrated in a single plane surface, whence $F_{r1} = C/(1 - C)$. Substitution into (16) indicates that the term $\epsilon_g^{-1} - 1$ in (17) must now be multiplied by $(1 - \epsilon_g)/(1 - C\epsilon_g)$. The total flux to the sink is the sum of the radiation and convective contributions, or

$$\dot{Q}_G = \overline{GS}_1(E_g - E_1) + h_1A_1(T_g - T_1). \tag{18}$$

Because the convective flux is small compared to radiation it may be linearized in T^4 to give

$$A_1h_1(T_g - T_1) \cong (A_1h_1/4\sigma T_{g1}^3)(E_g - E_1),$$

where T_{g1} is the arithmetic mean of T_g and T_1. Equation (18) may therefore be rewritten as

$$\dot{Q}_G = [\overline{GS}_1 + (h_1A_1/4\sigma T_{g1}^3)](E_g - E_1) \equiv (\overline{GS}_1)_{r,c}(E_g - E_1), \tag{19}$$

where $(\overline{GS_1})_{r,c}$ includes allowance for multiple reflection at all walls, reradiation by the refractory, and convection. Equation (19) may be combined with an energy balance. The mean heat capacity applicable to the interval T_g to the base temperature T_0 is used to define a kind of adiabatic-flame temperature T_F, and if H_F is the entering hourly enthalpy of the fuel and air above the base, the energy balance becomes

$$(H_F - \dot{Q}_G)/H_F = (T_g - T_0)/(T_F - T_0) \tag{20}$$

This may be combined with Eq. (19) to eliminate the unknown T_g and give the dimensionless relation

$$\dot{Q}'D' + \tau^4 = (1 - \dot{Q}')^4, \tag{21}$$

where $\dot{Q}'$ is the reduced furnace efficiency, the actual furnace efficiency times the temperature ratio $(T_F - T_0)/T_F$,

$$D' = H_F/\sigma(GS_1)_{r,c}T_F^4(1 - T_0/T_F)$$

and τ is the reduced sink temperature T_1/T_F. Equation (21) gives the efficiency of a combustion chamber as a function of firing density and heat sink temperature; and the firing density term makes due allowance for any operating variable such as fuel type, excess air, or air preheat which affect flame temperature or gas emissivity, for fractional occupancy of wall by sink surfaces, and for sink emissivity. This is the well-stirred speckled-furnace model (Hottel, 1961). The characteristics of Eq. (21) are shown in Fig. 5, together with shaded areas indicating the operating

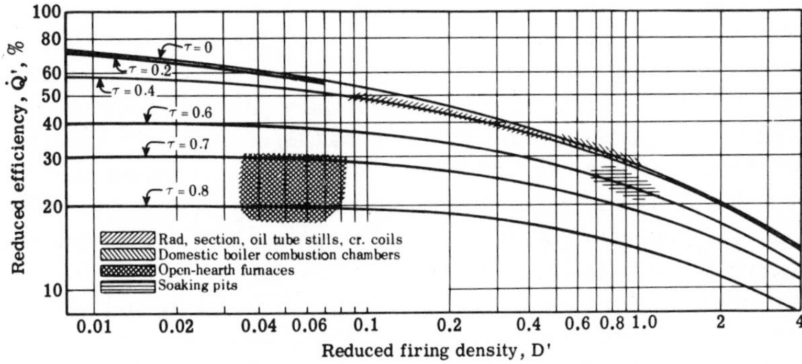

Fig. 5. The thermal performance of well-stirred furnace chambers. Reduced efficiency as a function of reduced firing density D' and reduced sink temperature τ. Gas turbine combustors are off scale, far right. [From "Radiative Transfer" by H. C. Hottel and A. F. Sarofim. Copyright 1967, McGraw-Hill, New York. Used with permission of the McGraw-Hill Book Company.]

regimes of various types of furnace. Note the significant properties of the function presented: (1) As firing rate D' goes down, the efficiency rises and approaches $1 - \tau$ in the limit; (2) changes in sink temperature have little effect if $\tau < 0.3$; (3) at very high firing rates, where $\dot{Q}'$ approaches inverse proportionality to D', efficiency of heat transfer varies directly with ϵ_g when C_{ϵ_1} approaches 1 (gas turbine combustor); but (4) at low firing rates, where $\dot{Q}'$ is insensitive to changes in D', ϵ_g has relatively little effect.

A. Illustration 2

Consider a gas-fired, 200-MW utility boiler with a radiant chamber 30×60 ft^2 in cross section and 80 ft high. The side walls are water cooled, the top surface consists of banks of superheater tubes, and the floor is lined with refractory insulation. The mean temperature of the water walls is 700°F. What is the efficiency of the heat transfer in the radiant chamber when it is fired with (i) 95,000 lb/hr and (ii) 133,000 lb/hr of natural gas (lower heating value 21,500 Btu/lb) with 10% excess air at 600°F air preheat? From the information just provided it is readily shown that $p_c = 0.088$ atm, $p_w = 0.175$ atm, $A_T = 18{,}000$ ft^2, $L_m = 0.88 \times 4 \times (144{,}000)/18{,}000 = 28$ ft, $C_s = 0.9$. In order to evaluate $(\overline{GS_1})_{r,c}$ and T_F it is assumed that the tube wall emissivity is 0.8, that the convective heat transfer coefficient is 2.0 Btu/hr ft^2 °F and the $\dot{m}C_p$ at the leaving gas temperature is 8.7 Btu/mole °F. $T_0 = 520$°R. Then $H_F = 2.27 \times 10^9$ and 3.18×10^9 Btu/hr for the two firing rates; the pseudo-adiabatic flame temperature is 4360°R; $\epsilon_g = 0.38$; $(\overline{GS_1})_{r,c} = 6190$ ft^2 of which convection contributes about 4%; $\tau = 0.266$; and $D' = 0.7$ or 0.98 depending on firing rate. The reduced efficiencies are 0.31 and 0.29 [the actual efficiencies equal $T_F/(T_F - T_0)$ times the reduced efficiency or 0.35 and 0.33] corresponding to flux densities of 46,000 Btu/hr ft^2 and 69,000 Btu/hr ft^2 for the two firing rates. It is to be remembered that efficiency here is based on combustion-chamber heat transfer; the convection section of course adds additional performance. The consequences of increasing the gas emissivity from 0.38 to 1.0 are to increase $(GS_1)_{r,c}$ to 12,920 ft^2 and to decrease D' to 0.32 and 0.45. The corresponding reduced efficiencies read from Fig. 5 for the two firing densities are 0.4 and 0.37. From these numerical examples it is concluded that the efficiency of the radiant chamber is insensitive to changes in firing density and emissivity, a change of 40% in firing rate producing a 7% change in efficiency, and an increase by a factor of 2.5 in emissivity producing an only 30% change in heat transfer. The predicted efficiencies are on the low side because the effective radiating temperature in a boiler is higher than the leaving gas temperature, allowance for which may be made by the procedures discussed in the next section.

B. Illustration 3

By contrast to the conclusions arrived at for a utility boiler, an increase in emissivity for a high-output combustor produces a proportional increase in the heat transfer to its walls. For purposes of illustration a combustor will be modeled by a cylinder 6 in. in internal diameter, 1 ft long, and operating at 20 atm. A kerosene fuel and air (at 1440°R) are fed at rates of 1000 and 50,000 lb/hr, respectively. For this system, $p_c = p_w = 0.087$ atm, $L_m = 0.35$ ft, $C_s = 1.0$, $\epsilon_1 \equiv 0.8$, $A_T = 2.0\ \text{ft}^2$, $(\overline{GS_1})_{r,c} \cong A_T/(1/\epsilon_g + 0.25) \cong A_T\epsilon_g$, $T_F = 4400°\text{R}$, $T_0 = 520°\text{R}$, and $D' = 26.5/\epsilon_g$. The wall temperature is about 1620°R and therefore $\tau = 0.37$. The value of D' is off-scale on the right-hand side of Fig. 5 in the region in which $\dot{Q}'$ is inversely proportional to D' and hence directly proportional to ϵ_g. The efficiency for heat transfer in a gas-turbine combustor is sufficiently low that the temperature profiles can be determined adequately from the combustion, flow, and mixing patterns in a system assumed to be adiabatic. Since the fuel–air ratio may vary by more than a factor of 5 as the combustion products are diluted by air bled through the liner, the assumption of a well-stirred chamber is a poor one. A reasonable first approximation for a thermal model of the chamber is a well-stirred chamber followed by a plug-flow section. With the temperature and concentration field established the flux incident on the combustion liner may be calculated along different lines of sight using the procedures described in Chapter VI. An alternative is to zone the enclosure and calculate the flux from the different zones to the wall. Approximation of the combustion tube by a right circular cylinder would permit the use of the cylindrical exchange areas developed by Erkku (1959). But the values tabulated by Erkku would need to be modified to allow for concentration gradient. The point to the illustrative calculation here is that the flux will be approximately proportional to gas emissivity and it is therefore important that this quantity be estimated accurately, including allowance for pressure broadening of H_2O and CO_2 and for soot luminosity.

Application of the well-stirred model to a number of different furnace types was reported at the International Flame Research Foundation's 2nd Members Conference at Ijmuiden, May 13–14, 1971, and the results are summarized in Table I, taken from the paper by J. Crowther (1971). The agreement between measured and calculated values is impressive, but

TABLE I

COMPARISON OF WELL-STIRRED FURNACE EFFICIENCIES WITH MEASURED VALUES

Furnace type		H_F (therms/hr)	A_T (ft²)	C	D'	T'	$\dot{Q}'$	Efficiency (%) Calc.	Meas.
OH steel		900	1500	0.32	0.69	0.56	0.28	39	35
Copper matte smelter		1500	7690	0.42	0.15	0.58	0.36	39	37.5
IFRF (various conditions)	(a)	32	626	0.217	0.17	0.58	0.35	25	23
	(b)	32	500	0.019	1.2	0.185	0.285	15.5	15
	(c)	38.6	626	0.217	0.49	0.62	0.270	23.5	23
	(d)	38.6	500	0.019	3.96	0.215	0.135	9.8	10
Aluminum melter	high flame	134	736	0.22	0.72	0.52	0.275	25.7[a]	24.5[a]
	low flame	60	736	0.22	1.22	0.73	0.172	—	—

[a] Represent mean values.

must depend to a certain extent on the assumptions made in defining the input conditions.

V. EFFECTIVE RADIATING TEMPERATURE

In practice, because of departures from the well-stirred furnace approximation, the effective radiating temperature of a gas will differ from the mean enthalpy temperature of the gases at the outlet of the combustion chamber by an amount Δ that will depend on firing density and flow patterns for a particular chamber. Allowance for the effect of Δ in the well-stirred furnace formulation [the subtraction of Δ from T_g in Eq. (20)] yields the modified expression:

$$\dot{Q}'D' + \tau^4 = (1 + \Delta' - \dot{Q}')^4, \tag{22}$$

where $\Delta' = \Delta/T_F$. Calculations performed for a number of different conditions suggest that Δ' is approximately proportional to $\dot{Q}'$. Let the proportionality constant be $(1 - 1/d)$. Then

$$\Delta' = (1 - 1/d)\dot{Q}' \qquad \text{or} \qquad \dot{Q}' - \Delta' = \dot{Q}'/d. \tag{23}$$

Substitution of this into (22) gives

$$[\dot{Q}'/d][D'd] + \tau^4 = [1 - \dot{Q}'/d]^4. \tag{24}$$

Equation (24) is seen to be identical to (21) when $\dot{Q}'$ is replaced by $\dot{Q}'/d$ and D' is replaced by $D'd$. Thus Fig. 5 includes the case of a difference Δ between the gas-radiating and gas-enthalpy temperatures. There is some evidence that d is about $\frac{4}{3}$. Thus for the conditions of Illustration 2, (Section IV,A), with $\dot{Q}' = 0.3$ and $T_F = 4360°\text{R}$, $\Delta' = 0.075$ and $\Delta = 320°\text{F}$. Inclusion of Δ in the calculation of $\dot{Q}'$ for the case of the lower firing density increased the computed value of $\dot{Q}'$ from 0.31 to 0.36. For a high-output system such as that in Illustration 3, Δ is small.

VI. ALLOWANCE FOR WALL LOSSES

Wall losses may justifiably be neglected for many furnace types. However, in the operation of high-temperature furnaces such as glass furnaces and ore-sintering furnaces losses through ports or refractories worn thin may become significant. The analysis in the preceding sections has been extended to include this case. The equation of flux, Eq. (18), modified to include allowance for loss through the refractory, becomes

$$\dot{Q}_G = (\overline{GS_1})_{r,c}(E_g - E_1) + UA_r(T_g - T_0), \tag{25}$$

where U is the overall coefficient of heat transfer through the refractory. Elimination of T_g between Eqs. (20) and (25) with Δ included as a subtraction from T_g in the former, yields, after substitution from (24),

$$\dot{Q}'D' + \tau^4 = (1 - \dot{Q}'/d)^4 + L(1 - T_0' - \dot{Q}'/d), \tag{26}$$

where two new dimensionless groups have been defined, the loss coefficient term $L = UA_r/(\overline{GS_1})_{r,c}\sigma T_F^3$ and the ambient temperature term $T_0' = T_0/T_F$. It should be recognized that $\dot{Q}'$ measures the energy $\dot{Q}_G$ lost by the gas [Eq. (25)] and that the energy loss term must be subtracted before a useful transfer term is obtained. The efficiency η of energy transfer for the sink can be shown to be given by

$$\eta \equiv \dot{Q}_{\text{useful}}/H_F = \dot{Q}'/(1 - T_0') - (L/D')\{1 - [(\dot{Q}'/d)/(1 - T_0')]\}. \tag{27}$$

Equations (26) and (27) can be used to eliminate $\dot{Q}'$ and obtain the efficiency η by substituting $(1 - \dot{Q}'/d) \equiv x$, with which substitution (26) and (27) become

$$(1 - x)(D'd) + \tau^4 = x^4 + L(x - T_0') \tag{28}$$

and

$$\eta/d = [(1 - x)/(1 - T_0')] - [L/(D'd)][1 - (1 - x)/(1 - T_0')]. \tag{29}$$

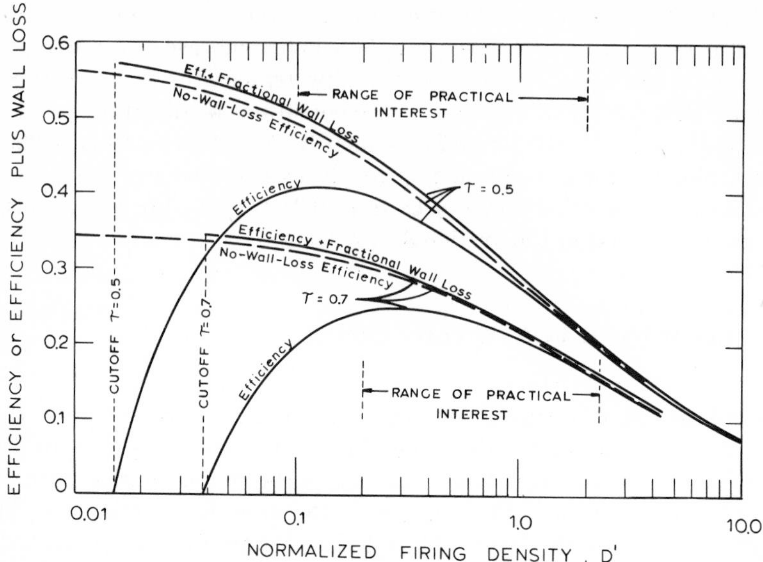

Fig. 6. Example of effect of wall loss on performance of well-stirred furnace chamber. $\tau = \frac{1}{2}$, $L = 0.02$, $T_0' = \frac{1}{8}$. Either $\Delta' = 0$ (curves as labeled) or, if $\Delta' = (1 - 1/d)\dot{Q}'$, efficiency η is changed to η/d, wall loss is changed to (wall loss)$/d$, and firing density D' is changed to $D'd$.

Equations (28) and (29) may be considered parametric equations in x, expressing the relation

$$\eta/d = f(D'd, \tau, L, T_0'). \tag{30}$$

Since T_0' normally varies negligibly (it is about $\frac{1}{8}$), Eq. (30), or (28) and (29), gives efficiency as a function of firing density, sink temperature, and wall-loss coefficient. The use of $(\overline{GS_1})_{r,c}$ from Eq. (15) in the calculations above restricts the validity of the derivation to the cases of a white refractory or that in which the internal convection balances the conduction through the wall. The derivation can readily be extended to the more general case in which the refractory surface is no longer a radiatively adiabatic surface but the solutions no longer assume simple algebraic form. For the restricted case of a white refractory the efficiency falls to zero when the gas temperature equals the sink temperature or

$$D'd = L(\tau - T_0')/(1 - \tau). \tag{31}$$

The consequence of allowing for wall losses is shown for the case of $L = 0.02$ in Fig. 6. The conditions selected for the calculated curves were a dimensionless sink temperature $\tau = \frac{1}{2}$, a dimensionless ambient temperature $T_0' = \frac{1}{8}$, and no difference between the effective radiating and the mean enthalpy temperatures ($\Delta = 0$). The top curve represents the energy flux $\dot{Q}_G$ to stock and walls normalized by the entering enthalpy H_F. The curve extending to $D' = 0$ is the corresponding value of $\dot{Q}_G$ for the case of no wall loss ($L = 0$); it has an asymptotic value of $(1 - \tau)/(1 - T_0')$ or 0.572 as $D' \to 0$. The curve exhibiting a maximum is the efficiency η, representing the flux to the stock normalized by H_F; it goes to zero when the transfer from the gas is balanced by the loss through the wall, at $D' =$ 0.15 in this case. The value of $L = 0.02$ is realistic. To place it in perspective, for a furnace with 50% refractory ($C = 0.5$), a gas emissivity of 0.3, a sink emissivity of 0.8, $T_F = 4000°F$, and $T_0 = 500°R$, the computed value of $(\overline{GS_1})_{r,c}$, neglecting convection and using the speckled-furnace model, is $0.203A_T$, and therefore L equals $0.0226U$ in English engineering units. The value of U allowing for an internal convective heat transfer coefficient of 3 Btu/hr ft² °F, conduction across a 2-in. thickness of brick with a thermal conductivity of 1 Btu/hr °F ft, and an external coefficient of 5 Btu/hr ft² °F to allow for convection and radiation from the outside surface, is found to be 1.5 Btu/hr ft² °F. Thus the value for L is found to be 0.032 for a case in which the insulation is not particularly good.

VII. ALLOWANCE FOR GAS TEMPERATURE GRADIENTS

The assumption of a well-stirred furnace neglects the effects of both temperature and concentration gradients. Allowance for temperature

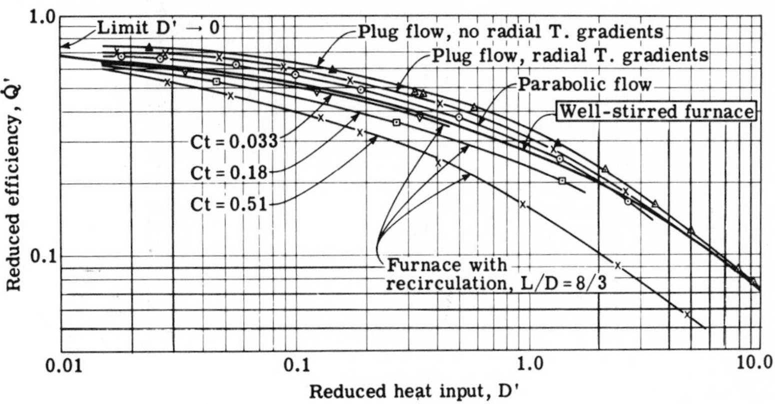

Fig. 7. Comparison of performance of plug-flow and recirculatory-flow furnaces with that of the "well-stirred" chamber. $\dot{Q}'$, τ as in Fig. 5. [From "Radiative Transfer" by H. C. Hottel and A. F. Sarofim. Copyright 1967, McGraw-Hill, New York. Used with permission of the McGraw-Hill Book Company.]

gradients may be made by zoning a combustion chamber and formulating energy balances about each zone following the methodology developed by Hottel and Cohen (1958). Application of the zone method by Hottel and Sarofim (1965) to a cylindrical combustion chamber provided the results summarized in Figs. 7–10. Other than assuming a uniform concentration and adopting flow patterns measured on cold systems the analysis was fairly rigorous. Figure 7 compares the reduced efficiency calculated for different flow models with the value for a well-stirred furnace (heavy line). The cases of plug and parabolic flow through the furnace as expected yield higher efficiencies than the well-stirred furnace model. The three bottom curves corresponding to the firing of a jet along the axis of the

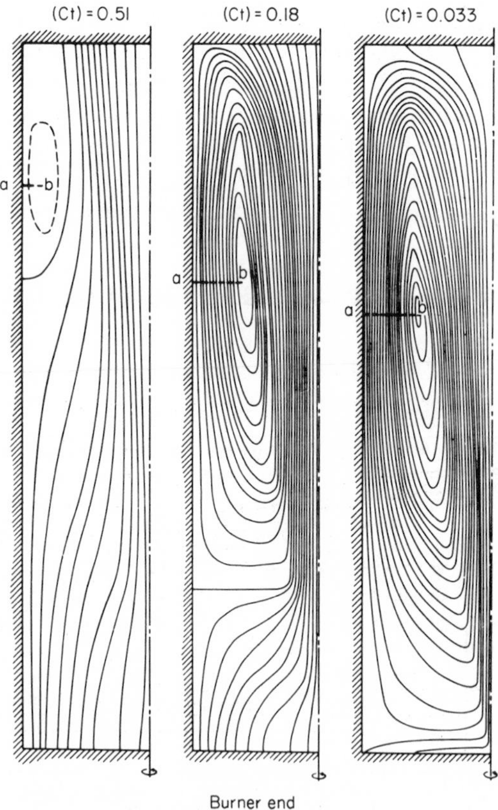

Fig. 8. Mean flow patterns for axially cylindrical furnace (flow between streamlines for Ct = 0.033 is five times the value at Ct's of 0.18 and 0.51). [From "Radiative Transfer" by H. C. Hottel and A. F. Sarofim. Copyright 1967, McGraw-Hill, New York. Used with permission of the McGraw-Hill Book Company.]

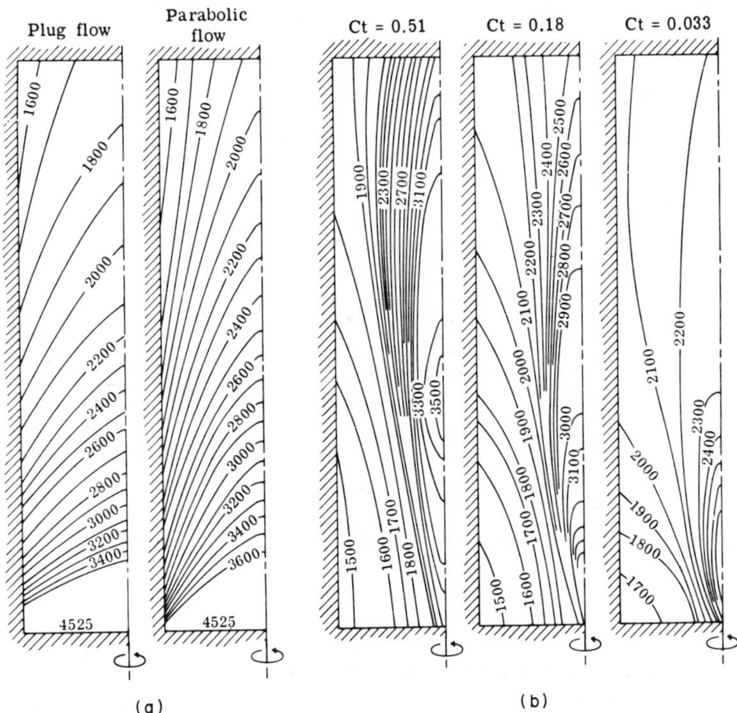

Fig. 9. Temperature profiles (°R) for a cylindrical furnace; diameter is 4 ft, firing density is 10^4 Btu/hr ft^2 of sink. Because of symmetry, the distribution to the left of the axis only is shown. Temperature is in degrees Rankine. (a) Furnace fired across entire burner end—instantaneous combustion. (b) Furnace firing axially—combustion limited by mixing of fuel and air. [From "Radiative Transfer" by H. C. Hottel and A. F. Sarofim. Copyright 1967, McGraw-Hill, New York. Used with permission of the McGraw-Hill Book Company.]

chamber, however, fall below the well-stirred furnace results. The mean stream lines for the ducted jet shown in Fig. 8 were taken from the measurements by Becker (Becker, 1961; Becker *et al.*, 1963). The flow is a function of the momenta of the fuel and air, characterized by the Craya–Curtet number Ct. [For discussion of the Craya–Curtet number and the closely related Thring–Newby number refer to Becker *et al.* (1963).] The three conditions selected correspond to a lazy flame (Ct = 0.51), a flame typical of industrial operation (Ct = 0.18), and one with an unrealistically high degree of recirculation (Ct = 0.033). The curves fall below the well-stirred result because the flame near the burner is narrow and consequently an ineffective radiator. Recirculation of combustion products improves

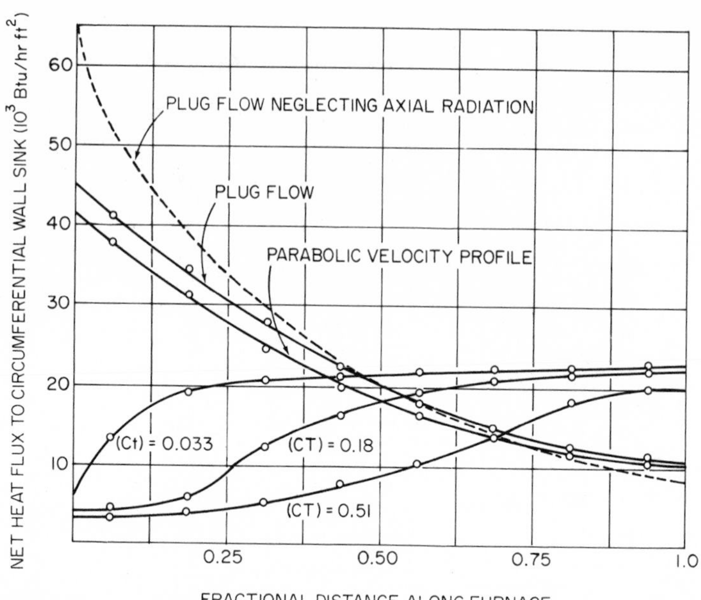

Fig. 10. Wall flux distribution along cylindrical furnace calculated for different flow models: length-to-diameter ratio is 8/3, diameter is 4 ft, firing density is 4×10^4 Btu/hr ft^2 of sink. [From 'Radiative Transfer'' by H. C. Hottel and A. F. Sarofim. Copyright 1967, McGraw-Hill, New York. Used with permission of the McGraw-Hill Book Company.]

efficiency and in the limit, as $Ct \rightarrow 0$, provides the well-stirred case. Calculated temperature and wall-flux distributions shown in Figs. 9 and 10 are consistent with those expected for the different postulated flow patterns.

VIII. ALLOWANCE FOR CONCENTRATION DISTRIBUTIONS

The zone method can be extended to the case of nonuniform concentration by proper allowance in the formulation of the exchange areas for the concentration distribution along the paths between two zones. The amount of effort required to evaluate the direct exchange areas increases enormously since relative position of two zones is no longer a sufficient specification of zone position for the formulation of direct exchange. The number of direct exchange areas is of the order of the square of the number of zones and imposes an upper limit on the number that can be handled with the present generation of computers.

Calculations by Pieri *et al.* (1973) for a 3 : 3 : 4 rectangular parallelepiped subdivided into 36 equal cubes demonstrate the effect of concentration gradients. The calculations were based on the hypothetical case of a jet fired through the central square of the 3 × 3 end of the chamber and spreading toward the edges of the far end wall (an exaggerated jet expansion forced by the restriction on the number of zones that could be handled). Entrainment into the jet was calculated by application of the Thring–Newby theory. Concentration within and outside the jet was assumed to be uniform and different. The calculations were repeated assuming the concentration throughout the enclosure was uniform and equal to that of the leaving combustion products. The results for nine different cases showed that allowance for concentration gradients reduced the overall

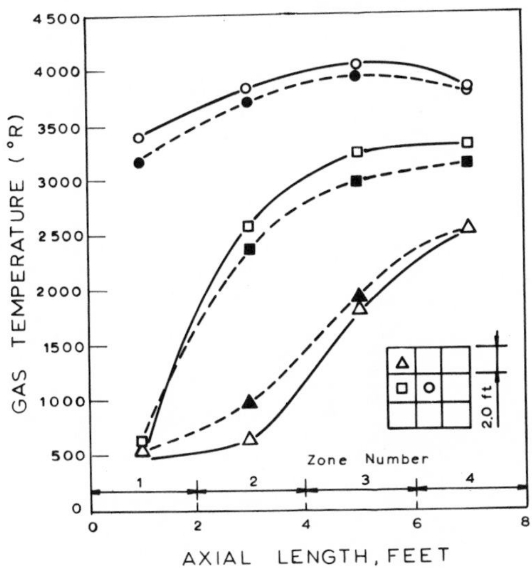

Fig. 11. Temperature distribution. Flow pattern: nonrecirculating jet. Firing density is 20,000 Btu/hr ft^2; air temperature, 492°R; sink temperature, 1460°R. Fuel $(C_2H_2)_n$, no excess air.

Constant composition	Allowance for compositional variations	Position
●	○	Axis
■	□	Wall
▲	△	Corner

efficiency by an average of 5%, a maximum of 10%. The effect of concentration gradients decreased with increasing recirculation, as expected. The temperature and flux distribution calculated with and without allowance for concentration are shown in Figs. 11 and 12, for the extreme case of no recirculation. Fortunately the effect of concentration gradients on the calculated fluxes is small, with the exception of that on the fluxes to the zone in which the burner is located. The fluxes calculated to the burner zone, however, are suspect since the zoning is too coarse adequately to resolve the effect of concentration gradients near the burner. The results

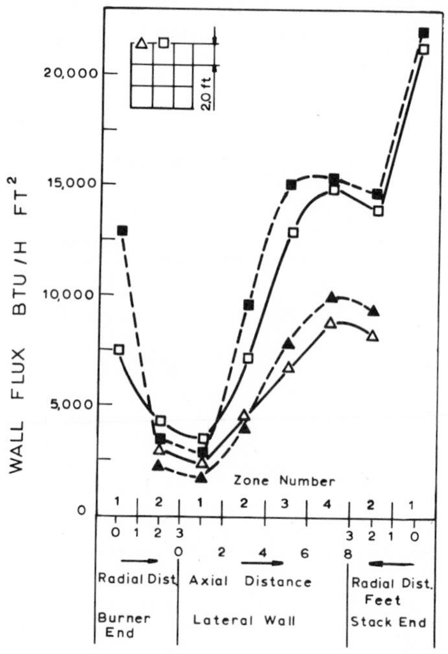

Fig. 12. Wall flux distribution. Flow pattern: nonrecirculating jet. Firing density is 20,000 Btu/hr ft²; air temperature, 492°R; sink temperature, 1460°R. Fuel $(C_2H_2)_n$, no excess air.

Constant composition	Allowance for compositional variations	Position
■	□	Wall
▲	△	Corner

presented here are supported by those of Cannon and Steward (1971) and suggest that the dependence of flux on the fourth power of temperature tends to minimize the effect of concentration gradients.

IX. COMPARISON OF THEORY AND EXPERIMENT

The major deficiency in the application of the zone method appears to be in the paucity of input data rather than in the inadequacies of the computations. Several groups are setting up parallel computational and experimental programs to check the method. Preliminary results by Osuwan (1971) on a 10-in.-diameter cylindrical furnace and by Johnson and Beér (1972) on the M-1 tests at Ijmuiden provide impressive agreement between theory and experiment.

List of Symbols

a	As subscript, absorption
A	Area
c	As subscript, carbon dioxide (in combustion gas radiation)
c_p, C_p	Specific heat
C_s	Cold-surface fraction of a furnace enclosure
Ct	Craya–Curtet number
D	Diameter, characteristic dimension
D'	Reduced firing density
E (E_B)	Hemispherical blackbody flux density at a surface, or hemispherical emissive power of a blackbody. Units, energy/area × time
E_λ (E_ω)	Monochromatic emissive power of a blackbody at wavelength λ (wave number ω). Units, energy/area × time wavelength or wave number
$E_{\lambda,g}$ ($E_{\lambda,s}$)	E_λ for a gas (surface)
F_{ij}	View factor, fraction of isotropic radiation from A_i intercepted directly by A_j
$\bar{F}$	Total view factor between black source and sink, with allowance for refractory surfaces
g (G)	Subscript, gas
$\overline{gs}$	Gas-surface exchange area [Eq. (3)]
$(G_iS_j)_R$	Total-exchange area with allowance for effect of surface zones in radiative equilibrium
h	Height; Planck's constant (units, momentum × length); coefficient of heat transfer
H	Incident radiant flux density; enthalpy flux in moving stream
i	Subscript usage: incident
i	General zone number. Subscript usage: integer; zone identification number
I	Intensity, radiant energy flux density per unit solid angle of divergence
I_B	Intensity of blackbody radiation
I_0	Intensity on entry into system of interest
j, k	Subscript usage: integers; zone identification numbers

k_λ (k_ω)	Monochromatic absorption coefficient
L	Path length; distance; characterizing dimension
L	Loss term, Eq. (26), dimensionless
L_e	Mean beam length
L_m	Average mean beam length
L_o	Mean beam length at vanishing optical thickness [Eq. (6)]
m	Subscript usage: measured in the medium
$\dot{m}$	Mass flow rate
n	An integer; number of source sink zones, plates, terms in a series, and so on
p	Partial pressure of radiating gas component
p_c (p_w)	Partial pressure of carbon dioxide (water vapor)
q	Flux density (energy/area × time)
$\dot{Q}$	Energy flux (energy/time)
$\dot{Q}'$	Reduced furnace efficiency, dimensionless
r, s, t	Subscript usage: reflected; refractory or radiative-equilibrium zones
r	Distance between two elements; radius
r_c	Center-to-center distance
R	Reflected flux density
s (S)	Subscript usage: surface
T (T_g, T_s)	Temperature (of a gas, of a surface)
T	Transmittance of a system
T_F	Adiabatic flame temperature
T_0'	Ambient temperature term [Eq. (26), dimensionless]
U	Overall coefficient of heat transfer
V	Volume
w	Subscript, water vapor (in combustion gas radiation)
W	Leaving-flux density (radiosity)
Z	Ratio of a/h, dimensionless
α	Absorptivity, absorptance
α_{gs} ($\alpha_{1,2}$)	Absorptivity of a gas for radiation from a surface (of surface 1 for radiation from 2)
Δ	Difference between radiating gas temperature and leaving-gas temperature
ϵ (ϵ_g, ϵ_s)	Emissivity or emittance (of a gas, of a surface)
ϵ_w (ϵ_c)	Emissivity of water vapor (carbon dioxide)
η	Efficiency of energy transfer to sink
θ	Polar angle
λ	Wavelength; thermal conductivity. Subscript usage: monochromatic value at wavelength λ
ν	Frequency. Subscript usage: monochromatic value at frequency ν
ρ	Reflectance
σ	Stefan–Boltzmann constant
τ	Reduced sink temperature, dimensionless
ω	Wave number λ^{-1}. Subscript usage: monochromatic value at wave number ω
Ω	Solid angle

References

Becker, H. A. (1961). Sc.D. Thesis in Chem. Eng. M.I.T., Cambridge, Massachusetts.

Becker, H. A., Hottel, H. C., and Williams, G. C. (1963). *Int. Symp. Combust., 9th*, p. 7. Academic Press, New York.

Cannon, P., and Steward, F. R. (1971). The Calculation of Radiative Heat Flux in a Cylindrical Furnace Using the Monte-Carlo Method, *Int. J. Heat Mass Transfer* **14,** 245.

Crowther, J. (1971). Preliminary Studies of Heat Transfer in Non-Ferrous Metal Melting Furnaces of the British Non-Ferrous Metal Research Association. Presented at *Int. Flame Res. Foundation's 2nd Members Conf., Ijmuiden, May 1971.*

Erkku, H. (1959). Sc.D. Thesis in Chem. Eng. M.I.T., Cambridge, Massachusetts.

Hottel, H. C. (1961). *J. Inst. Fuel.* **34,** 22.

Hottel, H. C., and Cohen, E. S. (1958). *A.I.Ch.E. J.* **4,** 3.

Hottel, H. C., and Sarofim, A. F. (1965). *Int. J. Heat Mass Transfer.* **8,** 1153.

Hottel, H. C., and Sarofim, A. F. (1967). "Radiative Transfer." McGraw-Hill, New York.

Johnson, T. R., and Beér, J. M. (1972). Radiative Heat Transfer in Furnaces: Further Development of the Zone Method. Paper presented at *Int. Symp. Combust., 14th.* Pennsylvania State University, August 20–25, 1972.

Osuwan, S. (1971). ScD. Thesis. Dept. of Chem. Eng. Univ. of New Brunswick, Canada.

Pieri, G., Sarofim, A. F., and Hottel, H. C. (1973). Radiant Heat Transfer in Enclosures: Extension of Hottel–Cohen Zone Method to Allow for Concentration Gradients. *J. Inst. Fuel* (to be published).

VIII

Radiation from Flames in Furnaces*

J. M. BEÉR

DEPARTMENT OF CHEMICAL ENGINEERING AND FUEL TECHNOLOGY
THE UNIVERSITY OF SHEFFIELD
SHEFFIELD, ENGLAND

Theoretical and experimental research on the radiative transport from flames in furnaces is reviewed in this chapter. A "flow chart" of the procedure of solving a radiative transfer problem is presented. Banded and continuous emission in furnace flames is considered. Emission from particulate clouds is discussed in three groups according to the value of the Mie parameter $X = \pi d/\lambda$: for $X \ll 1$ (soot in flames) where scattering is negligible, and calculated and measured optical constants are available; for $X \approx 1$ (cenospheres in flames) where scattering by individual particles is taken into account; and for $X \gg 1$ (pulverized coal flames) where the

* Much of the material in this chapter originally appeared in a paper contributed by J. M. Beér and C. R. Howarth to the *Int. Symp. Combust., 12th, 1969,* and is used with permission of the copyright owner, The Combustion Institute, Pittsburgh, Pennsylvania.

scattering is multiple and anisotropic. Simplified furnace calculations, the use of "comparative" design parameters, the well-stirred and the plug-flow furnace models, and the zone method of analysis are discussed.

The experimental information surveyed is presented in two groups:

(a) radiometric measurements on industrial-sized flames, and
(b) physical measurements of radiative properties of absorbing–emitting scattering media.

The experimental research reviewed includes data of radiant emittance and emissivity of flames as a function of design and operational input variables obtained at Ijmuiden and discussions on the correlation of values of absorption coefficients determined from flame measurements with results of theoretical studies.

Further research is required on methods of predicting temperature, gas-, and soot-concentration distributions in flames from input parameters; on the radiative properties of soots at flame temperatures; and on scattering characteristics of particles that are large compared with the wavelength of the incident radiation. Approximate methods for the solution of the transport equation for scattering media need checking experimentally.

I. INTRODUCTION

The purpose of this chapter is to review recent advances in the field of radiative transfer in high-temperature systems, such as industrial furnaces and combustors. Reference is made to various areas of theoretical and experimental research that made a contribution to predicting radiative heat transfer from flames. It is also intended to point to research areas in which further work is urgently required.

In solving problems of radiative heat transfer involving absorbing–emitting scattering media, it is usual to assume that any small gas volume in the system is in thermodynamic equilibrium. The significance of this is that, when this assumption is valid, the fundamental laws of radiation such as Planck's law of spectral energy distribution, Wien's displacement law, and the Stefan–Boltzmann law of blackbody radiation are applicable to the discussion of radiative transport. If thermodynamic equilibrium cannot be assumed, radiant transfer problems become considerably more complex as the actual distribution of radiant energy with wavelength and the gas absorption coefficients become functions of the population of the energy states in the gas, and detailed quantum statistical calculations are necessary for determining the emission and absorption of radiant energy. Conditions for radiative equilibrium are discussed by Gaydon and Wolfhard (1960) in detail. In small clear flames, for example, the loss of energy by radiation may not be fully compensated for by collision processes within the flame, with the effect that, due to deactivation, the proportion of excited atoms and molecules falls below that determined from the Maxwell–Boltzmann distribution law. Other practical cases where nonequilibrium effects can be significant include those in shocks, rarified gases exposed to a strong radiation field, or systems in which transients of incident radiation are so rapid that there are significant changes in incident radiation over periods of time commensurable with those required for electron and vibrational transition processes in the gas (Kulander, 1964; Oxenius, 1966; Ferrari and Clarke, 1964; Bond *et al.*, 1965; Howell, 1967). In combustion processes in industrial-sized flames, the colliding molecules are not excited or ionized because the ionization potentials of molecules or atoms are generally high and hence the chance of their excitation at normal combustion temperatures is low. Thus industrial-sized flames can be regarded as systems in equilibrium with good approximation, and the fundamental laws of radiation can therefore be applied to their discussion.

In furnaces fired by gaseous, liquid, or solid fuel the principal contributors to the radiation from the flames and from the fully burned combustion products are CO_2, H_2O, soots, and fly-ash particles. Figure 1 represents examples of infrared emission spectra of combustion products of various fuels compared with blackbody emission, illustrating the nature of these

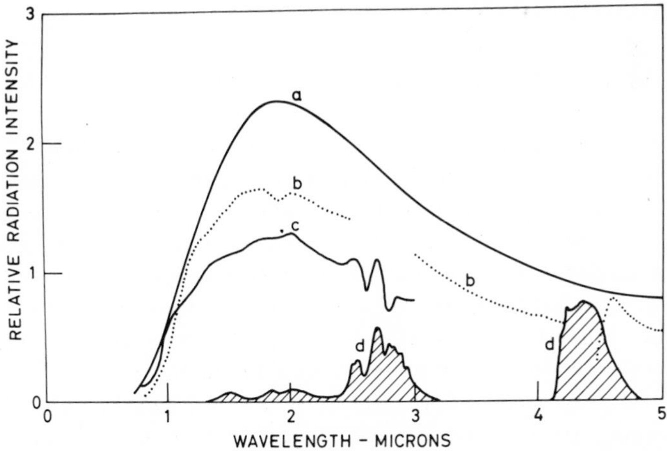

Fig. 1. Spectral emission of radiation from luminous and nonluminous flames at 1500°K. (a) Blackbody; (b) pulverized-coal flame [data from Penzias (1968)]; (c) liquid-fuel flame [data from Weeks and Saunders (1958)]; (d) nonluminous exhaust gas from jet engine combustor [data from Tourin (1966)].

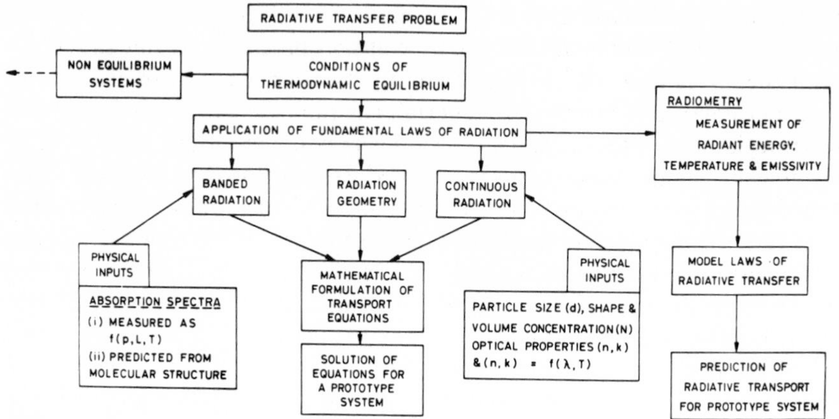

Fig. 2. Flow chart for the solution of a radiative transport problem for flame systems. [From Beér and Howarth (1969).]

emissions (banded or continuous). The complexity of the radiative transfer problem is increased by the fact that banded and continuous emission from gaseous and particulate media are superimposed in most practical flames. A "flow chart" of steps in the solution of radiative transport problems is presented in Fig. 2.

II. BANDED RADIATION

The radiation from flame gases of conventional fuels is mainly due to carbon dioxide and water vapor. Both CO_2 and H_2O emit in bands of the infrared (CO_2 at 2.7, 4.3, and 15 μ and H_2O at 2.7, 6.3, and 20 μ) (Tourin, 1966). Radiation from nonluminous gases is dependent on the number of molecules in the optical path and on temperature. Charts developed by Hottel (1954) enable calculation of total emissivity of the gas ε_g as a function of gas temperature T_g, and of the product of partial pressure and path length of the optical beam pL. These charts, covering ranges up to $T =$ 2800°K, $pL = 0.001–1.7$ m atm, and $pL = 0.002–6.0$ m atm for CO_2 and H_2O, respectively, are mainly based on measurement both for CO_2 and for H_2O (Hottel and Mangelsdorf, 1935; Hottel and Smith, 1951; Eckert, 1937; Schmidt, 1932). Radiation from nonluminous gases also can be computed from absorption spectra predicted from knowledge of molecular structure of the gas (Penner, 1959; Elsasser, 1942; Plass, 1958). For purposes of determining absorptance of CO_2 and H_2O use was made of infrared band models in which tractable mathematical presentations were made of infrared bands consisting of many hundreds of spectral lines. Penzias and co-workers (1962, 1963) have fitted the statistical random band model (Plass, 1958) to experimental absorptance data and have determined band-model parameters for CO_2 and H_2O in the temperature range of 1273 to 2400°K.

The significance of this semiempirical method is that it offers improved accuracy and also that it enables extrapolation of gas emissivities to high temperatures and pressures—conditions not favorable for accurate experimentation.

The application of theoretical and experimental data of infrared band spectra to the calculation of radiative performance of combustion chambers is complicated by the fact that the contribution of different bands to the gas emissivity varies differently with the path length L. Hottel (1961) suggested that the gas may be visualized as a mixture of gray gases and that the ε_g–L function may be represented in exponential series form:

$$\varepsilon_g = \sum_n a_n'[1 - \exp(-K_n L)], \tag{1}$$

where a_n' is the fractional amount of a component gas in the mixture having an absorption coefficient of K_n.

A comparison of emissivities of a typical furnace gas, calculated from the basic emissivity charts of CO_2 and H_2O, with those obtained from Eq. (1) showed that the consideration of two terms in the exponential series (a single gray gas plus a clear gas with $K = 0$) gives an adequate answer in most practical cases. When four terms of the series have been considered, the curve so obtained fits the data of the basic emissivity charts over a 2500-fold range of pL (Hottel, 1961).

III. CONTINUOUS RADIATION

The spectrum of radiation from solids such as furnace walls and particulate clouds in flames is continuous over a wide range of wavelengths.

A. Radiative Properties of Solid Particles in Flames

For any theoretical treatment of radiative transfer from luminous flames containing soot and larger solid particles, knowledge of radiative properties of the particles is necessary. The term radiative properties broadly covers the efficiency factors of absorption A, extinction E, and scattering S of a particle and, for purposes of this review, the angular distribution of the scattered intensity $S(\theta)$ is also included.

The Mie theory (Mie, 1908) that describes the interaction of an electromagnetic wave with a spherical particle is the one generally used for determining the efficiency factors A, E, and S for carbon particles (Thring *et al.*, 1961). The solutions of the Mie equations are given in terms of two parameters: the particle perimeter–wavelength ratio $X = d\pi/\lambda$, and the complex refractive index of the particle $m = n(1 - ik)$.

A concept generally referred to is the attenuation of a parallel beam after penetrating a distance dL into an absorbing–scattering medium. For a cloud of particles with varying sizes, the attenuation of a monochromatic parallel beam can be given as

$$I_{dL} = I_0 \exp(-\tau) = I_0 \exp[-C\, dL \sum_z E_z N_z (\pi d_z^2/4)]. \qquad (2)$$

Extinction, expressed by the factor E in Eq. (2), is due to absorption and scattering by the particles of the incident radiation. Because of the variation of the significance of scattering with the parameter $X = \pi d/\lambda$, three characteristic ranges of X are considered: $X \ll 1$, $X \approx 1$, and $X \gg 1$.

1. $X \ll 1$ (Very Small Particles Such as Soots in Flames)

Soot particles are very much smaller than the wavelengths important for thermal emission in flames (Gill, 1958; Kaskan, 1961; Lee *et al.*, 1962). For this case, scattering is negligible (Penner, 1959; Hawksley, 1952) and hence extinction equals absorption: $E = A$. The emissivity (absorptivity) of a cloud of particles can be calculated as a function of the mass concentration $\bar{C}$ and the optical constants of the particles: the real refractive index n and the absorption index k:

$$\varepsilon_\lambda = 1 - \exp[-\bar{C}L36(\pi/\rho)f(n, k)/\lambda], \qquad (3)$$

where

$$f(n, k) = n^2k/[(n^2 + n^2k^2) + 4(n^2 - n^2k^2 + 1)]. \qquad (4)$$

As shown by Eq. (3), the monochromatic emissivity in this size range is not a function of particle size.

The values of n and k generally used (Blokh, 1964; Ibiricu, 1962; Beér, 1962) for calculating ε_λ originate from experimental data on carbons in the visible range of the spectrum (Senftleben and Benedict, 1919) theoretically extended into the infrared (Stull and Plass, 1960). Experimental values of n and k obtained in the infrared at room temperature are now available for soots of varying carbon content (Howarth, 1966) and have been theoretically extended to temperatures up to 2000°C (Howarth *et al.*, 1966).

For determining the total emissivity ε_T we write Eq. (3) in the form

$$\varepsilon_\lambda = 1 - \exp[-\bar{C}LK_\lambda], \qquad (3a)$$

and integrate ε_λ over the whole spectrum as

$$\varepsilon_T = \int_0^\infty I(\lambda T)\varepsilon_\lambda \, d\lambda \bigg/ \int_0^\infty I(\lambda T)\, d\lambda = 1 - \exp[-\bar{C}LK_m]. \quad (5)$$

It was shown that substitution of the emission mean wavelength $\lambda_{0.5}$ into Eq. (3) yields good approximate values of ε_T. The emission mean wavelength is given by $\lambda_{0.5}T = 0.411$ cm °K (Hottel, 1954).

Correlations of theoretically predicted flame emissivities and experimental data obtained on industrial-sized flames are discussed in Section V, B.

2. $X \approx 1$ (Cenospheres, Chars, Fly Ash in Flames)

Scattering is no longer negligible. Calculations of the attenuation coefficient by Foster (1963) showed that neglecting scattering in optically thin clouds can cause errors up to 50%.

Because scattering for the $X \approx 1$ range can be considered isotropic (Mie, 1908), a simple correction for the effect of scattering on ε_λ can be made:

$$\varepsilon_\lambda = [1 - \exp(-\bar{C}LK_\lambda)][(E - S)/E]. \quad (6)$$

3. $X \gg 1$ (Pulverized Coal, Chars, Fly Ash in Flames)

Scattering is commensurable with absorption and is anisotropic (Blokh, 1967). Extinction E and scattering S can be calculated from simple relationships of geometrical optics which describe the scattering in terms of diffraction reflection and refraction. The value of E tends toward 2 and becomes independent of particle size; the absorption A may be taken as the emissivitt of a plane surface; and the scattering S then can be determined as $S = 2 = A$. While S is independent of particle size, the angular distribution of the scattered intensity $S(\theta)$ is dependent upon the size and shape of the particles and is forward directed. Because of this anisotropic nature of the scattering, measurements of attenuation coefficients in pulverized coal flames using narrow-angle instruments are open to error (Beér and Claus, 1964).

For predictions of radiative transfer, values of the scattering albedo ω (ratio of the scattering to total extinction) and the phase function $\rho(\theta)$ (normalized intensity distribution for single scatter) need to be known. Scattering intensity distributions of suspensions of coal particles have been measured by Hodkinson (1963) from which the phase functions can be directly determined.

IV. THE CALCULATION OF RADIATIVE TRANSFER

With the information on the radiative characteristics of combustion products and solid surfaces as considered in Section II and III, and by applying the laws of radiation geometry, the radiative heat transfer from the combustion products to heat sinks and walls of the combustion chamber can be calculated (Fig. 2).

The emission of radiation from a volume of emitting–absorbing fluid of specific shape to a specified portion of its bounding surface can be given as (Hottel, 1961)

$$\int_0^\infty \int_0^{2\pi} \int_V [4A_{z\lambda}c(l) \times I(\lambda, T)\, dV] \times [d\Omega] \times \left[\exp\left(-\int_0^L E_{z\lambda}c(l)\, dl\right)\right] d\lambda. \tag{7}$$

Here the first bracketed term is total emission from volume dV; the second is fraction toward element of surface; and the third is fraction of emission transmitted toward element of surface.

The integral above can be solved for various enclosure shapes and for cases when the values of the coefficients E_z and A_z, and also the spatial distribution of the concentrations, are known. It holds for cases of banded emission and emission from nonscattering particulate clouds, and for the case of a particulate cloud of small particles in which intermediate-size particles are present as a small fraction of the total mass.

A. Simplified Models of Furnace Heat Transfer

Because of the extensive computations required for a rigorous solution of a radiative heat transfer problem and the lack of physical input data, combustion engineers relied on methods of calculation based on simplifying assumptions.

The design engineer has in his work to rely on so-called *comparative design parameters,* for example, permissible maximum heat release rate (heat input per unit volume and unit time), or heat absorption rate (heat absorbed per unit surface area of heat sink and unit time), and so on. Although these parameters should be used only as tentative initial values in design it can be shown that simple relationships between them can give useful qualitative and in a comparative way quantitative information about some performance data.

The heat flow rate into a furnace can be given as the product of the volumetric heat release rate and the volume of the furnace as

$$Q = \dot{q}_v'V \tag{8}$$

and the heat absorption by the heat sink in the furnace as

$$Q_{tr} = \dot{q}_s'A_s. \tag{9}$$

We can write for the dimensionless ratio of heat transferred in the furnace to the heat input

$$\mu = \dot{q}_s'A_s/\dot{q}_v'V, \tag{10}$$

and by using the same value of the specific heat between a reference temperature and the adiabatic flame temperature on one hand and a reference temperature and the furnace exit gas temperature on the other, we can write for the exit gas temperature

$$T_E = T_{AF}(1 - \mu) = T_{AF}[1 - (\dot{q}_s'A_s/\dot{q}_v'V)], \tag{11}$$

or in the dimensionless form,

$$\theta_E = T_E/T_{AF} = 1 - (\dot{q}_s'A_s/\dot{q}_v'V). \tag{12}$$

The right-hand side of Eq. (12) contains the product of two ratios: the ratio of $\dot{q}_s'/\dot{q}_v'$ and that of the area of the heat sink and the volume of the furnace. For furnaces with a large proportion of their walls cooled (e.g., boiler combustion chambers), the latter ratio may be termed the "shape

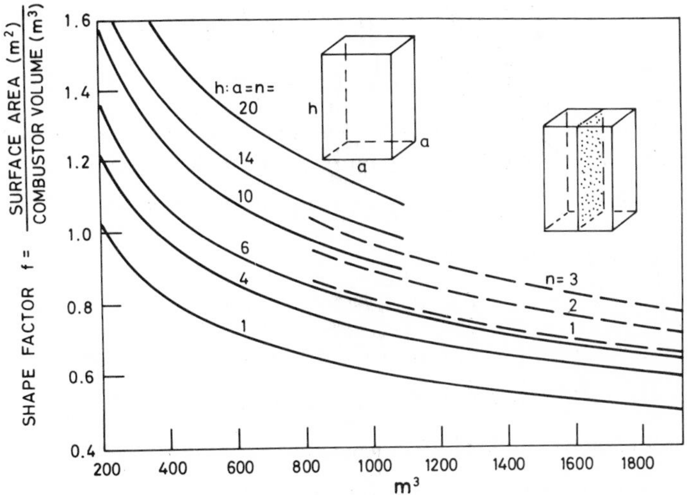

Fig. 3. Shape factor as a function of volume and relative dimensions of the furnace: – – with dividing wall; — without dividing wall. [From R. Doležal, "Large Boiler Furnaces." Elsevier, Amsterdam, 1967.]

factor." It can be seen from Eq. (12) that for a case when the specific heat absorption rate cannot be increased (because of the danger of burnout) an increase in the volumetric heat release rate will result in increased exit gas temperature unless it is compensated for by an increase in the shape factor, the surface-to-volume ratio of the furnace chamber. Figure 3 represents values of the shape factor as a function of the absolute size of the furnace and a parameter h which is the aspect ratio of a parallelepiped-shaped furnace.

1. The Well-Stirred Furnace Model

This model is based on the assumption that mixing in the furnace is so effective that the gas temperature and concentration are uniform throughout the furnace volume. It is assumed further that the gas is gray; the bounding walls of the total area A_T are divisible into a heat sink area A_1, gray at a single temperature T_1, and an adiabatic refractory area A_r. Convective heat transfer is neglected. Hottel in his Melchett lecture (1961) derived a relationship between nondimensional groups representing the heat transfer efficiency or the proportion of input heat transferred in the furnace as a function of the input enthalpy, the heat sink temperature, and a so-called total exchange area.

By using the same specific heat over the interval T_0 to T_{AF} as over T_0 to T_E we can write

$$(Q - Q_{tr})/Q = (T_E - T_0)/(T_{AF} - T_0), \tag{13}$$

and for

$$Q_{tr} = \overrightarrow{GS}\sigma(T_g^4 - T_s^4). \tag{14}$$

It has been assumed that the gas mass and flame transfer heat is at a mean temperature T_g and that $T_g = T_E$; T_g can thus be written as

$$T_g = [1 - (Q_{tr}/Q)](T_{AF} - T_0) + T_0 \tag{15}$$

and Eq. (14) can be rewritten as

$$(Q_{tr}/\sigma\overrightarrow{GS}) + T_s^4 = T_g^4, \tag{14a}$$

and by substituting T_g from Eq. (15) into Eq. (14a) we have

$$(Q_{tr}/\sigma\overrightarrow{GS}) + T_s^4 = T_{AF}^4[1 - Q_{tr}(T_{AF} - T_0)/QT_{AF}]^4. \tag{16}$$

By introducing the following dimensionless groups

μ' = nondimensional heat transfer efficiency
D' = nondimensional firing density
$= Q/[\sigma\overrightarrow{GS}T_{AF}^3(T_{AF} - T_0)]$
$\tau = (T_s/T_{AF})$ — ratio of heat sink and adiabatic flame temperatures,

and by rearranging Eq. (16) Hottel obtained the following relationship:

$$\underbrace{(T_s^4/T_{AF}^4)}_{\tau^4} + \underbrace{(Q_{tr}/Q)[(T_{AF} - T_0)/T_{AF}]}_{\mu'} \cdot \underbrace{Q/[\sigma GST_{AF}^3(T_{AF} - T_0)]}_{D'}$$

$$= \underbrace{\{1 - (Q_{tr}/Q)[(T_{AF} - T_0)/T_{AF}]\}^4}_{(1-\mu')^4}, \tag{17}$$

and hence we have

$$\mu' D' + \tau^4 = (1 - \mu')^4. \tag{18}$$

Equation (18) is represented in Fig. 4. The efficiency μ' of the heat transfer in the furnace is seen to be a function of the firing density and the heat sink temperature. The firing density—the ratio of heat input and radiating ability of the flame—makes due allowance for any operating variables such as fuel type or excess air or air preheat which affect flame temperature or gas emissivity, for fractional occupancy of the walls by sink surfaces, and for wall emissivity.

The total gas–surface interchange area $\overrightarrow{GS}$ can be given in the simple form

$$\overrightarrow{GS} = A_T/[(1/\varepsilon_g)(1/C_s\varepsilon_s) - 1]; \qquad C_s = A_s/A, \tag{19}$$

where A_T is the total surface area of the furnace chamber; A_s is the area of the sink; ε_g, ε_s are the gas and surface emissivities, respectively; and C_s is the fraction of the furnace wall covered by sink.

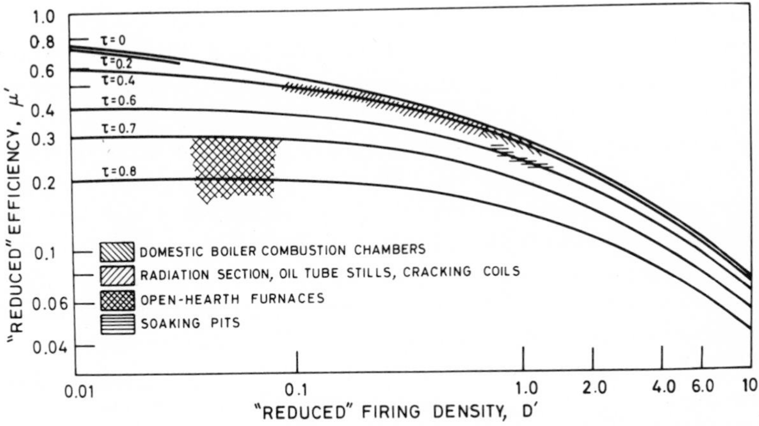

Fig. 4. Thermal performance of a "well-stirred" furnace. [From Hottel (1961).]

a. Discussion

Some of the consequences of the relationship [Eq. (18)] are as follows:

(a) At low firing density the efficiency increases with decreasing firing density and approaches the limit of $1 - \tau$ as $D \to 0$.
(b) At high values of D' the efficiency is inversely proportional to D'.
(c) Changes in sink temperature have little effect if $\tau < 0.3$.
(d) As the furnace wall approaches complete coverage by sink ($C_s\varepsilon_s \to 1$), D' becomes inversely proportional to ε_g. This means that at high firing density (e.g., gas turbine) the heat transfer is directly proportional to the gas emissivity, but that at low values of D' the value of ε_g is relatively unimportant.
(e) When $C_s\varepsilon_s \ll 1$ because of highly reflecting heat sink or small fractional coverage of the wall by sink, changes in ε_g produce little changes in heat transfer efficiency.

Application of the well-stirred model to a number of different furnace types was reported at the International Flame Research Foundation's 2nd Members Conference at Ijmuiden on May 13–14, 1971. Crowther (1971) reported comparisons between measured and calculated values of furnace efficiencies. Table I is taken from his paper.

2. The Long Furnace Model

The well-stirred furnace model yields the relationship between heat transfer efficiency and important design variables, but does not give information on the heat flux distribution in the furnace.

Thring (1962) has developed an analysis based on the long furnace model. It is assumed that the furnace is long compared with its dimensions normal to the direction of flow, and that the gas temperature varies only along the gas path, but is constant across any cross section of the furnace. By setting up a heat balance for a short element of length of the gas path and integrating it along the furnace, the exit heat losses have been determined. By considering the heat balance of the furnace, and by introducing dimensionless parameters, a relationship for the heat transfer efficiency as a function of varying heat input, of furnace wall, heat loss, and heat sink temperature was derived. The comparison of the predicted data with those measured in industrial furnaces showed good agreement and has yielded information on the effect of design variables on furnace efficiency.

A similar method for determining the gas temperature distribution was proposed by Gurwich and Blokh (1956) and quoted by Doležal (1967).

TABLE I

CALCULATION OF PERFORMANCE OF INDUSTRIAL FURNACES USING HOTTEL'S SINGLE GAS ZONE MODEL[a]

Furnace type	Fuel input (therms/hr) Q_{in}	Charge area (ft²) A_s	Total area A_T	Gas emissivity ε_g	Reduced firing density D'	Reduced load temp. τ	Reduced efficiency μ'	Efficiency (%) Calc.	Efficiency (%) Meas.
OH steel furnace	900	484	1500	0.11	0.69	0.56	27.5	38.7	35.4
Copper matte smelter	1500	3200	7690	0.4	0.15	0.58	36.2	39.0	37.5
IFRF (various conditions) (a)	32	136	626	0.14	0.17	0.58	35.0	($h=0$) 24.5 ($h=15$) 26.5	23
(b)	32	9.7	500	0.14	1.2	0.185	28.5	15.5	15
(c)	38.6	136	626	0.11	0.49	0.62	27.0	23.5	23
(d)	38.6	9.7	500	0.11	3.96	0.215	13.5	9.8	10
Aluminum melter high flame	134	160	736	0.185	0.72	0.52	27.5	25.7[b]	24.5[b]
low flame	60			0.157	1.22	0.73	17.2		

[a] After Crowther (1971).

[b] Mean value.

Allowing for the nature of temperature–distance curves—they rise to a peak at some distance from the furnace entry and then the temperature drops along the furnace—they represent the T^4–X function as

$$\theta^4 = T^4/T_{AF}{}^4 = \exp(-\alpha X) - \exp(-\beta X), \tag{20}$$

where X is the dimensionless length, and α and β are constants dependent upon furnace cooling and upon the variation of the heat-release rate along the furnace, respectively (Fig. 5). From Eq. (20), gas temperature at the furnace exit $(X = 1)$ can be given as

$$\theta_E = T_E/T_{AF} = [\exp(-\alpha) - \exp(-\beta)]^{1/4}, \tag{21}$$

and the position of the maximum flame temperature as

$$X_m - (\ln \alpha - \ln \beta)/(\alpha - \beta). \tag{22}$$

The long furnace model, often referred to as the plug-flow combustor model, was used by a number of research workers for both combustor design

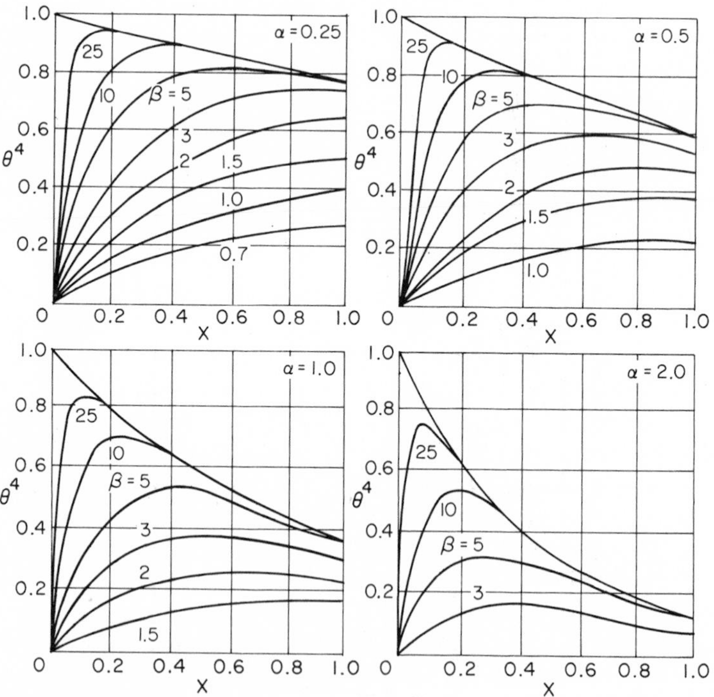

Fig. 5. Unidimensional model variation of normalized T^4 with distance along the furnace. [From R. Doležal, "Large Boiler Furnaces." Elsevier, Amsterdam, 1967.]

purposes (Hottel, 1961; Gumz, 1962; Beér *et al.*, 1956) and also for the development theories of thermal flame propagation in strongly absorbing dust clouds (Nusselt, 1924; Csaba, 1964; Essenhigh and Csaba, 1963).

B. The Zone Method of Analysis

With increasing unit capacity of industrial furnaces and combustors, there is great interest in the detailed heat-flux distribution in the furnaces, information that cannot be obtained from the well-stirred and long-furnace model calculations. Improved methods of predicting flow patterns and the progress of combustion, and, thus, the concentration of the absorbing–emitting species in the furnace allows the use of the zone-method analysis (Hottel and Cohen, 1958; Erkku, 1959; Yardley and Patrick, 1963; Hottel and Sarofim, 1965), the most useful procedure for calculating detailed furnace performance.

The volume and surface of the furnace are subdivided into zones assumed to be of uniform temperature and concentration, the number of which will depend on the accuracy required of the solution. Equations are then written for the energy balance of each zone. The setting up of these balances necessitates the description of emissivity and absorptivity of combustion products by the weighted sum of gray gases, and the calculation of so-called exchange-area coefficients, the latter of which represent the radiative interchange between two finite zones of gas and/or surface of any shape and relative disposition. The exchange-area factors are evaluated between all zone pairs.

The simultaneous solutions of the energy-balance equations containing unknown temperatures yield the temperature distribution in the furnace. Substitution of these temperatures into equations containing unknown heat fluxes permits the calculation of wall heat-flux distribution. A check on the convergence of the solution of heat-balance equations is provided by the comparison of the sum of fluxes to the wall with the decrease in total enthalpy along the gas path.

In the zone-method analysis a complicated integral equation is replaced by a series of algebraic equations. The solution usually requires the use of a high-speed digital computer.

The further development of the zone method will greatly depend upon the progress that can be made in predicting velocity and temperature patterns in furnaces from input and design parameters.

For the case of multiple scattering, the mathematical description of the heat transfer from an absorbing–emitting–scattering medium to its bounding surface is more complex. The optical depth may be considered to be a criterion of multiplicity of scatter: If $\tau < 0.3$, that is, $\varepsilon \geq 0.25$, multiple scattering theory has to be applied (Van Der Hulst, 1963).

The general integrodifferential equation for the radiant energy balance along a pencil beam can be written as

$$(dI/dr)(\tau, \theta, \phi) = -I(\tau, \theta, \phi) + (K\sigma T^4/\pi) + (\omega/4\pi) \int_0^{2\pi} \int_0^{\pi} I(\tau', \theta', \phi') p(\theta, \phi, \theta', \phi') \sin \theta' \, d\theta' \, d\phi'. \tag{23}$$

Here, the left-hand expression is the net change in intensity over an optical depth dr and the right-hand expressions are decrease in intensity due to attenuation, plus increase in intensity due to thermal emission, plus increase in intensity due to scatter into the direction ϕ from all other directions.

The mathematical task of solving the transport equation (23) is considered even when the physical input data are available. For simplifying the analysis, it is usual to assume that the scattering is isotropic and to treat the problem as one dimensional. Merits of some simplified solutions (Richards, 1955; Grosjean, 1963; Triplett, 1958; Chu *et al.*, 1963; Chu and Churchill, 1955) have been considered by Tien and Churchill (1965) and the effect of anisotropy on the solution has been discussed by Hottel *et al.* (1967) and by Evans *et al.* (1965). Approximate solutions closely relevant to the calculation of the effect of scattering in burning particulate clouds have been presented by Love and Grosh (1965) and by Spalding (1965). Calculating the scattering function from Mie theory for the unidimensional system of carbon particles between partially reflecting surfaces, Love and Grosh concluded that the assumption of isotropic scattering gave good approximate results for the overall heat transfer. Since the size parameter considered in these studies was between $X = \pi d/\lambda = 1 \div 6$, it remains to be determined whether the assumption of isotropic scattering can be made for pulverized-coal flames with a size parameter of $X \approx 40$.

C. Flux Methods*

The basis of the method is to replace the continuous variation of radiant intensity with direction by mean values representing various angle ranges. In this way the infinite number of radiant intensity balances necessary for a

* See Schuster (1905) and Hamaker (1947).

complete description of radiant energy transfer in all possible directions can be replaced by a small (n) number of balances corresponding to the same number of solid angles. Addition of a total energy balance at any point provides a complete system of $n + 1$ simultaneous, ordinary, nonlinear differential equations. Iterative solution of these equations leads to the required temperature and flux distributions. A special advantage of this method is that the general integrodifferential equations of transfer are reduced to a set of ordinary differential equations which can be readily cast in finite difference form. This then makes the flux method compatible with the conservation equations of momentum, mass, and energy also in finite difference form used in the prediction procedure for velocity, concentration, and temperature distributions.

V. EXPERIMENTAL RESULTS

Radiometric measurements are discussed in two groups: Section V,A, measurements on optically thick flames for determining radiative heat transfer parameters as a function of design and operation variables; and Section V,B, detailed measurements both outside and within flames with the objective of determining more fundamental flame properties by comparison of measurement data with theoretical predictions.

A. Radiometric Measurements on Industrial-Sized Flames

Since the beginning of the work of the International Flame Research Foundation at Ijmuiden (Holland) in 1948, systematic measurements have been made of radiant emittance and flame emissivity of gas, oil, and pulverized-coal flames. The measurement method most generally used in these trials is the "modified Schmidt method" (Tourin, 1966) an emission–absorption method originally developed by Schmidt (1909) for monochromatic radiation. The total absorptivity of the flame for a given blackbody background temperature can be determined from two readings taken with a narrow-angle total radiation pyrometer: radiation from the flame alone (cold background) R_1 and the combined radiation of flame and hot blackbody background R_2. The effective flame absorptivity (emissivity) can then be given as

$$\varepsilon_T = 1 - [(R_2 - R_1)/\sigma T_B^4]. \tag{24}$$

Observations of the radiation from industrial-sized turbulent diffusion flames were interpreted to provide empirical curves of flame emissivity and radiant emittance for the effect of variables, such as fuel-jet momentum, C/H ratio of the fuel, oxygen enrichment, and particle size and coal rank in pulverized-fuel flames. Figure 6 presents the results of an investigation in which coke oven gas was mixed with liquid hydrocarbon fuel to show the effect of the C/H ratio of the fuel on the radiation from and the emissivity of the flame (Rivière, 1956). These results are of particular interest to combustion engineers for their applications to the design and operation of gas-fired furnaces and boilers.

Another set of typical results by IFRF is shown in Fig. 7, illustrating the effect of fineness of pulverized anthracite and of the volatile combustion on the radiation from the flame. Radiant emittance R_1 is increasing with increasing fineness of the anthracite and the peak radiation is translated

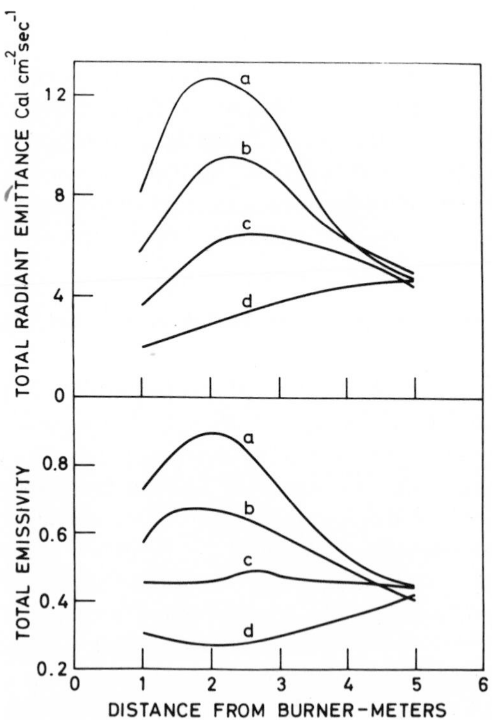

Fig. 6. The effect of carburation on the emissivity and radiation from oil–coke oven-gas flames of varying mixture concentrations: (a) 100% oil/0% coke–oven gas; (b) 40% oil/60% coke–oven gas; (d) 20% oil/80% coke–oven gas; (e) 0% oil/100% coke-oven gas. [From Rivière (1956).]

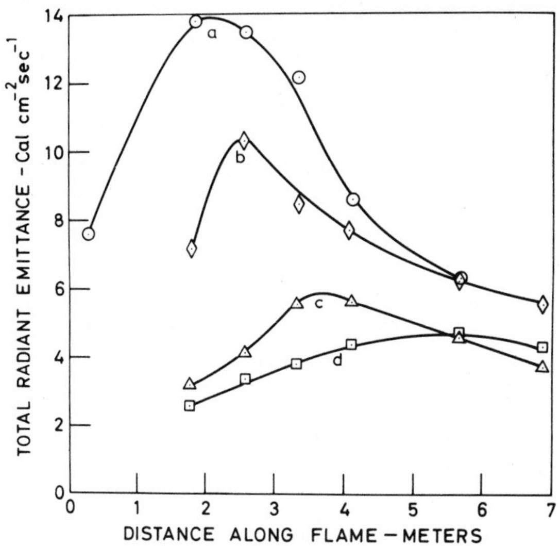

Fig. 7. The effect of input particle size and coal rank on the total radiation from pulverized-coal flames. (a) Bituminous coal 70% < 76 μ diameter; (b) anthracite coal sized 100% < 76 μ; (c) anthracite 60% < 76 μ; (d) anthracite 20% < 76 μ. [From Beér (1964).]

toward the burner. For a bituminous coal the volatile combustion associated with soot formation intensifies these trends (Beér, 1964). Detailed results of the Ijmuiden furnace trials, providing a wealth of data on the influence of input parameters on flame radiation, are available in published reports of the I.F.R.F.*

B. Measurements of Radiative Properties of Luminous Flames

The Schmidt method was used by a number of research workers (Ibiricu, 1962; Beér, 1962; Ashton, 1962) to determine the mean absorption coefficient K_m in sooty flames. Effective flame emissivities of liquid-fuel flames were correlated with soot-concentration distributions measured along the path of the optical measurements. The following relationship was used to

* Details may be found in articles by various authors appearing in the following issues of the *Journal of the Institute of Fuel*: November 1951, January 1952, January 1956, October 1957, July 1959, and August 1960. Certain more recent results from the IFRF are discussed by Chedaille and Braud (1971).

determine the value of K_m:

$$\varepsilon_T = 1 - \exp\left[-K_m \int_0^L C\, dl\right] (1 - \varepsilon_g). \tag{25}$$

Computed values of the parameter $36(\pi/\rho)f(n, k)$ [see Eqs. (3) and (4)] plotted as a function of n and k (Howarth, 1966) are presented in Fig. 8, and the ranges of n and k are shown where experimental or theoretical data are available.

Products of the experimentally determined absorption coefficient K_m and the emission wavelength $\lambda_{0.5}$ are compared with the theoretically predicted parameter in Fig. 8. Experimental values of K_m determined from pressure-jet oil flames at Ijmuiden (Beér, 1962) and Sheffield (Ibiricu, 1962) gave two distinct values for two different modes of flame stabilization. For the case without stabilization near the burner, high values of K_m were obtained: $K_m = 0.03$, $\lambda_m = 2.45$ (Beér, 1962), and $K_m = 0.032$, $\lambda_m = 2.79$ (Ibiricu, 1962), approaching the theoretical maximum for small

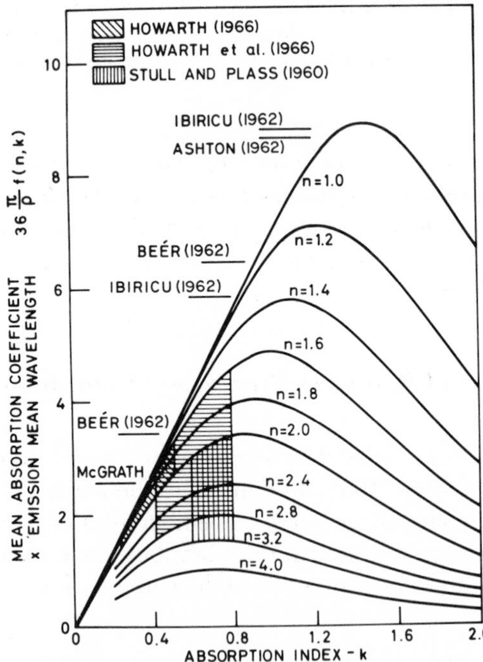

Fig. 8. A comparison of calculated and measured absorption coefficient parameters $36(\pi/\rho)f(n, k)$ for clouds of small particles. The shaded area refers to the regions of experimental values of optical constants of carbons. [From Howarth (1966).]

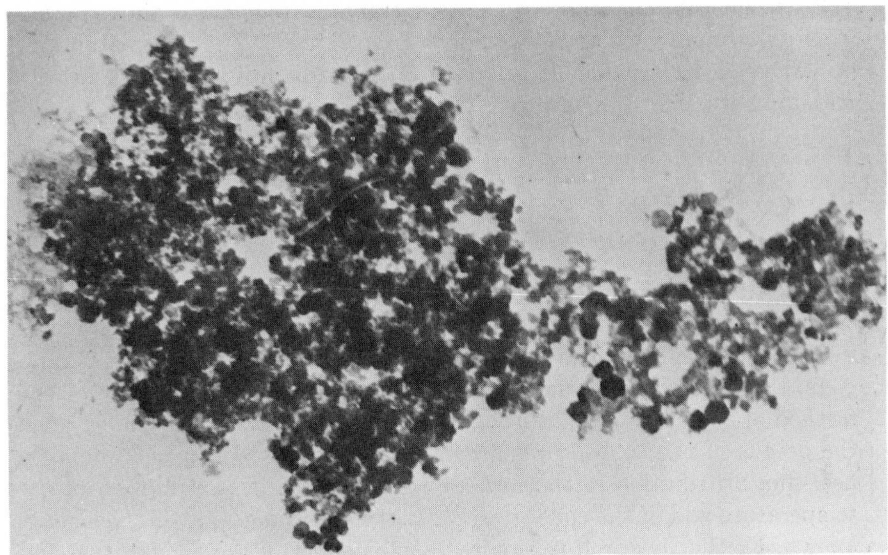

Fig. 9. An electron micrograph of graphitized soot particles from liquid-fuel flames at 1700°K. [From Ibiricu (1962).]

particles ($n = 1.0$, $k = 1.6$). The lower values of $K_m = 0.014$ and 0.021, obtained in stabilized flames, are close to the range of known values of n and k (Fig. 6).

Figure 9 is an electron micrograph showing soot particles from a plug-flow-type oil flame (without stabilization) (Ibiricu, 1962). The graphitic hexagonal structure of the soot particles may account for the apparent high values of n and k associated with the high value of K_m for this flame (Fig. 8). It is noteworthy that the graphitization of soot particles occurred here at temperatures very much lower (1600–1700°C) than the range of 2500 to 3000°C reported by Walker (1962).

Measurements traversing the flame by a water-cooled pyrometer probe were made at Ijmuiden to determine the contribution of radiation from finite layers along the optical beam. Comparison of results of these measurements with traverses predicted from known temperatures and concentration distributions, and by using the appropriate value of the absorption coefficient, showed good agreement (Beér and Claus, 1962).

Information on spectral absorptivity in industrial-sized flames is scarce. Weeks and Saunders (1958) have obtained spectral radiant emittance values and emissivities for liquid-fuel flames in a gas-turbine combustor. They warned that the application of the Schmidt method to flames with large temperature gradients can lead to erroneous values of emissivities.

Penzias (1968) has recently made spectral measurements of the radiation from pulverized-coal flames in the Ijmuiden furnace. The information can be expected to provide useful data on spectral absorption coefficients measured at flame temperature.

VI. CONCLUSIONS

Analytical solutions of radiative transfer problems are difficult even for systems of simple geometry. For many particular cases, however, simplified furnace model calculations such as the use of "comparative" design parameters, the well-stirred furnace model, the long furnace model, or the zone method of analysis suffice, the choice of type of the model depending upon the details of the solution required. The zone method can yield detailed heat-flux distribution in the furnace when the spatial distribution of gas temperature and of the concentration of the absorbing–emitting media are known. Further research is required on the effect of the composition and physical properties of soots (C/H ratio, agglomeration) on their optical properties at flame temperatures. Information is lacking also on the rate of formation of soot in hydrocarbon flames, which is required for the prediction of soot concentration distribution in luminous flames.

Acknowledgment

The discussion in this paper is largely based on a survey by Beér and Howarth (1969). Permission by the Combustion Institute to reprint part of the text and figures from this publication is gratefully acknowledged.

List of Symbols

$a'(\ )$	Weighting factor for calculating absorptivity (emissivity) of gases, dimensionless
A_T, A_s	Area of a surface zone; T refers to total and s to sink area
A	Absorption, the ratio of energy absorbed per unit time to the energy incident on the projected area of a particle per unit time
AF	Subscript, refers to adiabatic flame temperature
c	Dimensionless particulate concentration mass per unit volume, c refers to mean
C	Coverage factor A_s/A_T
d	Diameter of the particle
D'	Reduced firing density, Eq. (8)
E	Subscript, refers to exit gas

E	Extinction, the ratio of energy absorbed and scattered per unit time to the energy incident on the projected area of the particle per unit time, dimensionless
g	Subscript, refers to gas
$\overleftrightarrow{GS}$	Total interchange area between a gas and surface zone
I	Intensity of radiation (function of position)
$I(\lambda, T)$	Planck's law, distribution function
in	Subscript, refers to input
K	Gas, solid absorption coefficient, reciprocal length
L	Path length along a pencil beam
m	Subscript, refers to mean
m	Complex refractive index $m = n(1 - ik)$, where n is the real refractive index and k is the absorption index
n	Subscript, number of gas zones, also general (nth) term in a series expression of emissivity (absorptivity) of gas
N	Number of particles per unit volume
p	Partial pressure of radiating component of a gas
$p(\theta)$	Phase function, normalized intensity distribution for single scatter
Q	Heat flow rate
$S(\theta)$	Scattered intensity coefficient, the ratio of the light flux per unit solid angle in the direction θ to the flux geometrically incident on the particle
S	Scattering, the ratio of energy scattered per unit time to the energy incident on the projected area of the particle per unit time, dimensionless
T	Temperature
tr	Subscript, refers to transferred
V	Volume
X	Size parameter, $X = \pi d/\lambda$; dimensionless length, Eq. (19)
z	Subscript, refers to zth particle
α, β	Constants in Eq. (20)
ε	Emissivity
$\theta, \emptyset$	Polar and azimuthal angles defining pencil-beam position
θ	Dimensionless gas temperature T/T_{AF}
λ	Wavelength of radiation
μ'	"Reduced" heat transfer efficiency
ρ	Particle density, g/cm^3
τ	Reduced sink temperature T_s/T_{AF} in Eq. (8)
τ	Optical depth, product attenuation coefficient, and path length
ω	Albedo of scatter, ratio of scatter to extinction S/E
Ω	Solid angle

References

Ashton, J. S. (1962). Ph.D. Thesis. Sheffield Univ., Sheffield, England.

Beér, J. M. (1962). *J. Inst. Fuel* **35,** 3.

Beér, J. M. (1964). *J. Inst. Fuel* **37,** 286.

Beér, J. M., and Claus, J. (1962). *J. Inst. Fuel* **35,** 437.

Beér, J. M., and Claus, J. (1964). Doc. G04/a/2, Int. Flame Res. Foundation, Ijmuiden, Holland.

Beér, J. M., and Howarth, C. R. (1969). *Int. Symp. Combust., 12th,* p. 1205. Combustion Inst., Pittsburgh, Pennsylvania.

Beér, J. M., Csorba, T., and Csaba, J. (1956). *J. Magy. Energiagazdasag* **2,** 61 [Transl.: Ministry of Power (Great Britain) No. T.179].

Blokh, A. G. (1964). *Teploenergetika* **11,** 4, 26.

Blokh, A. G. (1967). "Heat Transfer in Boiler Furnaces." Energia, Leningrad.

Bond, J. W., Watson, K. M., and Welch, J. A. (1965). "Atomic Theory of Gas Dynamics." Addison-Wesley, Reading, Massachusetts.

Chedaille, J., and Braud, Y. (1971). "Industrial Flames," Vol. 1. Arnolds, London.

Chu, C. M., and Churchill, S. W. (1955). *J. Phys. Chem.* **59,** 855.

Chu, C. M., Churchill, S. W., and Pang, S. C. (1963). *Proc. Interdisciplinary Conf. Electromagnetic Scattering,* p. 507. Macmillan, New York.

Crowther, J. (1971). Preliminary Studies of Heat Transfer in Non-Ferrous Metal Melting Furnaces of the British Non-Ferrous Metal Research Association. Presented at the *Int. Flame Res. Foundation's 2nd Members Conf., Ijmuiden, May 1971.*

Csaba, J. (1964). Ph.D. Thesis. Sheffield Univ., Sheffield, England.

Doležal, R. (1967). "Large Boiler Furnaces." Elsevier, Amsterdam.

Eckert, E. R. G. (1937). *Forschungsheft* **387,** 1.

Elsasser, W. M. (1942). Heat Transfer by Infrared Radiation in the Atmosphere. *Harvard Meteorolog. Stud.* No. 6.

Erkku, H. (1959). Sc.D. Thesis in Chem. Eng. MIT, Cambridge, Massachusetts.

Essenhigh, R. H., and Csaba, J. (1963). *Int. Symp. Combust., 9th,* p. 111. Academic Press, New York.

Evans, L. B., Chu, C. M., and Churchill, S. W. (1965). *Trans. ASME Ser. C, J. Heat Transfer* **87,** 381.

Ferrari, C., and Clarke, J. H. (1964). New Determination of the Photoionisation Upstream of a Strong Shock Wave. *Arch. Mechan. Stosowanej* **2,** 223.

Field, M. A., Gill, D. W., Morgan, B. B., and Hawksley, P. G. W. (1967). Combustion of Pulverised Fuel. *BCURA Mon. Bull.* **31,** 3.

Foster, P. J. (1963). *Combust. Flame* **7,** 277.

Gaydon, A. G., and Wolfhard, H. G. (1960). "Flames." Chapman & Hall, London.

Gill, D. W. (1958). *BCURA Mon. Bull.* **12,** Part 2.

Grosjean, C. C. (1963). *Proc. Interdisciplinary Conf. Electromagnetic Scattering,* p. 485. Macmillan, New York.

Gumz, W. (1962). "Kurzes Handbuch der Brennstoff und Feuerungs-Technik," p. 380. Springer, Berlin.

Gurwich, A. M., and Blokh, A. G. (1956). *Energomashinostroenie* **1956,** 11.

Hamaker, H. C. (1947). *Philips Res. Rep.* **2,** 55.

Hawksley, P. G. W. (1952). *BCURA Mon. Bull.* **14,** Parts 4 and 5.

Hodkinson, J. R. (1963). *Proc. Interdisciplinary Conf. Electromagnetic Scattering,* p. 87. Macmillan, New York.

Hodkinson, J. R., and Greenleaves, I. (1963). *J. Opt. Soc. Amer.* **53,** 577.

Hottel, H. C. (1954). *In* "Heat Transmission" (W. H. McAdams, ed.), 3rd ed., Chapter 4. McGraw-Hill, New York.

Hottel, H. C. (1961). *J. Inst. Fuel* **34,** 220.

Hottel, H. C., and Cohen, E. S. (1958). *AIChE J.* **4,** 3.

Hottel, H. C., and Mangelsdorf, H. G. (1935). *Trans. AIChE* **31,** 517.

Hottel, H. C., and Sarofim, A. F. (1965). *Int. J. Heat Mass Transfer* **8,** 1153.

Hottel, H. C., and Smith, V. C. (195¹). *Trans ASME* **57,** 463.

Hottel, H. C., Sarofim, A. F., Evans, L. B., and Basalos, I. A. (1967). *ASME–AIChE Heat Transfer Conf., Seattle, Washington, August 1967.* Paper No. 67-HT-19.

Howarth, C. R. (1966). Ph.D. Thesis. Sheffield Univ., Sheffield, England.

Howarth, C. R., Foster, P. J., and Thring, M. W. (1966). *Proc. Heat Transfer Conf. ASME, 3rd, Chicago,* Part V, p. 122.
Howell, J. R. (1967). *ASME–AIChE Heat Transfer Conf., Seattle, Washington, August 1967.* Paper No. 67-HT-51.
Ibiricu, M. (1962). Ph.D. Thesis. Sheffield Univ., Sheffield, England.
Kaskan, W. E. (1961). *Combust. Flame* **5,** 93.
Kulander, J. L. (1964). Non-equilibrium Radiation. General Electric Space Sci. Lab. Rep. R64SD41, DDC No. AD-605827.
Lee, K. B., Thring, M. W., and Beér, J. M. (1962). *Combust. Flame* **6,** 137.
Love, T. J. and Grosh, R. J. (1965). *Trans. ASME Ser. C, J. Heat Transfer* **87,** 161.
McGrath, I. A. (1960). Ph.D. Thesis. Sheffield Univ., Sheffield, England.
Mie, G. (1908). *Ann. Phys.* **25,** 377.
Nusselt, W. (1924). *Z. Ver. Deut. Ing.* **68,** 124.
Oxenius, J. (1966). *J. Quant. Spectrosc. Radiative Transfer* **6,** 65.
Penner, S. S. (1959). "Quantitative Molecular Spectroscopy and Gas Emissivities." Addison-Wesley, Reading, Massachusetts.
Penzias, G. J. (1968). Private communication of spectral radiation measurements on pulverized coal flames at IFRF, Ijmuiden, Holland.
Penzias, G. J., and Maclay, G. J. (1963). NASA Rep. CR-54002. Lewis Res. Center, Cleveland, Ohio.
Penzias, G. J., Maclay, G. J., and Babrov, H. J. (1962). NASA Rep., Contract NAS 3-1542. Lewis Res. Center, Cleveland, Ohio.
Plass, G. N. (1958). *J. Opt. Soc. Amer.* **48,** 690.
Richards, P. I. (1955). *Phys. Rev.* **100,** 517.
Rivière, M. (1956). *J. Inst. Fuel* **29,** 9.
Sarofim, A. F. (1971). Principles and Applications of Radiative Heat Transfer. Short Lecture Course, Univ. of Sheffield, Sheffield, England.
Schmidt, E. (1932). *Forsch. Gebiete Ingenieurw.* **3,** 57.
Schmidt, H. (1909). *Ann. Phys.* **29,** 1027.
Schuster, A. (1905). *Astrophys. J.* **21,** 1.
Senftleben, H., and Benedict, E. (1919). *Ann. Phys.* **60,** 297.
Spalding, D. B. (1965). 14th Coal Sci. Lecture. *BCURA Gazette* No. 55.
Stull, V. R., and Plass, G. N. (1960). *J. Opt. Soc. Amer.* **50,** 121.
Tien, L. C., and Churchill, S. W. (1965). *Chem. Eng. Progr. Symp. Ser.* **61,** 135.
Thring, M. W. (1962). "The Science of Flames and Furnaces," 2nd ed., p. 591. Chapman & Hall, London.
Thring, M. W., Foster, P. J., McGrath, I. A., and Ashton, J. S. (1961). *Int. Conf. Heat Trans. ASME, Boulder, Colorado.* Paper 96.
Tourin, R. H. (1966). "Spectroscopic Gas Temperature Measurement." Fuel and Energy Sci. Monograph, Amer. Elsevier, New York.
Triplett, J. R. (1958). *Proc. Int. Conf. Peaceful Uses At. Energy, Geneva, 2nd,* p. 1869. United Nations, New York.
Usiskin, C. M., and Sparrow, E. M. (1960). *Int. J. Heat Mass Transfer* **1,** 28.
Van Der Hulst, H. C. (1963). *Proc. Interdisciplinary Conf. Electromagnetic Scattering,* p. 583. Macmillan, New York.
Walker, P. L., Jr. (1962). *Amer. Scientist* **50,** 259.
Weeks, D. J., and Saunders, O. A. (1958). *J. Inst. Fuel* **31,** 247.
Yardley, B. E., and Patrick, E. A. K. (1963). *Symp. Ind. Uses Towns Gas,* p. B1. Inst. Fuel, London.

IX

Effects of Electric Fields on Flames

ROBERT J. HEINSOHN

DEPARTMENT OF MECHANICAL ENGINEERING
THE PENNSYLVANIA STATE UNIVERSITY
UNIVERSITY PARK, PENNSYLVANIA

and

PHILIP M. BECKER

FUEL SCIENCE SECTION
DEPARTMENT OF MATERIAL SCIENCES
THE PENNSYLVANIA STATE UNIVERSITY
UNIVERSITY PARK, PENNSYLVANIA

I. INTRODUCTION

When an electric field is applied to a hydrocarbon–air flame, whether premixed or diffusion, the flame is deflected toward the cathode. Besides causing deflections, electric fields can also widen flammability limits of premixed flames, shorten flame length, increase stability, and affect temperature, heat release rates, noise, and sooting characteristics (and thus the color) of flames. There is some controversy about whether burning velocities are influenced. In addition, the flames can conduct measurable electric currents.

These effects are caused indirectly. In hydrocarbon–air flames and in other flames arising from certain types of fuel and oxidant mixtures, electrons and ions are produced in very small concentrations (although much larger than predicted by thermodynamic equilibrium) as by-products of the complex chemical kinetics of combustion. Charged particles are also produced in flames seeded with alkali metals and salts. The electric field affects the motion of the charged particles, which then interact strongly enough with the surrounding neutral gas molecules to affect the concentration and flow pattern of the neutral species.

As a consequence of the flame–electric field interaction, there exists an additional means by which one can manage the behavior of flames. The object of this chapter is to set forth the current understanding of the relationship between flame species and electric fields to identify conditions under which fields can be used to manage flames, and to summarize past, present, and future engineering applications.

II. HISTORICAL REVIEW

The electrical character of flames has been observed for several centuries. Indeed, interest in the electrical characteristics of flames seems to be as old as man's interest in science itself and in electricity in particular. Flames along with cats' skins and amber rods were among the implements of the early intrepid inquirers. Sir William Gilbert (1600), physician to Queen Elizabeth, described attempts to attract flames to electrically charged bodies. Bacon and Boyle (Priestley, 1775) continued these experiments. During the eighteenth century countless experiments were conducted in Europe (Desaguliers, 1739; Watson, 1744, 1747, 1748; Henly, 1777; Ingenhoufz, 1778; Volta, 1782; Bennet, 1787; Brande, 1814) which showed that flames were conductors of electricity and could be deflected by electric fields. In the American colonies the remarkable Benjamin Franklin (Cohen, 1941) included flames in his research. As an example of the type of experiments carried out, a contemporary of Franklin, William Watson (1747), observed that if some burning spirit of wine were held in any vessel in the hand of a man standing on a charged plate, ". . . and if the End of an iron Rod in the Hand of the second Man be held at the Top of the Flame, this second Man will kindle other warm Spirits held near his Finger."

During the nineteenth century the studies of the electrical and magnetic aspects of flames continued (Grove, 1854; Kellogg, 1903; Faraday, 1855). The concept of "ionic wind" (or Chattock electric wind) is credited to the late nineteenth century when Chattock (1899) proposed that charged particles possessing preferential speed and direction transferred momentum to the neutral gas in which the charges were submerged which in turn gave rise to a net force on the neutral gas. The general idea was, however, used by Priestley (1775) a century before.

Early in this century flames were actively studied to examine kinetic theories of gases (Thomson and Thomson, 1928; Wilson, 1912, 1931). The suggestion was made that by controlling the behavior of high-velocity electrons, flames and explosions could be influenced (Garner and Saunders, 1926). Research was actively conducted in this country, Europe, and the Soviet Union to seek to identify the chemical and aerodynamic factors which govern the interaction (Malinowski, 1925; Bone *et al.*, 1931; Thornton, 1930; Lewis, 1931; Lind, 1924, 1926; Guénault and Wheeler, 1931, 1932; Tufts, 1906; Wendt and Grimm, 1924). Additional references to Malinowski have been cited by Lawton (1971a) and Jaggers and von Engel (1971).

The beginning of the modern explorations can be dated with the paper by Calcote (1949) in which he measured the deflection of a Bunsen flame by a transverse field and showed that the concentration of positive ions must be of the order of 10^{18} ions/m^3, a figure far in excess of the value predicted by thermodynamic equilibrium. Consequently, he suggested that ions were produced by chemical reactions. Since that time there have been extensive efforts in the following areas of research. A representative sample of references is included.

1. The identity of the ions:
   (a) positive ions (DeJaegere *et al.*, 1962; Deckers and Van

Tiggelen, 1959; Calcote *et al.*, 1965; Fontijn *et al.*, 1965; Spokes and Evans, 1965),

(b) negative ions (Jensen and Miller, 1971; Calcote *et al.*, 1965; Feugier and Van Tiggelen, 1965).

2. The concentrations of ions and electrons (Bell *et al.*, 1971; Calcote, 1957, 1962, 1963; Kinbara *et al.*, 1959; Carabetta and Porter, 1969; Tsuji and Hirano, 1970; Borgers, 1965; Wortberg, 1965).
3. Chemical mechanisms for formation of ions (Kinbara and Nakamura, 1955; Kinbara and Noda, 1971; Peeters *et al.*, 1971; Green and Sugden, 1963; Porter *et al.*, 1967; Miller, 1967; Calcote, 1965; Nesterko and Tsikora, 1968).
4. Rate of ion generation (Van Tiggelen *et al.*, 1971; Boothman *et al.*, 1969; Peeters and Van Tiggelen, 1969).
5. Propagation and burning velocities (Bowser and Weinberg, 1972; Jaggers and von Engel, 1971; Fowler and Corrigan, 1966; Salamandra, 1969).
6. The body force or ionic wind (Jones *et al.*, 1972; Lawton and Mayo, 1971a, b; Lawton, 1971a, b; Calcote and Pease, 1951; Lawton and Weinberg, 1964; Payne and Weinberg, 1959).
7. Speculation on the role of electrons (Heinsohn *et al.*, 1967; von Engel and Cozens, 1964a).
8. The electron temperature (Silla and Dougherty, 1972; Bell *et al.*, 1971; von Engel and Cozens, 1964a; Bradley and Sheppard, 1970; Bell and Bradley, 1970; Porter, 1970; Bradley and Matthews, 1967; Travers and Williams, 1965).
9. Formation of soot (Ball and Howard, 1971; Howard, 1969; Place and Weinberg, 1967).
10. Seeding of flames with alkali metals and salts (Uhlherr and Walsh, 1971; Carabetta and Porter, 1969; Schofield and Sugden, 1965).
11. High-frequency fields (Jaggers and von Engel, 1971).
12. Measurement of flame area (Fox and Weinberg, 1971).
13. Ionization in rocket exhausts (Jensen, 1972; Jensen and Pergament, 1971; Pergament and Calcote, 1967; Balwanz, 1965).

Surveys of this work can be found in many books on combustion (Gaydon and Wolfhard, 1970; Lewis and von Elbe, 1961; Fristrom and Westenberg, 1965; Becker, 1929; Minkoff and Tipper, 1962) and review articles (Wilson, 1931; Miller, 1968, 1972; von Engel and Cozens, 1964b; Sugden, 1965). An excellent modern monograph (Lawton and Weinberg, 1969) should be consulted for a detailed assessment of the subject.

III. THE NATURE OF THE FLAME–ELECTRIC FIELD INTERACTION

Positive ions and electrons are produced in the fast reaction zone of hydrocarbon flames as by-products of the complex chemical kinetics of combustion. Their concentrations are governed by their rates of production, recombination, and diffusion.

When an electric field is applied to the flame, the positive ions are accelerated preferentially in the direction of the cathode. Each time they undergo a collision they give up a portion of their momentum to the molecules with which they collide. Since the ions are present only in very small concentrations, the collision partners of the ions must be neutral species. Because the mean free path is small and the ions are reaccelerated by the field after each collision (Payne and Weinberg, 1959), each ion is responsible for causing a large number of neutrals to be displaced toward the cathode. The net result is that the electric field produces a body force, or "ionic wind," which is the cause of the observed deflection of a flame toward the cathode. This is an example of the coupling that the charged species provide between the electric field and the neutral molecules and free radicals responsible for propagating the flame.

Generally, the momentum transfer from the electrons can be neglected because their mass is small and their kinetic energy is too low for efficient momentum transfer. The kinetic energy ought to be high enough, however, for certain types of chemical reactions to occur, such as excitation of neutral molecules and dissociative attachment. Momentum transfer from negative species could be a factor under certain conditions. When the electrons have traveled a long distance they will acquire enough kinetic energy for efficient transfer of momentum to neutral molecules. More likely, though, is that they will then be outside the flame region, where they will become attached to neutral species at the lower temperatures prevailing there. In that event, the momentum transfer will be from negative ions rather than electrons.

After decades of speculation about the possible role of electrons in the flame–field interaction, Jaggers and von Engel (1971) have provided what seems to be proof that electrons are involved in some manner. In their work, they measured increases in the propagation velocity of CH_4 and C_2H_4 flames in dc and ac fields and in very high frequency ac fields where the ions could have no influence. On the other hand, it must be recognized that Bowser and Weinberg (1972) and Jaggers *et al.* (1972) observed no appreciable increase in burning velocity for a flame subjected to a dc or high frequency field. [If there were an increase, it couldn't have been greater than about 4%. Such an increase could have been present, however (Weinberg, 1972).] In this laboratory, recent work has found that there may be a very small field induced in the direction transverse to the applied field which may be responsible for flame shifts in the direction of the applied field under certain conditions. If electrons are involved, it is because they can be accelerated by the field to velocities in certain regions of the flame where transfer of kinetic energy from electrons to internal energy of neutral species is likely. Further details will appear in Section V.

To summarize, in principle an electric field can influence a flame in any combination of three ways.

1. The migration of positive particles toward the cathode produces a net force or "ionic wind" on the neutral gas which is not counterbalanced by an opposing force from the negative species. On the macroscopic scale the wind distorts the flame geometry.
2. The ionic wind may alter the concentrations of different neutral species by different amounts and change relative rates of reaction.
3. Kinetics may be influenced by the initiation of new chemical reactions.

IV. IONIC WIND

In recent times, Calcote (1949) and Calcote and Pease (1951) found that the deflection, stability, blowoff, and dead space of Bunsen flames depended on the intensity and polarity of an electric field in ways that were explicable by the mechanical movement of the neutral gas brought on by the ionic wind. Kinbara *et al.* (1959) addressed themselves to the origin and intensity of flame ions. Nakamura (1959) explained the displacement of CH, OH, and C_2 emission spectra profiles of propane diffusion flames by the action of an ionic wind on a microscopic level. In a theoretical calculation, Jones *et al.* (1972) showed that a simplified version of the ionic wind can cause deflections of an opposed-jet diffusion flame comparable to observed deflections. Iya (1971) has also shown that concentration and temperature profiles of a premixed flame can be shifted by a body force arising from the positive ions. On the level of particulates, Asakawa (1957) found that the combustion of fuel droplets was influenced by electric fields.

A. Ionic Wind Velocity

When a dc field is applied to a flame and no discharge occurs, that is, the ratio of electric power to the rate of chemical heat release is of the order of

10^{-3}, many of the effects of the field can be explained in terms of the ionic wind. Under these conditions, charged particles acquire a velocity proportional to the electric field. The constant of proportionality is called the mobility.

$$\mathbf{v}_+ = k_+\mathbf{E}, \qquad \mathbf{v}_- = k_-\mathbf{E}. \tag{1}$$

For a first approximation the mobility may be taken as constant (note, $k_+ > 0$, $k_- < 0$). At field intensities where collisions result in secondary ionization, this mobility notion is invalid. When the charged particles (i.e., agglomerates) are large compared with neutral gas molecules, Stokes and Newtonian flow concepts can be used to define a mobility.

If the ion–molecule collision is elastic, ion momentum is conserved and transferred to the neutral gas as a body force $\mathbf{F}$ per unit volume,

$$\mathbf{F} = \mathbf{E}(q_+n_+ + q_-n_-), \tag{2}$$

where q is the charge per particle and n is the number of particles per unit volume (note, $q_+ > 0$, $q_- < 0$).

The motion of the ions results in a current I:

$$I = \iint \mathbf{J}\cdot d\mathbf{A}, \tag{3}$$

where $\mathbf{A}$ is an area and $\mathbf{J}$ the current density given by

$$\mathbf{J} = (q_+n_+\mathbf{v}_+ + q_-n_-\mathbf{v}_-) = \mathbf{E}(q_+n_+k_+ + q_-n_-k_-). \tag{4}$$

Since the spatial variation of the electric field intensity is not always known, one may express the body force $\mathbf{F}$ as

$$\mathbf{F} = \mathbf{J}[(q_+n_+ + q_-n_-)/(q_+n_+k_+ + q_-n_-k_-)]. \tag{5}$$

The usefulness of Eq. (5) was shown by Payne and Weinberg (1962) who applied a transverse electric field to a free-jet diffusion flame and measured the pressure induced by the ionic wind. The idealization of the flame used by Payne and Weinberg is shown in Fig. 1 and it was assumed that $n_- = 0$ in the region between the flame and cathode, and that the predominant ion has a molecular weight comparable to the neutral gas, $d > x > c$; $n_+ = 0$ in the region between the flame and anode, and the predominant ion is the electron, $a > x > 0$. Thus,

$$\begin{aligned} F_x &= J_x/k_-, \qquad 0 < x < a, \\ F_x &= J_x/k_+, \qquad c < x < d. \end{aligned} \tag{6}$$

The negative ion was assumed to be an electron, thus making $|k_e| : |k_+|$

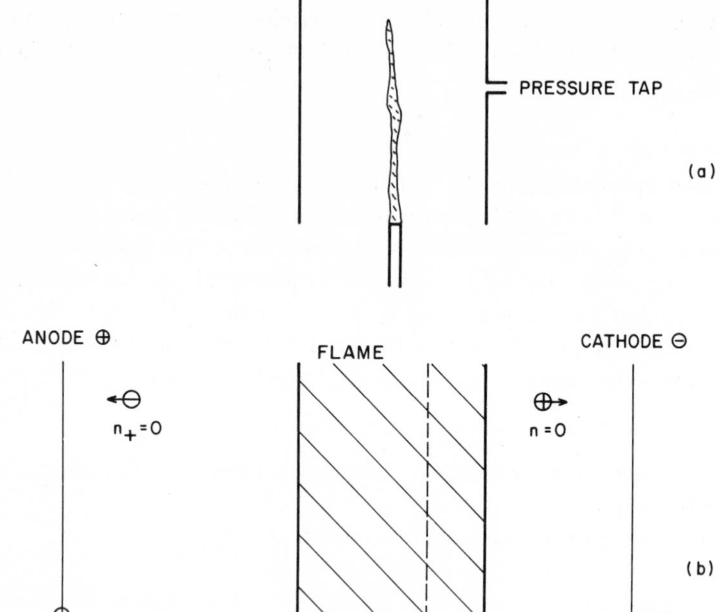

Fig. 1. (a) Free-jet diffusion flame and transverse electric field. (b) Idealization of flame used by Payne and Weinberg (1962).

of the order of $10^3 : 1$. Consequently,

$$F_x \simeq 0, \qquad 0 < x < a,$$
$$F_x = J_x/k_+, \qquad c < x < d. \tag{7}$$

Steady flow conditions were assumed to exist and the momentum equation in the field direction

$$\rho[(u\,\partial u/\partial x) + (v\,\partial u/\partial y)] = -(\partial p/\partial x) + \mu[(\partial^2 u/\partial x^2) + (\partial^2 u/\partial y^2)] + F_x \tag{8}$$

was solved.

On the anode side of the flame, $0 < x < a$, $F_x \simeq 0$, and the integration of the momentum equation (and other conservation equations) would yield the velocity field surrounding a conventional free jet in the absence of a field and was of no particular interest. On the other hand, on the cathode side of the flame, $c < x < d$, $F \neq 0$, and a different velocity field exists.

For a first approximation, the viscosity was neglected, the density was assumed constant, and the induced gas velocity **u** was assumed to be a function of x alone. Thus,

$$\rho\, d(\tfrac{1}{2}u^2)/dx = -(dp/dx) + (J_x/k_+). \tag{9}$$

Integrating between the flame, where the u velocity component is considered to be zero (the shear layer having been neglected) and the pressure is 1 atm, and a point x, $c < x < d$, where the ionic wind velocity is u and the pressure is also 1 atm, one obtains

$$\rho \int_0^u d(\tfrac{1}{2}u^2) = \int_c^x (J_x/k_+)\, dx. \tag{10}$$

If the current density J_x is also considered constant, the ionic wind velocity at any distance $(x - c)$ from the flame is

$$u = [2J_x(x - c)/\rho k_+]^{1/2}. \tag{11}$$

In terms of a stagnation pressure,

$$P_0 = p + \tfrac{1}{2}\rho u^2, \tag{12}$$

the stagnation pressure $P_0(x)$ at any value $c < x < d$ becomes

$$P_0(x) = p + J_x(x - c)/k_+. \tag{13}$$

From direct measurements of $P_0(x) - p$ (of the order of 0.1 N/m^2), Payne and Weinberg (1962) found that while the linear dependence of P_0 upon x was valid, the recorded pressures were approximately 70% of their expected values. Considering the uncertainty of the value of the mobility, the assumption of zero viscosity and u at the flame and the constancy of the current density J_x, the measurements support the contention that moving ions can entrain a neutral gas and produce a measurable bulk gas velocity, that is, the ionic wind. Additional evidence of an ionic wind can be found in research by Velkoff (1962, 1963) who found that an electric field and its resulting corona ion density (of the order of 10^{16} ions/m^3) produced alterations in the velocity field, friction factor, and convection heat transfer coefficient of fluids in various flow configurations.

B. Spatial Variations

The electric field intensity **E** is a function of both the applied voltage V and the space charge. In the absence of an electric field, the rates of ion generation and recombination are equal. When an electric field is applied, the ion concentrations are modified because ions migrate from the flame in addition to recombining. If the field strength does not result in secondary ionization, the ion generation rate will remain a function of combustion kinetics.

Figure 2 shows a typical variation of the current passing from a flame as a function of the applied voltage for the configuration shown in Fig. 1. As the applied voltage is increased, the ion current increases, presumably at the expense of recombination. The plateau portion of Fig. 2 corresponds to what will be referred to as "saturation." Lawton and Weinberg (1964) and Lawton *et al.* (1962) suggest that at saturation, the failure of the current to increase for further increases in voltage implies that charges are being removed from the flame before they can recombine. Thus the saturation current density is equal to the rate of charge generation per unit area of flame front. Even if the parent ion species withdrawn from the flame combine with the neutral gas species by the processes mentioned earlier, the charge is conserved, and its collection rate serves as a direct measure of the ion generation rate. Whether or not the value of the ion generation rate depends on the strength of the applied field is discussed in Section V. The rise in current with applied voltage beyond saturation values is due to secondary ionization.

In the space between the electrodes the field intensity and ion density

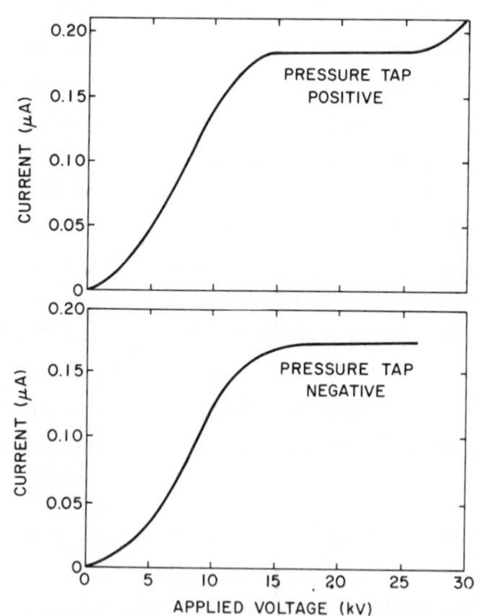

Fig. 2. Current through flame versus applied voltage. [After Payne and Weinberg (1962).]

are related by

$$\nabla \cdot \mathbf{E} = (n_+q_+ + n_-q_-)/\varepsilon. \tag{14}$$

The parameter ε is the permittivity of the gas and may often be taken as that of free space (ε_0).

If in addition, no magnetic field exists, or if one does exist which does not vary with time, the electric potential V is related as follows:

$$\nabla V = -\mathbf{E} \qquad \nabla^2 V = -(n_+q_+ + n_-q_-)/\varepsilon_0. \tag{15}$$

The conservation of charge requires that

$$\nabla \cdot \mathbf{J} = \partial(n_+q_+ + n_-q_-)/\partial t. \tag{16}$$

From Eqs. (14)–(16), the spatial distribution of the electric field strength, voltage, current density, and charge density can be estimated. An attempt to determine these spatial variations was performed by Lawton and Weinberg (1964). Three key assumptions characterize the Lawton–Weinberg hypothesis:

1. There is no ion generation or recombination between the flame and cathode, and the negative ion density is zero; that is, $n_- = 0$, $c < x < d$. Similarly, between the flame and anode, $n_+ = 0$, $0 < x < a$.
2. Since ions and electrons are known to exchange their charge with neutral gas species outside the flame, it will be assumed that in these regions the mobilities of the new charged species will be approximately the same; that is, $k_+ \simeq k$, $k_- \simeq k$.
3. Corresponding to the linear increase of current with applied voltage prior to saturation, there exists a region in the flame, that is, $b < x < c$, where the field is finite and ion recombination is negligible. In the remaining portion of the flame, that is, $a < x < b$, the field is zero and ion recombination matches generation. (See Fig. 3.)

From the conservation of charge [Eq. (16)], one finds that

$$\begin{aligned} J_x &= J_- = C_1, \qquad 0 < x < a, \\ J_x &= J_+ = C_2, \qquad c < x < d. \end{aligned} \tag{17}$$

The constants C_1 and C_2 are equal since the electrodes are connected through the external circuit. Thus at any applied voltage

$$J = J_x = \text{constant} = \begin{cases} J_-, & 0 < x < a, \\ (J_- + J_+), & a < x < c, \\ J_+, & c < x < d. \end{cases} \tag{18}$$

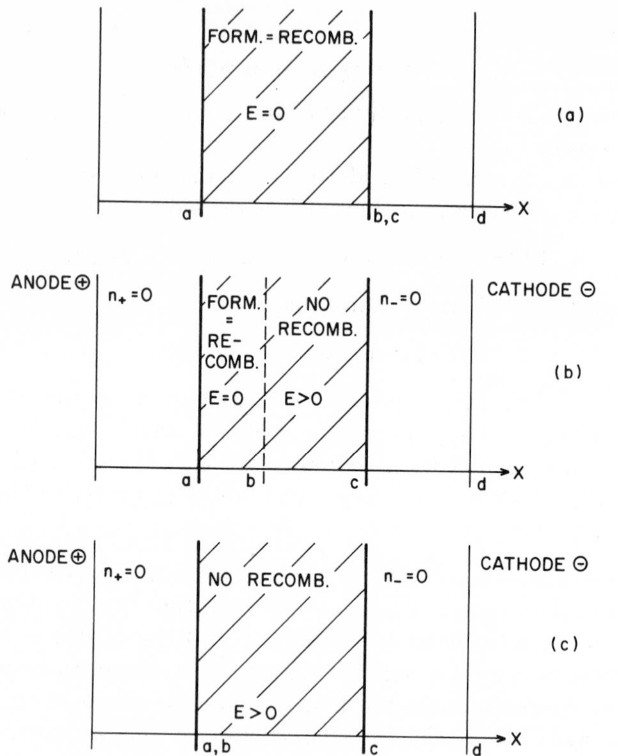

Fig. 3. Regions inside the flame. [After Lawton and Weinberg (1964).]

Application of the other field equations reveals that

$$\varepsilon_0 \, dE_x/dx = \begin{cases} n_- q_-, & 0 < x < a, \\ n_+ q_+, & c < x < d, \end{cases}$$

$$J_x = \begin{cases} -kE_x n_- q_-, & 0 < x < a, \\ kE_x n_+ q_+, & c < x < d. \end{cases} \tag{19}$$

By combining Eqs. (18) and (19) one obtains

$$E_x = [E(0)^2 - 2xJ_x/k\varepsilon_0]^{1/2}, \qquad 0 < x < a,$$

$$E_x = [2(x - c)J_x/k\varepsilon_0 + E(c)^2]^{1/2}, \qquad c < x < d, \tag{20}$$

where $E(c)$ is the field strength at the edge of the flame and can be evaluated from later equations, since the electric field strength is continuous at the flame's edge.

Inside the flame, positive ions may be thought to be generated at a rate r (ions/m^3 sec) and recombining in a second-order fashion with a recombination coefficient α. Thus the conservation of charge requires that

$$\nabla\cdot\mathbf{J}_+ = rq_+ - \alpha q_+ n_+ n_e, \qquad \nabla\cdot\mathbf{J}_- = rq_- - \alpha q_- n_+ n_e, \tag{21}$$

where it is assumed that the negative species is the electron and

$$\mathbf{J}_+ = n_+ q_+(\mathbf{u} + \mathbf{v}_+), \qquad \mathbf{J}_- = n_- q_-(\mathbf{u} + \mathbf{v}_-). \tag{22}$$

In the absence of a field and neglecting the transport of charge by bulk motion and diffusion, an equilibrium ion concentration is established where

$$r = \alpha n_+ n_e = \alpha n^2 \tag{23}$$

since the flame is electrically neutral ($n_+ = n_e = n$). Thus when the applied field is zero, *and* in the portion of the flame where the rate of ion formation equals the rate of recombination,

$$n_+ = n_e = [r/\alpha]^{1/2}, \qquad \mathbf{E} = 0, \qquad a < x < b. \tag{24}$$

Under these conditions the current densities satisfy

$$\nabla\cdot\mathbf{J}_+ = 0, \qquad \nabla\cdot\mathbf{J}_- = 0, \qquad \mathbf{J}_+ = C_3, \qquad \mathbf{J}_- = C_4. \tag{25}$$

The constants C_3 and C_4 must satisfy boundary conditions, which at $x = a$ require from Eq. (19) that

$$C_3 = 0, \qquad C_4 = J_- = -kE_x n_- q_- \qquad \text{at} \quad x = a. \tag{26}$$

In that portion of the flame where the ion formation rate exceeds the recombination rate,

$$\nabla\cdot\mathbf{J}_+ = rq_+, \qquad \nabla\cdot\mathbf{J}_- = rq_-, \qquad b < x < c. \tag{27}$$

Neglecting the transport of charge by bulk velocity $\mathbf{u}$, and the variation of r with position, and assuming charge moves as a consequence of the transverse field,

$$q_+ n_+ k_+ E_x = q_- n_e k_e E_x = rq_+(x - b) \tag{28}$$

since $E_x(b)$ is zero by assumption 3.

The ion densities in $b < x < c$ are

$$n_+ = [r(x-b)/E_x]k_+^{-1}, \qquad n_e = -[r(x-b)/E_x]\,k_e^{-1}. \tag{29}$$

Since the mobility of the electrons is some three orders of magnitude larger than those of the positive ion,

$$n_+ \gg n_e = \text{very small}, \qquad b < x < c. \tag{30}$$

(Note, while n_e is very small, it is *not* identically zero.)

By combining Eqs. (14) and (29), one finds that

$$E_x = (rq_+/k_+\varepsilon_0)^{1/2}(x - b). \tag{31}$$

Thus for a particular value of b, E_x increases linearly with $(x - b)$.

The current densities in $b < x < c$ may be expressed as

$$J_+ = n_+q_+E_xk_+ = q_+n_+k_+(rq_+/k_+\varepsilon_0)^{1/2}(x - b), \tag{32}$$

$$J_- = n_eq_-E_xk_e = q_-n_ek_e(rq_+/k_+\varepsilon_0)^{1/2}(x - b). \tag{33}$$

The significance of Eq. (33) may be lost in the algebra and it is wise to remember that in $b < x < c$,

$$\mathbf{J} = \mathbf{J}_+ + \mathbf{J}_-, \tag{34}$$

and

$$\nabla\cdot\mathbf{J} = \nabla\cdot(\mathbf{J}_+ + \mathbf{J}_-) = rq_+ + rq_- = 0 \tag{35}$$

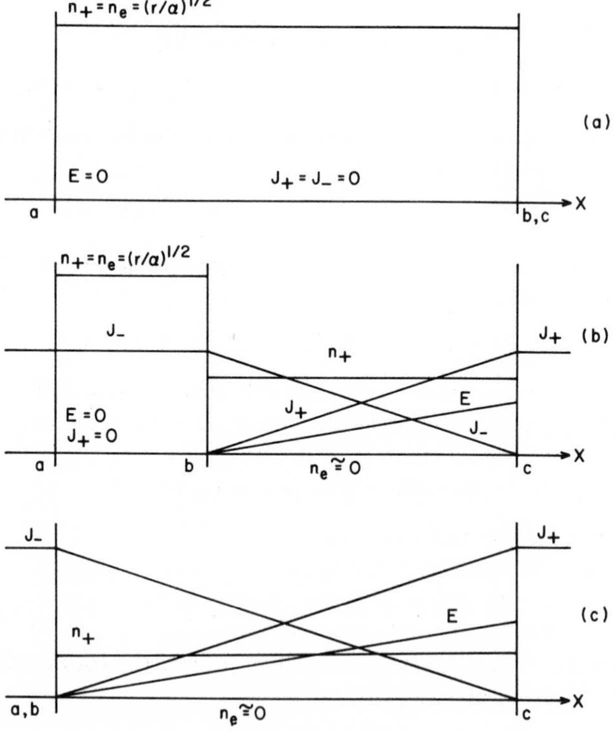

Fig. 4. Variation of electric field, ion density, and current density inside a flame. [After Lawton and Weinberg (1964).]

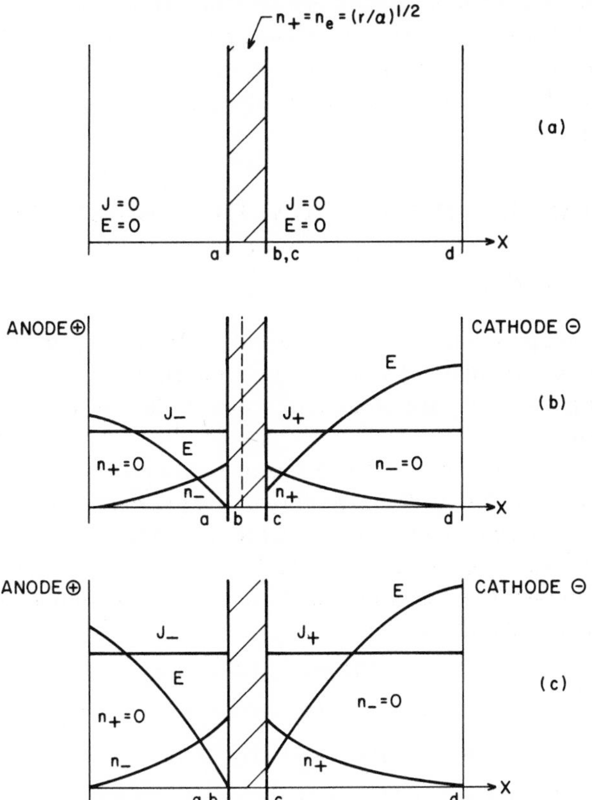

Fig. 5. Variation of electric field, ion density, and current density outside the flame. [After Lawton and Weinberg (1964).]

or

$$\mathbf{J} = \text{constant}. \tag{36}$$

Thus in $b < x < c$, a linear increase in J_+ with $(x - b)$ is matched with a linear decrease in J_-.

Values of r are not as yet known with certainty and the previous equations are difficult to visualize. Figure 4 graphically portrays the current and charge densities and field distribution inside the flame. The flame sheet is very thin in comparison with the distance between the flame and the electrodes and the ion generation rates are not immense; thus the electric field throughout the flame is considerably smaller than it is on the outside of the flame. Figure 5 graphically portrays conditions outside the flame.

Conclusions about the behavior of an ionized gas such as a flame in the presence of an electric field must also be judged in light of the portion of physics called gaseous electronics, plasma physics, and so on. Contributions in this field are numerous, but three of relevance to this chapter are the texts by von Engel (1965) and Loeb (1958, 1961), and the paper of Klein *et al.* (1968).

V. ORIGINS OF IONS AND KINETIC MECHANISMS

A. Ion Production

Table I compares the observed concentrations of ions in various flames with the values predicted by thermodynamic equilibrium. The high value for the hot gas seeded with potassium is due to the low ionization potential of potassium.

In discussing the fact that the concentration of ions in hydrocarbon–air flames is far greater than the value predicted by thermodynamics, Calcote (1957) gave convincing evidence that the ions must be generated initially by chemical reactions involving neutral species. Later, Green and Sugden (1963) proposed the following mechanism for the important reactions involving ionization in flames:

$$CH + O \rightarrow CHO^+ + e, \tag{37}$$

$$CHO^+ + H_2O \rightarrow H_3O^+ + CO, \tag{38}$$

$$H_3O^+ + e \rightarrow H_2O + H. \tag{39}$$

The principal reactants, CH and O, are produced by decomposition and other reactions of the fuel and oxidant. The CO and H species will continue to react and eventually will be incorporated into final products like CO_2 and H_2O. Excessively large concentrations of electrons and ions are pre-

TABLE I
Comparison of Observed and Predicted Ion Densities in Various Flames

Flame	Typical ion density (ions/m³)	
	Observed	Predicted
Hydrocarbon–air	10^{18}	10^{12}
CO or H_2–air	10^{12}	10^{12}
Gas seeded with K or KCl	10^{18}	10^{18}

vented from occurring by recombination reaction (39) and by diffusion from regions of high concentrations.

Later Calcote (1965) developed a classic proof in support of the assumption that reaction (37) is responsible for ion generation. The participation of excited species such as CH* and C_2* has been eliminated by Miller (1966) and Peeters *et al.* (1969). Although Porter *et al.* (1967) disagree, there is now general agreement that reaction (37) is the major contributor (with the possible exception of C_2H_2–air flames). It should also be noted that Klemm and Blades (1966) suggested a pair of reactions which have not yet been eliminated as contributing to ionization.

In most flames, H_3O^+ is the most abundant ion, being formed by the charge exchange reaction [Eq. (38)]. There are many secondary ions formed by similar reactions, such as CH_3O^+, $C_3H_3^+$, and $C_2H_3O^+$. Many of the concentration profiles of ions were obtained by Calcote and his coworkers by sampling ions with a microprobe and identifying them with a radio-frequency (rf) mass spectrometer. For the profiles to be valid, it must be determined whether the microprobe disturbs the flame and whether ion concentration ratios are changed to a significant degree by chemical reactions within the probe. Miller (1966), for example, has argued that these effects are small. Typical profiles are shown in Fig. 6 (Calcote *et al.*, 1965). Note that like the free radicals which are intermediates in the sequence of chemical reactions in the flame, the ions are localized to a very

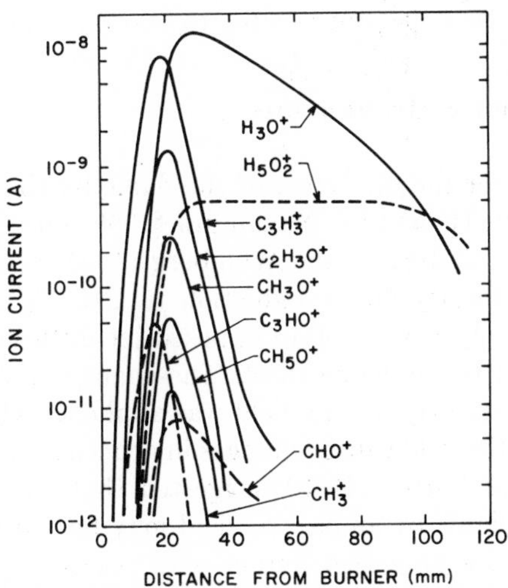

Fig. 6. Typical positive-ion profiles. [From Calcote *et al.* (1965).]

thin slab in the flame. The surveys by Miller (1968, 1973) should be consulted for more complete information.

Peeters *et al.* (1969) measured the current from charged particles striking the electrodes when an electric field is imposed on a flame. Under conditions of current saturation, the recombination of ions and electrons will be negligible and the number of ions impinging on the electrode per unit of surface area per unit time multiplied by the thickness of ion generation zone is equal to the rate of ion generation. On the assumption that the thickness of the ion production zone is equal to the thickness of the free radical reaction zone, they found the rate to depend on composition. The maximum value of 4×10^{23} ions/m^3 sec was observed in flames with low N_2 concentration and with the CH_4/O_2 ratio slightly richer than stoichiometric. Using the same method, Boothman *et al.* (1969) investigated the variation of saturation current for premixed and diffusion hydrocarbon flames with various additives, and with the use of O_2 and air. They did not estimate the thickness of the reaction zone producing ions and left their results in current per square meter of electrode surface area. There is one limitation in this method which must be pointed out. In principle the rate of ion generation could depend on the magnitude of the applied electric field because the ionic wind can cause distortion of the concentration profiles of the reactants in Eq. (37), but there have been no estimates of whether or not the effect is significant. Theoretical calculations by Iya (1971) predict that the ion generation rate does change by a measurable amount, but not more than a factor of 2. Thus the results of Peeters *et al.* (1969) and Boothman *et al.* (1969) need not be modified appreciably to take account of this effect.

B. Chemical Kinetic Mechanisms

The mechanism for the distortion of the flame by the electric wind was discussed in Sections III and IV. Nakamura (1959) observed such an effect in a diffusion flame, and Place and Weinberg (1965) observed an increase in OH emission intensity and a reduction in CH, C_2, and solid particle emission in the pyrolysis zone of an opposed-jet diffusion flame as a field was applied normal to the flame sheet. The change in sampling efficiency observed by Miller (1966) when a field was applied to the sampling probe also indicates that the field causes changes in concentration profiles of some species. Harker and Porter (1968) suggested that the stabilization of a methane diffusion flame by an electric field may be caused by drifting H^+ ions moving toward a grounded burner with reaction of H^+ to form hydrogen and with consequent increase of burning velocity. With reversed polarity, they could not reproduce the stabilization. [On the other hand,

Heinsohn *et al.* (1970a) observed that a detached diffusion flame could be caused to return to the burner by fields of either polarity.] The above-mentioned experiments are quoted to give justification to the belief that chemical kinetics might be influenced by a field. Obviously, the arguments are not at all conclusive. In fact Popov and Sheklein (1965) did not observe any microscopic changes in flame structure. The important point is that further work is needed. The following discussion will comment on the ways in which changes in chemical kinetics could occur, if they do.

If the electric wind does cause selective distortion of species concentration profiles, then the rates of various chemical reactions can be affected, perhaps significantly. It is possible that the entire mechanism for the reactions which are involved in flame propagation can be altered, although there is no evidence to indicate that changes are that large. The theoretical predictions of Iya (1971) suggest that the changes are detectable but not drastic. They have considered a one-dimensional model of a premixed methane–oxygen flame and solved the conservation equations of mass, momentum, and energy for a reasonably realistic set of chemical reactions.

It is also possible for the field to initiate chemical reactions which do not occur without the field. Reactions between high-velocity charged particles and neutrals are primary candidates. For example, under the influence of an electric field an electron may well be accelerated to velocities sufficient to allow reactions to occur during the collisions with a neutral species. At 2000°K and 10^5 N/m^2 pressure the mean free path of particles (ions, electrons, neutrals) is of the order of 10^{-5} m. Thus between collisions, a charged particle will acquire an energy of the order of 1 eV (which corresponds to a velocity of 5×10^5 m/sec for an electron) in a region where the field strength $E = 100$ kV/m. Such a region exists on the boundary of the flame, although the field strength is almost zero within the flame. Because of the distribution of distances between collisions, an appreciable fraction of electrons will achieve a kinetic energy of 2 or 3 eV, which is the value needed for the reaction (Henderson *et al.*, 1969)

$$e + O_2 \rightarrow O + O^- \tag{40}$$

to proceed with a reasonable rate.

Jaggers and von Engel (1971) have suggested that OH may be vibrationally excited by fast electrons at the boundary of the flame. Since the reaction of CH_4 with vibrationally excited $OH^\dagger$ would have a rate constant different from the reaction with ground state OH, the decay of CH_4 and subsequent flame kinetics could be affected by appreciable concentrations of $OH^\dagger$. Jaggers and von Engel (1971) also observed that the effects of high-frequency fields on C_2H_4 flames were different from the effects on CH_4

flames. Since the decay of C_2H_4 is primarily by reaction with O atoms, it is possible that reaction (40) could be a significant factor in C_2H_4 flames. This leads to the speculation that different mechanisms for the effects of electrons may be dominant for different fuels or under different operating conditions.

The suggestion that electron–molecule reactions may occur is not new. Thomson (Garner and Saunders, 1926) made the suggestion in 1910, and this has been repeated by von Engel and Cozens (1964a), Heinsohn *et al.* (1967), and Klein *et al.* (1968). However, to ascertain properly what reactions between neutrals and charged species are occurring, detailed knowledge is needed of the specific species present and of the reactions that may occur. Such knowledge is not always available.

In opposed-jet burners, jets of fuel diluted with an inert gas and oxygen (sometimes diluted with an inert gas) are allowed to impinge and form a reasonably flat flame midway between the burners (Jones *et al.*, 1972). By grounding one of the burners and applying a voltage to the other burner, an electric field is created in the direction of the jets. Nsakala (1973) observed that a methane/nitrogen/oxygen flame in such a burner shifted toward the fuel jet when fields of either polarity were applied to the burner. On the other hand Dayal and Pandya (1972) observed that an ethyl alcohol/nitrogen/oxygen flame moved toward the oxygen jet when that was positive. In a computer model of the methane/nitrogen/oxygen opposed-jet flame, Alavi-Naini (1973) found it impossible to cause a shift in the position of the flame toward either electrode without assuming the existence of a very small field induced in the direction transverse to the direction of the applied field. The existence of small transverse fields is obvious, because there will be slight variations in ion density in the transverse direction. That such fields must be invoked in order to explain deflections in the jet direction was very surprising. In related work on a one-dimensional computer model of a premixed methane/oxygen flame, Cordeiro (1973) found that the burning velocity did not increase when a dc field was applied to the flame in the direction of flow because the electron concentration in the flame zone was well below the value necessary to cause a significant increase in the $O_2^{\dagger}$ concentration. With the assumption that a high frequency field would not reduce the electron density below the value which occurs in the absence of a field, Cordeiro predicted an increase in burning velocity for that case. In the light of the above results, future work will have to consider the possibilities of transverse variations of charge density, of high electron and vibrational temperatures, and of the formation of negative ions.

This section concludes with the mention of two interesting sidelights.

Weinberg (1966) has used photographic emulsions to record flame ions, and Feugier and Van Tiggelen (1967) have accelerated the electrons naturally present in flames to velocities such that neutrals are ionized. By sampling the ions so produced, they were able to identify free radicals normally present in flames which may not be detectable by other methods. This provides an additional reason for the presence of high-energy electrons to be considered in any complete explanation of the flame–field interaction.

VI. ENGINEERING APPLICATIONS

At the present, the design of devices employing flames requires decisions about two general classes of parameters, that is, aerothermodynamic and geometrical. Aerothermodynamic parameters include the inlet fuel–air ratio, and the inlet and outlet velocities, temperatures, and pressures. The geometrical parameters include the length, breadth, and shape of the combustion volume. While this categorization masks important interrelationships between parameters, it is nonetheless useful, since manipulation of the variables in each group is the only means by which engineers design and control the combustion process.

The following characteristics of the flame–field interaction can be summarized from the preceding material and should be borne in mind for engineering purposes.

1. The electric field (voltage gradient) is small in the region where ions are generated and where CH and O are large, but at the boundaries of the region of ion generation, it is comparable to the overall applied voltage gradient.
2. It is possible, but not yet confirmed, that charged species may participate in flame reactions which do not occur in the absence of a field.
3. Momentum transfer from electrons may be neglected unless they have been accelerated over a long distance, of the order of 100 mean free path lengths.
4. Outside the flame, electrons become attached to neutral molecules, creating negative ions with mobility values comparable to those of positive ions. Thus forces may exist in the anode side of the flame comparable to those usually seen on the cathode side of the flame.
5. Ionization depends on the type of fuel selected as well as the fuel–air ratio.
6. Forces accruing from the ionic wind are small and are significant only if strategically located in the combustion flow field. Further, it appears

necessary to produce or maintain a high electron and/or ion density in the flame.

As a result of the known influence of electric fields on flames, it is suggested that the engineer now has an additional parameter in the design and control of flames—an electrical field. However, the effects are subtle, and unless the factors listed above are important to the combustor's performance, there is little expectation that the electric field will have technological importance.

Some practical applications have been proposed, and some of merit have progressed to the operational level. The following applications warrant consideration:

(1) electrically boosted (i.e., augmented) flames;
(2) flame temperatures and heat transfer;
(3) aerosols, particulates, and flame carbon;
(4) ducted diffusion flames;
(5) extinction limits;
(6) control of solid propellant burning;
(7) noise generation and suppression;
(8) air pollution.

A. Electrically Boosted (i.e., Augmented) Flames

In recent years there has been a renewed interest in the interaction of an electrical discharge and flames. The objective of this effort has been to increase gas enthalpies, temperatures, and thermal conductivities above values obtainable through combustion alone. Similar increases in these quantities cannot be achieved by merely preheating reactants since material limitations prevent it. Southgate (1924) produced a fuel–air flame through which an ac arc was passed which increased the output energy by 15 to 20%. Heating was achieved from a high-intensity concentrated arc produced between electrodes emersed in an open flame used in smelting.

In 1961 Richardson and Karlovitz (1961) [see also Karlovitz (1961)] reported on a device using a high-voltage, low-current ac discharge and a natural gas–air flame seeded with KCl. Measurements revealed that the flames could be stabilized at air–fuel ratios beyond the rich blowoff limit and that the burned gas enthalpy increased by roughly the input electrical energy. Similar results were also reported by Lawton *et al.* (1962), Hirt *et al.* (1965), Montes and Cushing (1967), Fells and Harker (1968, 1969), Davies (1969), and Harker (1969). Measured flame temperatures and rates of heat transfer to bodies immersed in the flame also increased with the application of the discharge. Harker and Allen (1969) have shown that the costs of electrically augmented methane–air flames are less than those of methane–oxygen flames both in attaining the same temperature and heat transferred per unit area.

To obtain uniform heating of the gas, it is necessary to distribute the arc uniformly throughout the flame. A concentrated discharge, that is, arc, envelopes only a minute gas volume and the associated energy enhancement is small with respect to the normal combustion energy. Furthermore, the remaining flow tends to avoid the heated channel. To spread the discharge uniformly, the following schemes have been suggested (Chen *et al.*, 1965):

(1) organized or disorganized (i.e., turbulence) alteration of the flow field,
(2) increasing the electrical conductivity by seeding,
(3) rf discharge,
(4) induced rapid oscillation or rotation of arcs by magnetic fields.

An excellent discussion of the merits of these schemes is given by Lawton and Weinberg (1969). Forcing the gas to follow the natural path of the discharge is far superior to using the flow to move the discharge. Small-scale turbulence, for example, of the width of the discharge column, slightly alters the discharge path, whereas large-scale turbulence does not. Seeding only succeeds in spreading the discharge in the burned gas region of the flame but alters little the path of the discharge in the luminous zone. Combustion tends to extinguish stable rf discharges and this scheme shows little promise. The passage of gas through a collimated discharge rotating (by means of an external magnetic field) in a plane perpendicular to the flow heats the gas stream uniformly and produces an augmented flame to high stability and combustion intensity. In conclusion, it should be borne in mind that the methods mentioned above are additive and may be tailored to fit specific applications.

A significant feature of boosted flames lies in the fact that at their high temperatures, the dissociation–recombination reactions in the combustion products cause an effective increase in thermal conductivity above that of a nonreacting system at the same temperature. Davies (1965) found that a 50% increase in heat content, that is, augmentation or boost ratio of 0.5, was sufficient to double the heat flux at constant mass flow. In addition Davies demonstrated that the heat flux from an electrically boosted gas–air flame seeded with NaCl or KCl can be approximately 35% greater than an equivalent gas–oxygen flame. Chen *et al.* (1965) and Fells and co-workers (Gawen *et al.*, 1966; Fells *et al.*, 1967; Harker and Fells, 1968) successfully obtained a diffuse discharge and eliminated seeding by a rotating magnetic field and a dc discharge. Engineering applications which have been patented include a method and apparatus for rock drilling (Karlovitz, 1964), a method for reducing metal oxides (Karlovitz, 1966), and a unique cutting tool for medical surgery (Anon., 1970a).

B. Flame Temperatures and Heat Transfer

The feasibility of influencing heat transfer from unboosted flames has been recognized for years. Payne and Weinberg (1959) measured the heat transferred to a cylindrical, constant-flow calorimeter surrounding a coaxial burner and found it to increase appreciably, the increase being at least two orders of magnitude greater than the dissipated electrical power. Pejack and Velkoff (1967) and Pejack and Jones (1968) studied the heat transfer from a diffuse flame in a parallel flow rectangular duct subjected to a transverse electric field. The flame profile, originally analyzed by Burke and Schumann (1928), experienced a marked displacement toward the cathode. While the heat transfer to the anode decreased, the heat transfer to the cathode increased. Pejack found no clear change in the total heat transfer nor was the exhaust temperature significantly altered.

Evidence indicating an alteration in volumetric heat release rate and temperatures of a flame by an electric field is incomplete and contradictory. Lewis and Kreutz (1933) reported a modest reduction in the flame temperature over the entire length of premixed flames subjected to a transverse field. The reduction increased along the flame length and was dependent on the inlet composition. Heinsohn *et al.* (1969) observed slight reductions in the maximum temperatures of stable opposed-jet diffusion flames subjected to an electric field of either polarity which were in general agreement with Lewis and Kreutz. Klein *et al.* (1968) measured flame temperatures within a premixed flame subjected to a voltage difference between a downstream annular electrode surrounding the flame and the burner port. The temperature decreased when the downstream electrode was positive and increased with opposite polarity. They concluded [as suggested by von Engel and Cozens (1964a)] that high-energy electrons created new species. Theoretical calculations by Jones *et al.* (1972) predict shifts in the temperature and heat release rate profiles, and small changes in the maximum value of the heat release rate.

C. Aerosols, Particulates, and Flame Carbon

Particulates such as soot, or fuel droplets, acquire a charge when they pass through an electric field of suitable strength (Lawton and Weinberg,

1969; Gugan *et al.*, 1965). Thus the designer has at his disposal a means to control, within limits, particle speed and trajectory and in the case of reacting particles, their formation or disappearance. The uses of this principle in electrostatic dust precipitation, spray painting, and so on, are well-known examples and will not be discussed.

Molecular carbon formed in combustion and later deposited on surfaces as soot is strongly influenced by electric fields. All three phases of the process, that is, nucleation, growth, and deposition, are affected. It is strongly suspected that ions act as centers of carbon growth, although uncharged particles may also act in this fashion. Particle growth and charge depend on its chronology (Place and Weinberg, 1965, 1967; Lawton and Weinberg, 1969). The ultimate deposition rate depends on the polarity of the electrode and particle. Each one of these three phases is the subject of active current study and the current literature should be consulted. Thus any or all three phases of the soot process can be influenced by an electric field; however, the phase in which the designer has the widest latitude is the growth process. All carbon particles acquire charge either by attachment of or transfer from another charge carrier, or by electron emission resulting from carbon's low work function. While electron emission results in positively charged carbon particles, the polarity, length of time, and character of the space charge surrounding the particle determine its ultimate charge, mass, and density.

The yellow luminosity of flames can be reduced by an electric field by reducing the carbon particle residence time in the flame and increasing the removal of growth centers. Depending on its size, carbon particles can either follow the flow field or move as charged particulate matter. In any event, luminosity is inherently bound to the flame's aerodynamics. For this reason it is not surprising for Weinberg (1968) and Place and Weinberg (1965) to report a reduction in luminosity of low-speed opposed-jet diffusion flames, but others (Rezy and Heinsohn, 1966; Wulfhorst, 1966; Heinsohn *et al.*, 1969) to report an increase in similar high-speed flames and ducted diffusion flames (Witt, 1965; Mitchell, 1969; Heinsohn *et al.*, 1970a).

D. Ducted Diffusion Flames

Laminar and turbulent diffusion flames, confined and unconfined, are strongly influenced by electric fields. The luminosity, geometry, and recirculation aerodynamics are flame characteristics which respond to electric fields. The effects are brought on by the influence of the ionic wind on the flow field and the motion of carbon particles. Asakawa (1957) reported that the length of a variety of flames burning at a wick fed by liquid fuels decreased when subjected to ac and dc fields. Witt (1965) observed a reduction in the length of a 17-in. free city gas diffusion flame to 6 or 7 in. with a radial field of either polarity and the consumption of only a few watts of electrical power. Heinsohn *et al.*, (1970a) found that confined laminar and turbulent diffusion flames all assume shorter lengths when subjected to axial fields. Mitchell (1969) and Mitchell and Wright (1969) found that radial electric fields caused off-port diffusion flames to become shorter, broader, and less luminous. In addition, the point at which burning commenced was displaced upward so that the region of isothermal mixing of the air and fuel was lengthened.

Lawton and Weinberg (1969) and Lawton *et al.* (1968) have shown that by confining entrainment produced by ionic winds to specified regions, large flow velocities can be induced at the flame. Their experiments with axial fields confirm that confined entrainment can be used to create diffusion flames in an accurately controllable manner, to reduce the flame lengths substantially and remove all traces of sooting.

One of the most exhaustive studies of diffusion flames was conducted by Pejack and Velkoff (1967) and Pejack and Jones (1968) who examined the effects of a transverse electric field on a confined diffusion flame of rectangular geometry of the type studied (but in the absence of fields) by Burke and Schumann (1928). Using the identical assumptions as these authors, Pejack showed that the location of the flame in an electric field was achieved by expressing the continuity of a species at any point in the chamber by

$$\partial C/\partial y = (D/U)(\partial^2 C/\partial x^2) - (1/U)[\partial(Cw)/\partial x], \tag{41}$$

where C is the concentration of the fuel or air, D is a diffusion coefficient U is the uniform inlet velocity of gas and air in the y direction, and x and w are transverse displacement and velocity, respectively. The last term in Eq. (41) represents the transverse velocity away from the flame brought on by the electric field. This velocity opposes the diffusion velocity of one species toward the flame and displaces the flame sheet toward the cathode. Pejack showed that the transverse drift velocity w is a function of $\partial J/\partial y$ (the y gradient of the transverse current density), that it is significant and produces first-order effects on the diffusion flame, particularly in the lower portions of the flame near the inlet plane.

E. Extinction Limits

Electric fields have pronounced effect on the extinction limits of many types of combustion systems. It is possible to support combustion with an electric field at velocities and mixture compositions where without the field there would be no combustion. Richardson and Karlovitz (1961) reported that an electrical discharge in a premixed gas–air boosted burner allowed the air–fuel ratio to be extended beyond the rich blowoff limit. Heinsohn and Lay (1964) observed a substantial shift in the lean extinction limits of seeded (KOH) premixed city gas and air flat flames. Putnam and Smith (1953) found that the blowoff limits of a Bunsen flame and bluff-body flames were extended by dc electric fields. Similarly, Parker and Heinsohn (1964) observed a 5% reduction in the lean extinction limits of bluff-body flames. Guest (1930) reported that the ignition of combustible mixtures by heated wires occurred at lower temperature if an electric field was present. Rezy and Heinsohn (1966) observed a large increase in the apparent flame strength of opposed-jet diffusion flames, and deduced an increase in the maximum volumetric heat release rate. Heinsohn *et al.* (1967) interpreted such data as a change in the overall order of reaction, which signals changes in the reaction mechanism. Enhancement of flame extinction limits has also been reported by Harker and Porter (1968), Fox (1965), and Harrison and Weinberg (1971). In summary, it is clear that electric fields provide valuable assistance in the combustion of mixtures nominally too lean or rich to burn satisfactorily. An engineering application which has been patented by Wright and Levine (1967) is a device using an ac field to assist in the stabilization of flare flames.

F. Control of Solid Propellant Burning

The possibility of varying the burning rate of solid propellants, after ignition, by the use of electric fields had been considered by Mayo *et al.* (1965). It was concluded that an ionic wind can be used to increase flame spreading by making the propellant one of the electrodes, or to decrease the flame spread by placing the electrode in the flame. Theory indicated that larger effects should be possible at higher pressures.

The application of an electric field to a cesium-enriched low-pressure flame plasma was investigated by Dimmock and Kineyko (1963). In these studies an arc discharge **J** was established across the flame and an axial magnetic field **B** applied parallel to the direction of flow. It was found that the deflection of the flame from the direction of flow increased with the Lorentz force $\mathbf{J} \times \mathbf{B}$, and decreased linearly with flame momentum at a constant value of the Lorentz force.

G. Noise Generation and Suppression

If the hot combustion gas column between electrodes is caused to vibrate by a time-varying field, sound is generated. The sound produced was found by Babcock *et al.* (1967) to be of high fidelity over the entire audible range. For satisfactory sound intensity they found that the conductivity of the flame had to be larger than normal and required an oxygen–acetylene flame seeded with an alkali salt. Anonymous (1970b) indicated that any plasma of comparable electrical conductivity can be used in place of the flame. By modeling the flame as a long cylindrical column of homogeneous, weakly ionized gas subjected to small periodic perturbations, Burchard (1969) found that

(a) for a flame of a particular size, the sound pressure level was proportional to the square root of the frequency and inversely proportional to the square root of the distance from the flame; and

(b) for a particular frequency and distance from the flame, the sound pressure level varied linearly with the product of the dc biasing voltage and the ac driving voltage and inversely with the product of the flame impedance and the electrode separation.

The impedance of the flame, as seen by an electrical driving circuit, was found by Russell (1970) to be characteristic of a parallel *RC* circuit. The resistance component varied inversely with the flame's conductivity, hence its seeding, while the capacitance depends on the electrode configuration.

The ability to generate sound suggests a chance to suppress undesirable and inherent noise in flames by destructive interference with an externally applied electric field. Abrukov *et al.* (1966, 1967) suppressed acoustic vibrations (of fundamental frequency 120 Hz) in "singing flames" with both constant and variable (50 Hz) transverse electric fields. By using the audio output of a "singing flame" as the driving source for an electric field but changing its phase by 180°, Wenaas and McChesney (1970) were able to suppress combustion oscillation completely. By aligning the phase of the electric field and the audio output, Wenaas caused an enhancement of the oscillation.

H. Air Pollution

Iya (1971) has shown theoretically that electric fields may change O concentrations by a factor of 2. The oxides of nitrogen, that is, NO, NO_2, produced in flames originate from reactions of N and O with their molecular counterparts. For this reason the production of NO and NO_2 should be affected. Experiments (Heinsohn *et al.*, 1970b) indicate that NO_2 concentrations are reduced slightly in premixed flat flames subjected to fields opposite to the direction of flow.

VII. CONCLUSIONS

In summary, it has been shown that electric fields can selectively act upon flame ions and in turn bring about large-scale changes in the flame. The motion of ions induces motion of the neutral gas in and near the flame. Inside the flame, the motion alters the rates of certain reactions in the kinetic mechanism and may bring about changes in the overall reaction. Outside the flame, the motion alters the shape and entrainment capacity of flames. While current conceptualizations are imperfect, it has been demonstrated that the luminosity, heat transfer, extinction, flame speed, and geometry of flames are significantly affected.

Acknowledgments

The preparation of this paper has been supported by National Science Foundation Grants GK-1905 and GK-19077, and National Air Pollution Control Administration of the Public Health Service Grant No. AP-00643 through the Center for Air Environment Studies at the Pennsylvania State University.

List of Symbols

- **A** Cross-sectional area
- a Position of boundary of flame on anode side
- **B** Axial magnetic field
- b Position within flame which is the boundary of the region where ion recombination is negligible
- C Fuel or air concentration
- C_j Constants of integration, $j = 1$ to 4
- c Position of boundary of flame on cathode side
- D Diffusion coefficient
- d Position of cathode
- e Subscript: electrons
- **E** Electric field strength
- **F** Body force per unit volume acting on the neutral gas
- I Current
- **J** Current density
- k_i Mobility of species i
- n_i Number density of species i
- P_0 Stagnation pressure
- p Pressure
- q_i Charge of species i
- r Rate of ion generation per unit volume
- U Uniform inlet velocity of gas or air in y direction
- u Component of neutral gas velocity in x direction
- **V** Velocity of charged particles
- V Voltage
- v Component of neutral gas velocity in y direction
- w Transverse gas velocity
- x Spatial coordinate; subscript usage: force or current in x direction
- y Spatial coordinate
- α Ion–electron recombination coefficient
- ε Permittivity of gas
- ε_0 Permittivity of free space
- μ Viscosity
- ρ Density of neutral gas
- ∇ Gradient operator
- $+$ Subscript, refers to positive ions
- $-$ Subscript, refers to negative ions or electrons

References

Abrukov, S. A., Kurzhunov, V. V., and Mezdrikov, V. N. (1966). *Combust. Explos. Shock Waves* (*USSR*) **2,** No. 2, p. 43 [Transl. publ. by Faraday Press, New York.]

Abrukov, S. A., Kurzhunov, V. V., and Mezdrikov, V. N. (1967). *Combust. Explos. Shock Waves* (*USSR*) **3,** No. 1, p. 99. [Transl. publ. by Faraday Press, New York.]

Alavi-Naini, S. (1973). Ph.D. Thesis in Fuel Science, Pennsylvania State University, University Park, Pennsylvania (in progress).

Anonymous (1970a). Plasma Scalpel 'Cuts' Surgical Hemorrhaging. *Ind. Res.* **12,** No. 6, p. 34.

Anonymous (1970b). Plasma, Not Flame, Produces Sound Waves. *Ind. Res.* **12,** No. 13, p. 26. (Sequel to Babcock *et al.*, 1967.)

Asakawa, Y. (1957). *Proc. Int. Symp. Combust., 6th,* p. 923. Reinhold, New York.

Babcock, W. R., Baker, K. L., and Cattaneo, A. G. (1967). *Nature* (*London*) **216,** 676.

Ball, R. T., and Howard, J. B. (1971). *Proc. Int. Symp. Combust., 13th,* p. 353. Combust. Inst., Pittsburgh, Pennsylvania.

Balwanz, W. W. (1965). *Proc. Int. Symp. Combust., 10th,* p. 685. Combust. Inst., Pittsburgh, Pennsylvania.

Becker, A. (1929). *In* "Handbuch der Experimentalphysik" (W. Wien and F. Harms, eds.), Vol. 13, part 1, p. 109. Akad. Verlagsges., Leipzig.

Bell, J. C., and Bradley, D. (1970). *Combust. Flame* **14,** 225.

Bell, J. C., Bradley, D., and Jesch, L. F. (1971). *Proc. Int. Symp. Combust., 13th,* p. 345. Combust. Inst., Pittsburgh, Pennsylvania.

Bennet, A. (1787). *Phil. Trans. Roy. Soc. London* **77,** 26.

Bone, W. A., Fraser, R. P., and Wheeler, W. H. (1931). *Proc. Roy. Soc., Ser. A* **132,** 1.

Boothman, D., Lawton, J., Melinek, S. J., and Weinberg, F. J. (1969). *Proc. Int. Symp. Combust., 12th,* p. 969. Combust. Inst., Pittsburgh, Pennsylvania.

Borgers, A. J. (1965). *Proc. Int. Symp. Combust., 10th,* p. 627. Combust. Inst., Pittsburgh, Pennsylvania.

Bowser, R. J., and Weinberg, F. J. (1972). *Combust. Flame* **18,** 296.

Bradley, D., and Matthews, K. J. (1967). *Proc. Int. Symp. Combust., 11th,* p. 359. Combust. Inst., Pittsburgh, Pennsylvania.

Bradley, D., and Sheppard, C. G. W. (1970). *Combust. Flame* **15,** 323.

Brande, W. T. (1814). *Phil. Trans. Roy. Soc. London* **104,** 51.

Burchard, J. K. (1969). *Combust. Flame* **13,** 82.

Burke, S., and Schumann, E. (1928). *Ind. Eng. Chem.* **20,** 998.

Calcote, H. F. (1949). *Symp. Combust. Flame Explos. Phenomena, 3rd,* p. 245. Williams & Wilkins, Baltimore, Maryland.

Calcote, H. F. (1957). *Combust. Flame* **1,** 385.

Calcote, H. F. (1962). *Proc. Int. Symp. Combust., 8th,* p. 184. Williams & Wilkins, Baltimore, Maryland.

Calcote, H. F. (1963). *Proc. Int. Symp. Combust., 9th,* p. 622. Academic Press, New York.

Calcote, H. F. (1965). Ionization in Hydrocarbon Flames. 26th Meeting of Propulsion and Energetics Panel, AGARD, Pisa, Italy, September 1965.

Calcote, H. F., and Pease, R. N. (1951). *Ind. Eng. Chem.* **43,** 2726.

Calcote, H. F., Kurzius, S. C., and Miller, W. J. (1965). *Proc. Int. Symp. Combust., 10th,* p. 605. Combust. Inst., Pittsburgh, Pennsylvania.

Carabetta, R., and Porter, R. P. (1969). *Proc. Int. Symp. Combust., 12th,* p. 423. Combust. Inst., Pittsburgh, Pennsylvania.

Chattock. A. P. (1899). *Phil. Mag.* **48,** 401.

Chen, D. C. C., Lawton, J., and Weinberg, F. J. (1965). *Proc. Int. Symp. Combust., 10th,* p. 743. Combust. Inst., Pittsburgh, Pennsylvania.

Cohen, I. B. (1941). "Benjamin Franklin's Experiments." Harvard Univ. Press, Cambridge, Massachusetts.

Cordeiro, A. A. (1973). Ph.D. Thesis in Fuel Science, Pennsylvania State University, University Park, Pennsylvania (in progress).

Davies, R. M. (1965). *Proc. Int. Symp. Combust., 10th,* p. 755. Combust. Inst., Pittsburgh, Pennsylvania.

Davies, R. M. (1969). *Combust. Flame* **13,** 332.

Dayal, S. K., and Pandya, T. P. (1972). *Combust. Flame* **19,** 113.

Deckers, J., and Van Tiggelen, A. (1959). *Proc. Int. Symp. Combust., 7th,* p. 254. Butterworths, London.

DeJaegere, S., Deckers, J., and Van Tiggelen, A. (1962). *Proc. Int. Symp. Combust., 8th,* p. 155. Williams & Wilkins, Baltimore, Maryland.

Desaguliers, J. T. (1739). *Phil. Trans. Roy. Soc. London* **41,** 186, 634.

Dimmock, T. H., and Kineyko, W. R. (1963). *Combust. Flame* **7,** 283.

Faraday, M. (1855). "Experimental Researches in Electricity," Vol. 3, pp. 467, 490. Taylor and Francis, London. [An unabridged edition has been publ. by Dover, New York (1965).]

Fells, I., and Harker, J. H. (1968). *Combust. Flame* **12,** 587.

Fells, I., and Harker, J. H. (1969). *Combust. Flame* **13,** 334.

Fells, I., Gawen, J. C., and Harker, J. H. (1967). *Combust. Flame* **11,** 309.

Feugier, A., and Van Tiggelen, A. (1965). *Proc. Int. Symp. Combust., 10th,* p. 621. Combust. Inst., Pittsburgh, Pennsylvania.

Feugier, A., and Van Tiggelen, A. (1967). *Combust. Flame* **11,** 234.

Fontijn, A., Miller, W. J., and Hogan, J. M. (1965). *Proc. Int. Symp. Combust., 10th,* p. 545. Combust. Inst., Pittsburgh, Pennsylvania.

Fowler, R. G., and Corrigan, S. J. B. (1966). *Phys. Fluids* **9,** 2073.

Fox, J. S. (1965). *Combust. Flame* **9,** 422.

Fox, M. D., and Weinberg, F. J. (1971). *Proc. Int. Symp. Combust., 13th,* p. 641. Combust. Inst., Pittsburgh, Pennsylvania.

Fristrom, R. M., and Westenberg, A. A. (1965). "Flame Structure," p. 219. McGraw-Hill, New York.

Garner, W. E., and Saunders, S. W. (1926). *Trans. Faraday Soc.* **22,** 281, 324.

Gawen, J. C., Harker, J. H., and Fells, I. (1966). *Nature (London)* **210,** 1149.

Gaydon, A. G., and Wolfhard, H. G. (1970). "Flames," 3rd ed., p. 189, and Chapter 13. Chapman & Hall, London.

Gilbert, W. (1600). "De Magnete." Short, London. [Translated into English by P. F. Mottelay, "On the Loadstone and Magnetic Bodies." Wiley, New York, 1893; and reprinted by Dover, New York, 1958 and by Encyclopedia Britannica as part of the series "Great Books of the Western World" (R. M. Hutchins, ed. in chief), Benton, Chicago, Illinois, 1952. A translation which captured the flavor of the original Latin and which contains notes was prepared by S. P. Thompson, "On the Magnet," and published by Chiswick, London, 1900, and reprinted as a part of "The Collector's Series in Science" (D. J. de S. Price, ed.). Basic Books, New York, 1958.]

Green, J. A., and Sugden, T. M. (1963). *Proc. Int. Symp. Combust., 9th,* p. 607. Academic Press, New York.

Grove, W. R. (1854). *Phil. Mag.* **7,** 47. [See also "Encyclopedia Britannica" (D. O. Kellogg, ed.), 9th ed. ,Vol. 9, p. 285. Werner, New York, 1903.]

Guénault, E. M., and Wheeler, R. V. (1931). *J. Chem. Soc.* **1931,** 195.

Guénault, E. M., and Wheeler, R. V. (1932). *J. Chem. Soc.* **1932,** 2788.

Guest, P. G. (1930). Ignition of Natural Gas–Air Mixtures by Heated Surfaces. Tech. Paper 475, U. S. Bur. of Mines.

Gugan, K., Lawton, J., and Weinberg, F. J. (1965). *Proc. Int. Symp. Combust., 10th,* p. 709. Combust. Inst., Pittsburgh, Pennsylvania.

Harker, J. H. (1969). *Combust. Flame* **13,** 661.

Harker, J. H., and Allen, D. (1969). *Chem. Eng.* (*London*) No. 231, p. 160.

Harker, J. H., and Fells, I. (1968). *Combust. Flame* **12,** 286.

Harker, J. H., and Porter, J. E. (1968). *J. Inst. Fuel* **41,** 264.

Harrison, A. J., and Weinberg, F. J. (1971). *Proc. Roy. Soc., Ser. A* **321,** 95.

Heinsohn, R. J., and Lay, J. E. (1964). Studies of a Flat Flame under Impressed Electric and Magnetic Fields. Paper 64-WA/ENER-2 presented at Ann. ASME meeting.

Heinsohn, R. J., Wulfhorst, D. E., and Becker, P. M. (1967). *Combust. Flame* **11,** 288.

Heinsohn, R. J., Thillard, S. V., and Becker, P. M. (1969). *Combust. Flame* **13,** 442.

Heinsohn, R. J., Wilhelm, C. F., and Becker, P. M. (1970a). *Combust. Flame* **14,** 341.

Heinsohn, R. J., Becker, P. M., Elwinger, G. F., and Margle, J. M. (1970b). Effect of Electric Fields on Nitrogen Dioxide in Flames. Paper presented at *Int. Air Pollut. Conf. Int. Un. Air Pollut. Prevention Ass., 2nd, Washington, D. C., December 1970.*

Henderson, W. R., Fite, W. L., and Brackman, R. T. (1969). *Phys. Rev.* **183,** 157.

Henly, W. (1777). *Phil. Trans. Roy. Soc. London* **67,** 85.

Hirt, T. J., Marynowski, C. W., and Karlovitz, B. (1965). Electrical Augmentation of Natural Gas Flames. Paper presented at ACS meeting, September 1965.

Howard, J. B. (1969). *Proc. Int. Symp. Combust., 12th,* p. 877. Combust. Inst., Pittsburgh, Pennsylvania.

Ingenhoufz, J. (1778). *Phil. Trans. Roy. Soc. London* **68,** 1022.

Iya, K. S. (1971). Influence of an Electric Field on the Species Concentrations of a One-Dimensional Flame. Ph.D. Thesis in Fuel Sci., Pennsylvania State Univ., University Park, Pennsylvania.

Jaggers, H. C., and von Engel, A. (1971). *Combust. Flame* **16,** 275.

Jaggers, H. C., Bowser, R. J., and Weinberg, F. J. (1972). *Combust. Flame* **19,** 135.

Jensen, D. E. (1972). *Combust. Flame* **18,** 217.

Jensen, D. E., and Miller, W. J. (1971). *Proc. Int. Symp. Combust., 13th,* p. 363. Combust. Inst., Pittsburgh, Pennsylvania.

Jensen, D. E., and Pergament, H. S. (1971). *Combust. Flame* **17,** 115.

Jones, F. L., Becker, P. M., and Heinsohn, R. J. (1972). *Combust. Flame* **19,** 351.

Karlovitz, B. (1961). Method and Apparatus for the Production of High Gas Temperatures. U.S. Patent No. 3,004,137.

Karlovitz, B. (1964). Method and Apparatus for the Drilling of Rock. U.S. Patent No. 3,122,212.

Karlovitz, B. (1966). Reducing Metal Oxides in Arc-Heated Flames from High Temperature Gas Burners. U.S. Patent No. 3,232,746. [See *Chem. Abstr.* **64,** No. 13, p. 19052 (1966).]

Kellogg, D. O., Ed. (1903). *In* "Encyclopedia Britannica," 9th ed., Vol. 9, p. 285. Werner, New York.

Kinbara, T., and Nakamura, J. (1955). *Proc. Int. Symp. Combust., 5th,* p. 285. Van Nostrand-Reinhold, Princeton, New Jersey.

Kinbara, T., and Noda, K. (1971). *Proc. Int. Symp. Combust., 13th,* p. 333. Combust. Inst., Pittsburgh, Pennsylvania.

Kinbara, T., Nakamura, J., and Ikegami, H. (1959). *Proc. Int. Symp. Combust., 7th,* p. 263. Butterworths, London.

Klein, S., Guttinger, B., and Sahni, O. (1968). *C. R. Acad. Sci. Paris, Ser. B* **267,** 605.
Klemm, R. F., and Blades, A. T. (1966). *Nature (London)* **212,** 920.
Lawton, J. (1971a). *Combust. Flame* **17,** 1.
Lawton, J. (1971b). *Combust. Flame* **17,** 7.
Lawton, J., and Mayo, P. J. (1971a). *Combust. Flame* **16,** 253.
Lawton, J., and Mayo, P. J. (1971b). *Combust. Flame* **17,** 243.
Lawton, J., and Weinberg, F. J. (1964). *Proc. Roy. Soc., Ser. A* **277,** 468.
Lawton, J., and Weinberg, F. J. (1969). "Electrical Aspects of Combustion." Oxford Univ. Press, London and New York.
Lawton, J., Payne, K. G., and Weinberg, F. J. (1962). *Nature (London)* **193,** 736.
Lawton, J., Mayo, P. J., and Weinberg, F. J. (1968). *Proc. Roy. Soc., Ser. A* **303,** 275.
Lewis, B. (1931). *J. Amer. Chem. Soc.* **53,** 1304.
Lewis, B., and Kreutz, C. D. (1933). *J. Amer. Chem. Soc.* **55,** 934.
Lewis, B., and von Elbe, G. (1961). "Combustion, Flames, and Explosions of Gases," 2nd ed., p. 558. Academic Press, New York.
Lind, S. C. (1924). *J. Phys. Chem.* **28,** 57.
Lind, S. C. (1926). *Trans. Faraday Soc.* **22,** 289.
Loeb, L. B. (1958). "Static Electrification." Springer-Verlag, Berlin.
Loeb, L. B. (1961). "Basic Processes of Gaseous Electronics." Univ. of California Press, Berkeley.
Malinowski, A. E. (1925). *Chem. Abstr.* **19,** 1630.
Mayo, P. J., Watermeier, L. A., and Weinberg, F. J. (1965). *Proc. Roy. Soc., Ser. A* **284,** 488.
Miller, W. J. (1966). Ion Sampling from Hydrocarbon Combustion Plasma. *Ann. ASTM Committee E-14 Conf. Mass Spectrom., 14th, Dallas, Texas, May 1966.*
Miller, W. J. (1967). *Proc. Int. Symp. Combust., 11th,* p. 311. Combust. Inst., Pittsburgh, Pennsylvania.
Miller, W. J. (1968). *Oxid. Combust. Rev.* **3,** 97.
Miller, W. J. (1973). *Proc. Int. Symp. Combust., 14th.* In Press.
Minkoff, G. J., and Tipper, C. F. H. (1962). "Chemistry of Combustion Reactions," p. 305. Butterworths, London.
Mitchell, J. E. (1969). *Combust. Flame* **13,** 605.
Mitchell, J. E., and Wright, F. J. (1969). *Combust. Flame* **13,** 413.
Montes, G. E., and Cushing, R. H. (1967). Development of the Northern Electrically Augmented Burner. Paper presented at Central States Sect. of the Combust. Inst., March 1967.
Nakamura, J. (1959). *Combust. Flame* **3,** 277.
Nesterko, N. A., and Tsikora, I. L. (1968). *Combust. Explos. Shock Waves (USSR)* **4,** 219. [Transl. publ. by Faraday Press, New York.]
Nsakala, N. Ya. (1973). An Experimental Study of the Effects of an Electric Field on Methane/Nitrogen/Oxygen and Methane/Argon/Oxygen Opposed-Jet Diffusion Flames. M.S. Thesis in Fuel Science, Pennsylvania State University, University Park, Pennsylvania.
Parker, T. A., and Heinsohn, R. J. (1964). *In Proc. Conf. Performance High Temp. Syst., 3rd* (G. Bahn, ed.), Vol. 1. Gordon & Breach, New York.
Payne, K. G., and Weinberg, F. J. (1959). *Proc. Roy. Soc., Ser. A* **250,** 316.
Payne, K. G., and Weinberg, F. J. (1962). *Proc. Int. Symp. Combust., 8th,* p. 207. Williams & Wilkins, Baltimore, Maryland.
Peeters, J., and Van Tiggelen, A. (1969). *Proc. Int. Symp. Combust., 12th,* p. 437. Combust. Inst., Pittsburgh, Pennsylvania.
Peeters, J., Vinckier, C., and Van Tiggelen, A. (1969). *Oxid. Combust. Rev.* **4,** 93.

Peeters, J., Lambert, J. F., Hertoghe, P., and Van Tiggelen, A. (1971). *Proc. Int. Symp. Combust., 13th*, p. 321. Combust. Inst., Pittsburgh, Pennsylvania.

Pejack, E. R., and Jones, C. D. (1968). *Combust. Flame* **12,** 509.

Pejack, E. R., and Velkoff, H. R. (1967). Effects of a Transverse Electric Field on the Characteristics and Heat Release Transfer of a Diffusion Flame. Tech. Rep. No. 8, US Army Res. Office, Durham, North Carolina, Contract No. DA-31-124-ARO-D-246, November 1967.

Pergament, H. S., and Calcote, H. F. (1967). *Proc. Int. Symp. Combust., 11th*, p. 597. Combust. Inst., Pittsburgh, Pennsylvania.

Place, E. R., and Weinberg, F. J. (1965). *Proc. Roy. Soc., Ser. A* **289,** 192.

Place, E. R., and Weinberg, F. J. (1967). *Proc. Int. Symp. Combust., 11th*, p. 245. Combust. Inst., Pittsburgh, Pennsylvania.

Popov, V. A., and Sheklein, A. V. (1965). *Combust. Explos. Shock Waves* (*USSR*) **1,** No. 1, p. 58. [Transl. publ. by Faraday Press, New York.]

Porter, R. P. (1970). *Combust. Flame* **14,** 275.

Porter, R. P., Clark, A. H., Kaskan, W. E., and Browne, W. G. (1967). *Proc. Int. Symp. Combust., 11th*, p. 907. Combust. Inst., Pittsburgh, Pennsylvania.

Priestley, J. (1775). "The History and Present State of Electricity," 3rd ed. Bathurst and Lowndes, London.

Putnam, A. A., and Smith, L. R. (1953). *Proc. Int. Symp. Combust.* (*Combust. Detonation Waves*) [*Pap.*], *4th*, p. 708. Williams & Wilkins, Baltimore, Maryland.

Rezy, B. J., and Heinsohn, R. J. (1966). *Trans. ASME, Ser. A* **88,** 157.

Richardson, D. L., and Karlovitz, B. (1961). A Burner with an Electrical Discharge Superimposed on the Combustion Flame. Paper No. 61–WA–251, Ann. meeting ASME.

Russell, G. A. (1970). *J. Acoust. Soc. Amer.* **47,** No. 6 (Part 1), 1482.

Salamandra, G. D. (1969). *Combust. Explos. Shock Waves* (*USSR*) **5,** 129. [Transl. publ. by Faraday Press, New York.]

Schofield, K., and Sugden, T. M. (1965). *Proc. Int. Symp. Combust., 10th*, p. 589. Combust. Inst., Pittsburgh, Pennsylvania.

Silla, H., and Dougherty, T. J. (1972). *Combust. Flame* **18,** 65.

Southgate, G. T. (1924). *Chem. Met. Eng.* **31,** 16.

Spokes, G. N., and Evans, B. E. (1965). *Proc. Int. Symp. Combust., 10th*, p. 639. Combust. Inst., Pittsburgh, Pennsylvania.

Sugden, T. M. (1965). *Proc. Int. Symp. Combust., 10th*, p. 539. Combust. Inst., Pittsburgh, Pennsylvania.

Thomson, J. J., and Thomson, G. P. (1928). "Conduction of Electricity through Gases," 3rd ed., Vol. 1. Cambridge Univ. Press, London and New York [Reprinted by Dover, New York, 1969.]

Thornton, W. M. (1930). *Phil. Mag.* **9,** 260.

Travers, B. E. L., and Williams, H. (1965). *Proc. Int. Symp. Combust., 10th*, p. 657. Combust. Inst., Pittsburgh, Pennsylvania.

Tsuji, H., and Hirano, T. (1970). *Combust. Flame* **15,** 47.

Tufts, F. L. (1906). *Phys. Rev.* **22,** 193.

Uhlherr, M. B., and Walsh, B. W. (1971). *Combust. Flame* **17,** 45.

Van Tiggelen, A., Peeters, J., and Vinckier, C. (1971). *Proc. Int. Symp. Combust., 13th*, p. 311. Combust. Inst., Pittsburgh, Pennsylvania.

Velkoff, H. R. (1962). Electric Fluid Mechanics: Investigation of the Effect of Electrostatic Fields on Heat Transfer and Boundary Layers. Tech. Doc. Rep. No. ASD-TDR-62-650. U.S. Dept. of Commerce Office of Tech. Services.

Velkoff, H. R. (1963). An Exploratory Investigation of the Effects of Ionization on the Flow and Heat Transfer with a Dense Gas. Tech. Doc. Rep. No. ASD-TDR-63-842. U.S. Dept. of Commerce Office of Tech. Services.
Volta, A. (1782). *Phil. Trans. Roy. Soc. London* **72,** 237.
von Engel, A. (1965). "Ionized Gases," 2nd ed. Oxford Univ. Press, London and New York.
von Engel, A., and Cozens, J. R. (1964a). *Nature (London)* **202,** 480.
von Engel, A., and Cozens, J. R. (1964b). *Advan. Electron. Electron Phys.* **20,** 99.
Watson, W. (1744). *Phil. Trans. Roy. Soc. London* **43,** 481.
Watson, W. (1747). *Phil. Trans. Roy. Soc. London* **44,** 41, 695.
Watson, W. (1748). *Phil. Trans. Roy. Soc. London* **45,** 93.
Weinberg, F. J. (1966). *Combust. Flame* **10,** 267.
Weinberg, F. J. (1968). *Proc. Roy. Soc., Ser. A* **307,** 195.
Weinberg, F. J. (1972). Private communication.
Wenaas, E. P., and McChesney, J. (1970). *Combust. Flame* **15,** 85.
Wendt, G. L., and Grimm, F. V. (1924). *Ind. Eng. Chem.* **16,** 890.
Wilson, H. A. (1912). "The Electrical Properties of Flames and Incandescent Solids." London Univ. Press, London.
Wilson, H. A. (1931). *Rev. Mod. Phys.* **2,** 156.
Witt, A. E. (1965). A Study of a City Gas Diffusion Flame, A. Interactions with an Electric Field; B. Properties of the Contained Soot. M.S. Thesis, Chem. Eng. Dept., Massachusetts Inst. of Technol., Cambridge, Massachusetts.
Wortberg, G. (1965). *Proc. Int. Symp. Combust., 10th,* p. 651. Combust. Inst., Pittsburgh, Pennsylvania.
Wright, F. J., and Levine, D. G. (1967). Apparatus for the Application of Insulated A. C. Fields to Flames. U.S. Patent No. 3,306,338.
Wulfhorst, D. E. (1966). The Influence of an Electric Field on the Over-all Kinetics of an Opposed-Jet Diffusion Flame. M.S. Thesis in Mech. Eng., The Pennsylvania State Univ., University Park, Pennsylvania.

Magnetohydrodynamics (MHD) and Electrogasdynamics (EGD) of Combustion Systems

J. SWITHENBANK

DEPARTMENT OF CHEMICAL ENGINEERING AND FUEL TECHNOLOGY
THE UNIVERSITY OF SHEFFIELD
SHEFFIELD, ENGLAND

I. INTRODUCTION

Conventional nuclear or fossil fuel power stations use rotating machinery to transform the internal energy of a gas into mechanical energy before it is converted into electrical energy. The mechanical components limit the maximum temperature and hence the Carnot efficiency of the cycle; and both MHD and EGD generators represent attempts to eliminate these highly stressed moving parts from at least part of the system.

Electricity may be generated by inducing current to flow in a conductor by moving it through a magnetic field in accordance with Faraday's laws of induction. This approach is followed in an MHD generator in which the gas itself is a conductor and its movement (or velocity) is attained by expanding it through a nozzle. Alternatively, electricity may be generated by transporting electrical charges through a voltage gradient and depositing them at a high-voltage electrode. The charges may be carried on dust particles suspended in a gas, and in this case the gas should be electrically nonconducting. In both cases a braking force is applied to the gas, and direct conversion of energy of a flowing gas to electricity is obtained. Although both systems accomplish the same overall objective, they are so different that the combustion aspects of each are discussed separately below.

II. MAGNETOHYDRODYNAMICS OF COMBUSTION SYSTEMS

Since the products of combustion are gaseous, it is of interest to see whether it is possible to use these gases directly in the MHD generator or whether a heat exchanger must be used to transfer the energy to a more suitable gas. Before discussing the combustion itself, we must therefore consider the *requirements* of the generator duct (Rosa, 1968). A schematic diagram of the duct is shown in Fig. 1 and it can be seen that this consists essentially of a pair of electrodes arranged orthogonally to a magnetic field and the gas flow.

Since the gas conductivity is small compared to that of metals, the interaction between the gas and the magnetic field is limited, and a certain length is required to extract a given proportion of the gas energy. The wall friction loss and heat loss to the walls depend on wall area, and these losses therefore represent a larger proportion of the energy as the size is decreased. We therefore find that MHD generators must be large (e.g., several

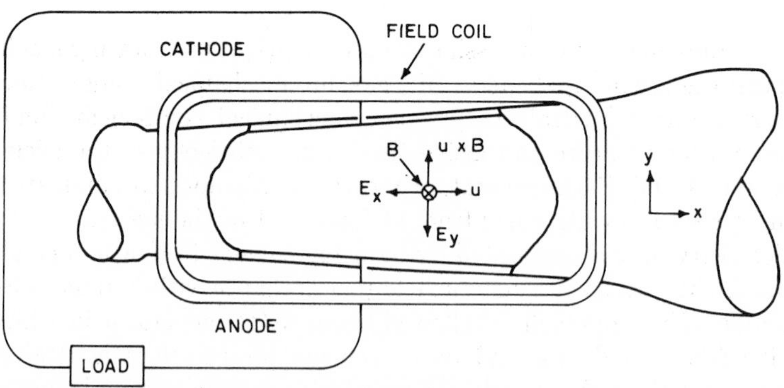

Fig. 1. Schematic diagram of a dc magnetohydrodynamic generator. [From "Magnetohydrodynamic Energy Conversion," by R. J. Rosa. Copyright 1968, McGraw-Hill, New York. Used with permission of McGraw-Hill Book Company.]

hundred megawatts) to be efficient. Because of the dependence of electrical conductivity on temperature, the higher the temperature, the smaller the generator can be and still be efficient. The other important losses are the power required to excite the magnet and the joule dissipation. The latter is recovered as heat although it represents an undesirable departure from thermodynamic reversibility. The current per unit area induced by the gas motion is

$$j = \sigma(uB - E), \tag{1}$$

where σ is the scalar conductivity and E is the electric field due to load.

If we define a loading parameter $K = E/uB$, then the power output per unit volume P is given by

$$P = jE = K(1 - K)\sigma u^2B^2. \tag{2}$$

Thus output power is proportional to gas conductivity, and is at a maximum when $K = 0.5$.

The force on the gas opposes the motion and is given by

$$F = jB = (1 - K)\sigma uB^2 \qquad \text{per unit cube,} \tag{3}$$

and the pressure difference along the duct is

$$\Delta p = FL = (1 - K)\sigma uB^2L, \tag{4}$$

where L is the generator length. The extent to which the gas flow distorts the magnetic field in any plasma flow problem is governed by the magnetic Reynolds number R_m where

$$R_m = \mu_0\sigma uL = \text{work done by gas } (FL)/\text{energy stored in field } (B^2/2\mu_0). \tag{5}$$

Since the value is less than unity, the mathematical treatment of MHD is greatly simplified because we do not need to solve both Maxwell's equations and the gas dynamic equations simultaneously.

Since the gas must be an electrical conductor, we are interested in the attainable level of conductivity.

Table I shows that the ionization potentials of gases are relatively high, and seeding with potassium or cesium is used to increase the conductivity. In common with simple chemical reactions, thermal ionization of seeded gases follows a mass action law (Saha equation)

$$n_e n_i/n_s = [(2\pi m_e kT)^{3/2}/h^3](2g_i/g_0) \exp(-eE_i/kT), \tag{6}$$

where n_e is the electron concentration, n_i is the ion concentration, n_s is the neutral seed atom concentration, h is Planck's constant, E_i is the ionization

TABLE I

IONIZATION POTENTIALS AND STATISTICAL WEIGHTS

Species	Ionization potential E_i (eV)	Statistical weights	
		g_o	g_i
Li	5.39	2	1
Na	5.14	2	1
K	4.34	2	1
Cs	3.89	2	1
He	24.58	1	2
Ne	21.56	1	6
A	15.76	1	6
H_2	15.6	—	—
O_2	12.05	3	4
O	13.61	9	4
N_2	15.6	1	2
NO	9.26	8	1
CO	14.1	1	2
CO_2	14.4	—	—
H_2O	12.6	—	—
OH	13.8	—	—
U	6.1	—	—

potential of the seed atom, g_i is the statistical weight of the ground state of the ion, and g_0 is the statistical weight of the ground state of the neutral atom.

In simple gases the electron density calculated from this equation is reliable; however, in combustion gases the attachment of electrons to the OH radical and the reduction in effective seed concentration by the formation of alkali oxides and hydroxides complicate the situation. Reduction of conductivity by a factor of 3 is suggested in this case (Rosa, 1968).

Although the overall results are uncertain, experiments (Freck, 1964) suggests that the formation of OH is unimportant, and the rate of formation of the oxide and hydroxide may not be fast enough to keep up with the rapid expansion of the gas. Typically about 1% of seed element is required for optimum conductivity.

The electrical conductivity of the gas is given by (Lin *et al.*, 1955)

$$\sigma = (n_e e^2/m_e c_e)\{\sum_k n_k Q_k + 3.9 n_i (e^2/8\pi\varepsilon_0 kT)^2 \ln[(12\pi/n_e)(\varepsilon_0 kT/e^2)^{3/2}]\}^{-1}, \tag{7}$$

where e is electron charge, m_e is electron mass, c_e is electron mean thermal velocity, Q is momentum transfer cross section, ε_0 is permittivity of free space, and summation over k includes all particles (atoms and molecules) except ions and electrons. (N.B.: The electron and ion concentrations in the bulk of the gas will generally be equal.)

At low temperatures, collisions with neutral particles dominate the process and σ varies exponentially with temperature. At high temperatures (>3000°K) the ion collision effects predominate and the conductivity varies approximately as $T^{3/2}$ [see Rosa (1968)].

The conductivity of the combustion products of JP-4-O_2 as a function of temperature is illustrated in Fig. 2. If we insert $u = 10^3$ m/sec, $B = 1$ Wb/m², and $K = 0.5$ in the equation for power density, we obtain $P = \sigma/4$ MW/m³ (i.e., power density is proportional to conductivity).

Referring now to Fig. 2, we see that a temperature between 2500 and 3000°K is desirable for a good power density, and the power density doubles for every 100°C increase in temperature.

Since the local temperature (and σ) decreases as the velocity is increased,

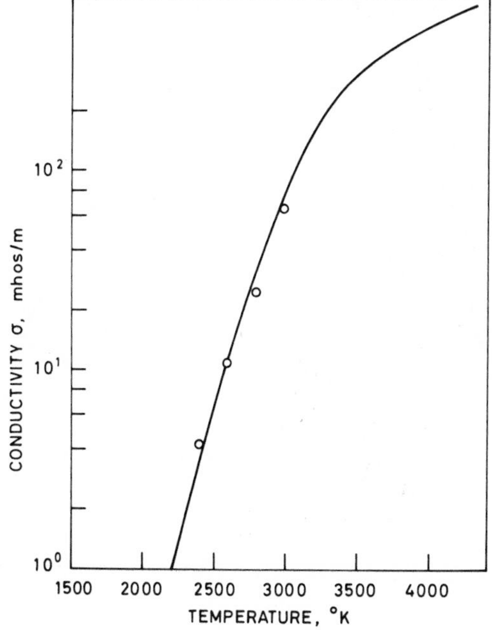

Fig. 2. Conductivity of JP-4-O_2 + 1% K combustion products. $P = 1$ atm. [From Brogan (1963).]

the optimum duct velocity may be obtained from the energy equation

$$\rho u\, \partial(\tfrac{1}{2}u^2 + h)/\partial x = \mathbf{j}\cdot\mathbf{E}. \tag{8}$$

Therefore

$$\partial(\tfrac{1}{2}u^2 + h)/\partial x = K(1 - K)\sigma u B^2/\rho. \tag{9}$$

For given K and B, $\sigma u/\rho$ may be plotted against duct Mach number as shown in Fig. 3. In the case of combustion products the optimum velocity lies in the low supersonic regime; however, there is a flat maximum and other factors can modify the Mach number chosen.

A further constraint imposed on the system is the effect of duct length-to-diameter ratio on the heat transfer and friction losses in the duct.

For a square duct of width D, the total heat loss is

$$Q = 4LDN_{st}\rho u C_p(T_g - T_w), \tag{10}$$

where N_{st} is the Stanton number. The enthalpy flux entering the duct is $P_t = D^2\rho u C_p T_g$; therefore the fraction of heat lost to the walls is

$$Q/P_t = 4N_{st}[1 - (T_w/T_g)](L/D). \tag{11}$$

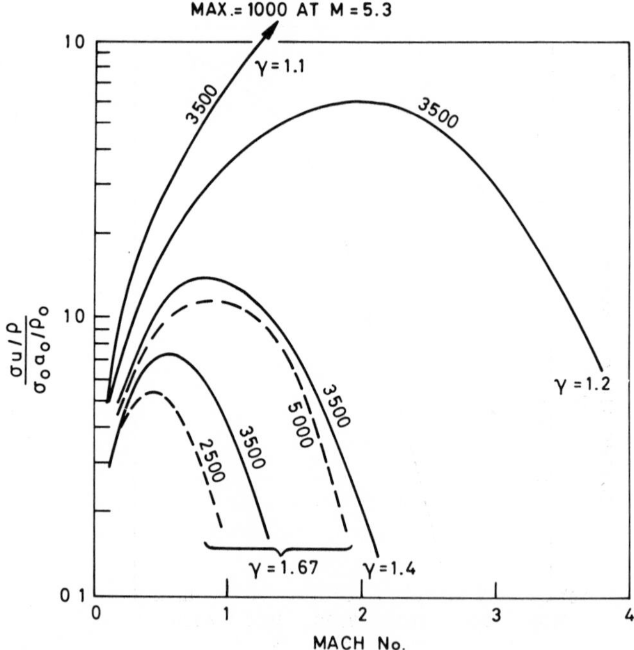

Fig. 3. Variation of $\sigma u/\rho$ versus Mach number. [From "Magnetohydrodynamic Energy Conversion," by R. J. Rosa. Copyright 1968, McGraw-Hill, New York. Used with permission of McGraw-Hill Book Company.]

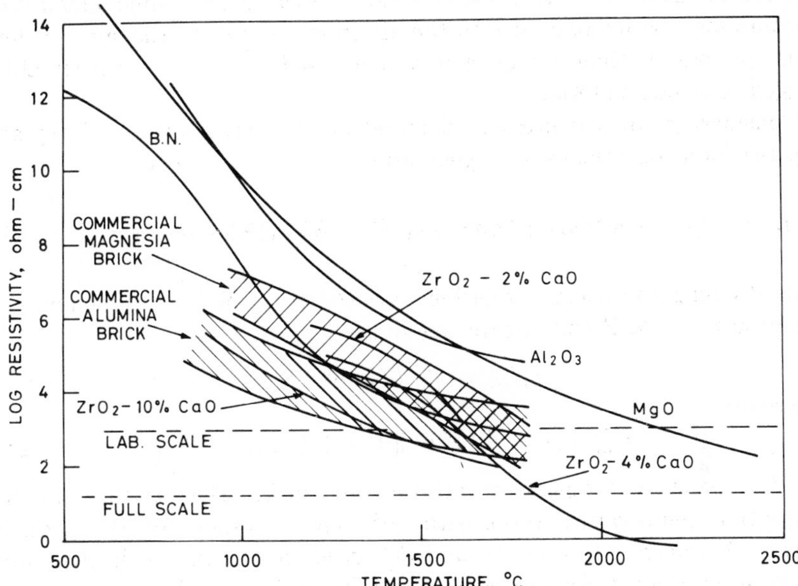

Fig. 4. Resistivity of ceramics. [From "Magnetohydrodynamic Energy Conversion," by R. J. Rosa. Copyright 1968, McGraw-Hill, New York. Used with permission of McGraw-Hill Book Company.]

Thus for $N_{st} \approx 0.0025$, $T_w/T_g \approx 0.5$, and heat loss to walls less than 10%, the L/D ratio must not exceed 20.

Similarly for the friction loss,

$$L/D \leq (2\gamma M^2 C_f)^{-1}(\Delta p_f/p), \tag{12}$$

where C_f is the friction coefficient for the wall. If $C_f \approx 0.003$, $M \approx 1$, and $\Delta p_f/p < 10\%$, the L/D ratio must not exceed 10!

Turning now to the duct wall material, without discussing the details of the optimum electrode configuration, it is apparent that the walls must contain both conducting and insulating regions. The permissible level of resistivity of the insulator depends on the scale of the apparatus. Figure 4 shows that while it is not too difficult to obtain a satisfactory leakage with a large generator, it is much more difficult to do so with a small generator, especially when the insulator is contaminated with seed material and slag.

Passage of electric current between the electrodes and the gas will take place by the formation of arc spots unless the electrode is sufficiently hot to provide thermionic emission. A thin coating of a material such as stabilized zirconia (which has a low work function) will achieve this in the presence of combustion gases; however, since it tends to wear away, a small amount of zirconia may be introduced into the airstream to restore the surface. For short periods of time carbon can also be used as a cheap, consumable, rugged electrode material.

Considering now the pressure required for the MHD duct, we integrate Eq. (4) for a constant velocity generator:

$$L = \int [(1 - K)\sigma u B^2]^{-1}\, dp = [p_1/(1 - K)\sigma u B^2][1 - (p_2/p_1)]. \quad (13)$$

Now dividing the energy equation $\rho u(dh/dx) = \mathbf{j}\cdot\mathbf{E}$ by the momentum equation $p = \mathbf{j} \times \mathbf{B}$ and integrating gives

$$p_1/p_2 = (T_1/T_2)^{\gamma/(\gamma-1)K}.$$

Therefore

$$L = [p_1/(1 - K)\sigma u B^2][1 - (T_2/T_1)^{\gamma/(\gamma-1)K}]. \quad (14)$$

(N.B.: The static temperature ratio above can be taken as equal to the stagnation temperature ratio with sufficient accuracy for the present purpose.) Referring to Eqs. (6) and (7), it can be shown that in the lower temperature range the electrical conductivity of the gas is inversely proportional to the square root of the gas pressure. (This may be more complicated in the case of combustion gases as explained previously.) Equation (14) may therefore be written as

$$L = C_1(p^{3/2}/B^2), \quad (15)$$

where the constant C_1 contains the various factors. Now the total power output depends on pressure and duct diameter:

$$P = C_2 p D^2 = C_3 p L^2 \qquad \text{for} \quad L/D = \text{constant}. \quad (16)$$

From Eqs. (15) and (16),

$$p = C_4 B P^{1/4}.$$

The maximum pressure thus depends slightly on the power rating of the generator, all other things being equal. For a typical system ($T_0 = 2500°$K, $B = 3$ Wb/m^2), the maximum permissible operating pressure would vary from 1.5 atm at 1 MW rating to 8 atm at 1000 MW rating. The operating pressure also depends on the temperature, and a typical envelope of allowed operating conditions is illustrated in Fig. 5.

Summing up the requirements of the MHD duct we see that the combustor must provide

(1) combustion products at temperatures of 2500°K or more,
(2) products at pressures in the vicinity of 10 atm,

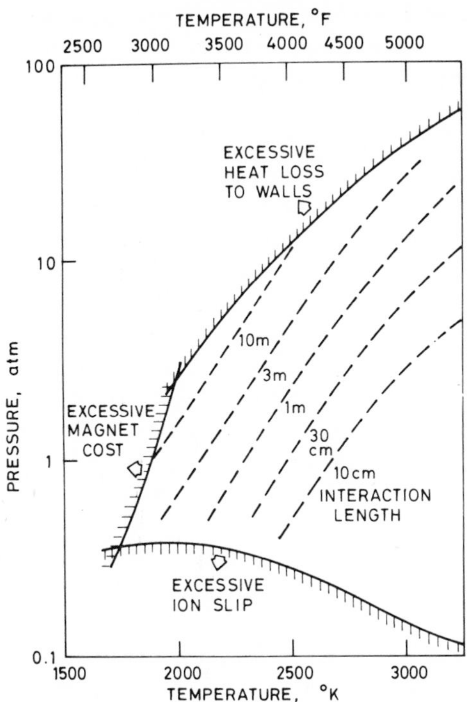

Fig. 5. Typical envelope of allowed operating conditions and the variation of interaction length. He + 1% Cs; 500 MW; B = 60 kg. [From "Magnetohydrodynamic Energy Conversion," by R. J. Rosa. Copyright 1968, McGraw-Hill, New York. Used with permission of McGraw-Hill Book Company.]

(3) a thermal output approaching 1000 MW,

(4) products seeded with about 1% of potassium (cesium is precluded on the grounds of expense),

(5) an output velocity in the low supersonic regime.

It is immediately apparent that these requirements can be readily attained with a conventional rocket motor, and indeed many duct test facilities have used such a motor. The rocket produces a high exhaust stagnation temperature since combustion of the fuel takes place with oxygen rather than with air. In the case of natural gas, the flame temperature with pure oxygen is 3200°K, whereas with air at room temperature it is only 2100°K. The disadvantage of such an MHD system is of course the cost of the oxygen, and alternative solutions must be sought. In order to attain a high enough combustion temperature with air, the highest possible air preheat is required, and should certainly be well in excess of 1000°K. For long life, this rules out conventional metallic heat exchangers and refractory regenerators appear to be most suitable. Both stationary and moving bed types have been used, and new systems using fluidic valves have also been conceived (Swithenbank and Taylor, 1968). Although the heat exchanger can be fired separately, this involves a severe loss of cycle efficiency and direct firing must be preferred. Unfortunately the presence of seed material in the MHD generator exhaust contaminates the heat exchanger, and to avoid condensation of the seed, the gas exit temperature must be maintained above 1500°K. Tests with different refractories have shown that high-purity magnesia is best for this application. Air preheat temperatures up to 1670°K have been attained with such a system.

The combustion takes place at such a high pressure and temperature that a very high combustion intensity is obtained. Although regenerative cooling of the combustion chamber could be used in some circumstances, an insulated wall is generally preferable, and commercial grades of zirconium oxide refractory have proved satisfactory in 1000-hr tests. This lining may be used at surface temperatures up to 2600°K with natural gas fuel. Satisfactory operation of combustors using both coal and oil have also been reported.

The seed material in the form of a salt of potassium may be injected as a powder, aqueous solution, or organic solution (e.g., potassium naphthanate dissolved in kerosene). Very strong aqueous solutions of K_2CO_3 are generally the more convenient form for injection and the detrimental effect of the water on conductivity is very small. In the combustor, generator duct, and heat exchanger system, the potassium combines with the sulfur and vanadium present in residual oil fuel to form sulfates, vanadates, and complexes thereof, which are highly corrosive in the liquid state. The binary mixture K_2SO_4–V_2O_5 has a very low melting point of 705°K. The seed is usually collected from the cooled (430°K) stack gases as K_2SO_4 with either a bag filter or electrostatic precipitator. The particle sizes are between 1 and 6 μ and therefore cyclone separators are not very efficient.

As mentioned earlier, in some cases it is desirable to introduce powdered zirconia into the combustor exit in addition to the seed so that a saturated zirconia vapor is produced. By careful control of the generator electrode surface temperature, losses of zirconia can be made up by condensation of this vapor.

Shock tubes or shock tunnels provide a convenient method of studying MHD interactions. At the University of Sheffield the open circuit voltage in an MHD duct is being used to determine the instantaneous velocity of flow in the region of Mach 4.

While successful operation of MHD systems has been obtained with clean fuels, there is currently a need to demonstrate satisfactory performance in the presence of slag and the corrosive chemistry associated with the impurities contained in coal. The alternative of producing a clean fuel from coal has not yet been evaluated, but could also be advantageous from the pollution aspect. Further regarding pollution, the production of large quantities of oxides of nitrogen by nonequilibrium cooling of high-temperature gas can be anticipated. This fixation of atmospheric nitrogen may represent a significant by-product if it can be removed economically from the exhaust.

One novel problem concerning the region between the combustor and MHD duct is worthy of mention. This section consists of a convergent–divergent Laval nozzle to accelerate the flow to supersonic velocities. When

TABLE II

PERFORMANCE OF PROJECTED OPEN-CYCLE MHD POWER PLANTS[a]

Date of initial operation	Pilot plant (U-25) 1969	First generation plant 1980	Advanced plant 1995?
Electrical output (MW),			
Total	80	1000	1500
MHD	25	500	1200
Steam cycle	55	500	300
Oxidizer conditions and preheat temperature	With oxygen enrichment to 1200°C	(a) With oxygen enrichment and metallic heat exchanger to 850°C or (b) With air preheat only and ceramic heat exchanger to 1100°C	Air preheat only to 1800°C
Fuel	Natural gas	Natural gas or coal	Coal
Magnetic field (tesla)	2	4	6
Net efficiency of overall plant (%)	33	50	58–60
Fuel saving over conventional plant by weight (%)	—	20	30

[a] From "Magnetohydrodynamic Energy Conversion," by R. J. Rosa. Copyright 1968, McGraw-Hill, New York. Used with permission of McGraw-Hill Book Company.

potassium seed is present, K-line radiation constitutes about 20% of the local energy loss in this region.

The current and predicted state of the art on MHD generators is outlined in Table II (Rosa, 1968). Although the road is long and slow, steady progress is being made toward cheaper electricity costs.

III. ELECTROGASDYNAMICS OF COMBUSTION SYSTEMS

The EGD generator is basically a Van de Graaf electrostatic generator in which the moving belt is replaced by a gas stream containing charged particles. The EGD generator therefore differs from the MHD generator in that the former produces high-voltage, low-current power, whereas the latter produces low-voltage, high-current power. As shown above, the *net* charge in the gas in an MHD generator is zero; that is, there are an equal number of positively and negatively charged species. However, in the EGD system the temperatures are low ($<1700°K$) and the pressure is high (>30 atm). Therefore the gas conductivity is low, and the gas may be seeded with a low concentration of unipolar charged species. Unfortunately the mere presence of the charges results in an electric field which can easily exceed the breakdown strength of the gas. It follows that a given size of chamber can only contain a certain number of charged particles at a given pressure, and the particle density can be made larger for smaller chambers. In the case of an EGD channel, the power density increases with the particle density, and in order to obtain a reasonable power density the diameter of the channel should be only a few millimeters! There is still a radial electrostatic field due to the particles, and this causes charged particles flowing down the tube to drift to the wall.

If the particles are simply ions, the drift speed in the axial and radial fields would be about 300 m/sec. Thus very high velocities would be required to carry them through the field. Using dust particles of 1 to 20 μ the drift velocity reduces to about 3 m/sec, and a gas flow of about 100 m/sec is sufficient for neglect of the axial drift velocity. The radial drift will limit the L/D ratio of the channel to $100/(2 \times 3) = 17$, and only a small fraction of the total power can be removed in such a stage. A multiple-stage system is therefore envisaged in practical generators. Informative discussions of EGD power generation have been presented in reports by Rentell (1968), Gourdine Systems, Inc. (1968), and Dynateck Corporation (1968).

Since coal-fired combustion systems contain particles in the form of fly ash, it has been thought that they may be suitable for EGD generators in spite of the difficulty of attaining high-power densities.

The required number of particles per unit volume may be evaluated as follows: The body force per unit volume is

$$f = en_iE,$$

where e is the electronic charge (1.6×10^{-19} C), E is the electric field, and n_i is the charge density (per cubic meter). The pressure drop in length L is then

$$\Delta p = fL.$$

To attain a reasonable power density we require a pressure gradient of about 1 atm/m (i.e., $\Delta p/L = 10^5$ N/m³). The maximum attainable electric field in the generator depends on the pressure; however, at STP, $E_{(\text{breakdown})} = 3 \times 10^6$ V/m; therefore

$$n_i = (\Delta p/L)(eE)^{-1} = 10^5/(3 \times 10^6 \times 1.6 \times 10^{-19}) = 2 \times 10^{17} \text{ ions/m}^3.$$

We now see the reason for the very high pressure; the Paschen curve shows that the breakdown strength of most gases varies linearly with density,

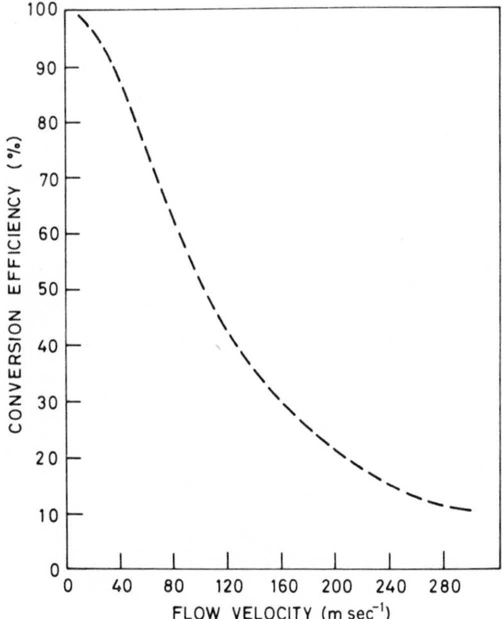

Fig. 6. Velocity dependence of conversion efficiency. $D = 10^{-2}$ m; $\sigma = 10^{-5}$ m; M = 0.1. [From Cowen *et al.* (1969).]

that is,

$$E_{\mathrm{b},\rho} = E_{\mathrm{b},\rho(0)}\rho/\rho(0).$$

If we define the stage efficiency η as the ratio of useful electrical energy produced L_{e} to the reduction in energy of the working fluid, then it can be shown that

$$\eta = L_{\mathrm{e}}/(L_{\mathrm{e}} + L_{\mathrm{d}}) = \{1 + [\xi\rho(V/E_{\mathrm{b},\rho(0)})^2][\rho(0)/\rho]\}^{-1},$$

where L_{d} is the energy loss due to drag and ξ is the overall drag loss coefficient. This equation shows that the conversion efficiency also depends strongly on velocity and Fig. 6 shows the dependence for a 10% by weight loading of 10-μ-diameter particles passing down a 1-cm duct (Cowen *et al.*, 1969).

The theoretical dependence of stage conversion efficiency on solids loading is shown in Fig. 7, and that on particle size is shown in Fig. 8 (Cowen *et al.*, 1969). Since about ten stages are necessary to remove a reasonable proportion of the overall energy, it can be seen that overall conversion efficiencies of about 50% could be attainable.

The practical results obtained to date have not been very promising. By far the major problem has been the excessive deposition of particles and

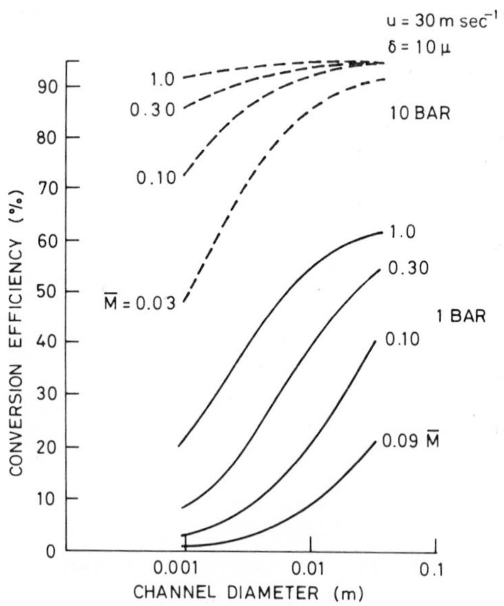

Fig. 7. Solids-loading dependence of conversion efficiency. [From Cowen *et al.* (1969).]

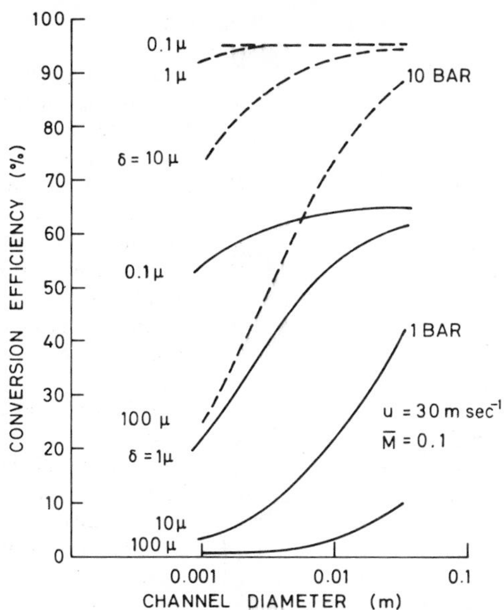

Fig. 8. Particle-size dependence of conversion efficiency. [From Cowen *et al.* (1969).]

hence low efficiency. The deposition is due to radial drift of the particles described above, and the generator behaves in many respects like an electrostatic precipitator. Recent work by Professor Quack (1968) on electrostatic precipitators has shown that particle deposition does not depend simply on the drift velocity, but is strongly dependent on the ionic wind effects from the electrodes. It therefore seems that extensive and detailed laboratory-scale investigation of this problem is required before any breakthrough in EGD generation is obtained.

Turning now to the combustion, we find that the associated problems are small compared to the generator duct problem. The normal density of the boiler ash varies from 0.7 to 30 gm/m^3, which represents a solids loading of about 1%. It follows that a large proportion of the ash must be recycled to maintain a 10% solids loading unless a very low grade of coal is used.

At the comparatively low exit temperatures and high pressure proposed, the combustor requirement is similar to that of a gas turbine combustor, and a similar technology would be used (i.e., a perforated metal combustion can would be fitted in an outer metal case). Whereas an MHD generator is used between the combustor and a boiler, the EGD generator could conveniently be fitted between the combustor and a gas turbine.

Until the EGD duct problems have been solved, no further work on the associated combustor seems warranted.

List of Symbols*

b	As subscript, breakdown
B	Magnetic field
C	Constant
C_p	Specific heat
c	Mean thermal velocity
D	Duct width
E	Electric field or potential
e	As subscript, electron
e	Electron charge
F	Force
f	Force per unit volume
g	As subscript, gas
g	Statistical weight
h	Planck's constant or enthalpy
i	As subscript, ion or ionization
j	Current per unit area
K	Loading parameter
k	Boltzmann constant
L	Length
L_e	Electrical energy produced
L_d	Energy loss due to drag
m	Mass
N_{st}	Stanton number
n	As subscript, neutral
n	Concentration
P_t	Enthalpy flux
P	Power output per unit volume or power output
p	Pressure
Q	Momentum transfer cross section or heat loss
R_m	Magnetic Reynolds number
T	Temperature
u	Velocity
w	As subscript, wall
x	Distance
γ	Ratio of specific heats
ϵ_0	Permittivity of free space
μ_0	Permeability of space
ξ	Drag loss coefficient
ρ	Density
σ	Scalar conductivity
0	Subscript, ground state
1	Subscript, duct entry
2	Subscript, duct exit

* SI units are used throughout.

References

Brogan, T. R. (1963). Electrical Properties of Seeded Combustion Gases. *Prog. Astronaut. Aeronaut.* **12,** 319–345.

Cowen, P. L., Gunzler, T., Kulka, R., and Gourdine, M. C. (1969). "Electrogasdynamic Power Generation. Final Rep., Gourdine Systems Inc., Livingston, New Jersey.

Dynateck Corp. (1968). Separate Evaluations of Electrogasdynamic (EGD) Power Generation. I.I.T. Res. Inst., Iowa State Univ., U.S. Dept. of Interior OCR Res. and Develop. Rep. No. 35.

Freck, D. V. (1964). On the Electrical Conductivity of Seeded Air Combustion Products. *Brit. J. Appl. Phys.* **15,** 301.

Gourdine Systems, Inc. (1968). EGD Generator Research May 1966 to September 1968. U.S. Dept. of Interior OCR Res. and Develop. Rep. No. 48.

Jackson, W. D. (ed.) (1969). MHD Electrical Power Generation. The 1969 Status Rep. Eur. Nucl. Energy Agency Organization for Econ. Co-operation and Develop.

Lin, S. C., Resler, E. L., and Kantrowitz, A. J. (1955). *J. Appl. Phys.* **26,** 95.

Quack, R. (1968). Private communication, Technical University, Stuttgart, Germany.

Rentell, T. D. (1968). Electrogasdynamics—A Survey. Brit. Coal Utilization Res. Ass. (Unpubl. Rep.).

Rosa, R. J. (1968). "Magnetohydrodynamic Energy Conversion." McGraw-Hill, New York.

Swithenbank, J., and Taylor, D. S. (1968). Improvements in or Relating to Fluid Flow Control Arrangements. U.K. Patent application No. 26813/68.

XI

Combustion Aspects of MHD Power Generation

STEWART WAY

WESTINGHOUSE RESEARCH LABORATORIES
PITTSBURGH, PENNSYLVANIA

This chapter discusses the various approaches to carrying out the combustion process to meet the high-temperature requirements of MHD generators. Subjects treated include laboratory-scale combustion chambers, combustors for liquid or gaseous fuels for pilot or prototype MHD plants, and coal-fired combustion chambers such as those that might be used in MHD central stations. Vortex-type chambers are given primary attention, but other types that have exhibited good performance are also described. For coal combustion the two-stage cyclone burner is of particular interest in affording a means of meeting the high-temperature requirements with minimal ash carryover. Excess air is used in the first stage and ash-free fuel, such as gas from a coal pyrolysis unit, is fed to the second stage. Developments in England and the U.S.S.R. are briefly described.

I. INTRODUCTION

In the magnetohydrodynamic process of energy conversion an electrically conducting fluid moving in a magnetic field is used in place of the solid metallic conductors of the usual turbine-driven generator. Either a liquid metal or a high-temperature gas may be considered as the working medium. We are concerned here with the use of the gaseous fluid; in this case, higher thermal efficiencies may be realized owing to the higher temperatures employed. To render the gaseous medium electrically conducting a slight addition of easily ionizable material such as potassium or cesium is used; if this were not done, the required temperatures would be too high. By the use of about 0.3 to 1.0% cesium or potassium adequate conductivities can be realized in the 2300–2800°K range.

One can use either a closed cycle system, with a gas such as helium or argon as working fluid and with heating in a gas-cooled nuclear reactor, or an open cycle arrangement with the gaseous products of combustion of a fossil fuel. The open cycle presents fewer problems and is probably better suited for early realization. We are concerned in this discussion with the problems of preparing the hot gas stream by carrying out a combustion process at high temperature.

The reasons for the interest in MHD power generation are that the method holds promise of quite high efficiencies, that thermal rejection to the environment is reduced, that our large coal resources can be effectively used for power generation, and that several of the more pressing air pollution problems are reduced.

The gaseous MHD method may be viewed not only as an efficient power plant system but also as an entirely new approach to the generation of electrical energy. The working fluid of the thermal cycle becomes identical with the working medium of the electrical generator. Enthalpy drop in the gas stream reappears as electrical work. The boiler, the steam turbine, and the generator of the conventional plant are replaced by one machine, the MHD generator.

As has been mentioned the MHD generator operates at high temperature; the leaving gases therefore possess considerable thermal energy, and this energy would be further utilized in a gas turbine or steam bottom plant (Way, 1971a,b). The bottom plant used may in itself be a combined cycle arrangement of high efficiency, such as a gas turbine–steam turbine combination. In this way it will be seen that the MHD generator is not only viewed as a competitor to the high-efficiency plant combinations but it is also an important additional thermodynamic component, promising overall efficiencies eventually in the 60–65% range.

II. CONDITIONS OF OPERATION

The combustion process for MHD open cycle power generation is characterized chiefly by the necessity for producing high temperatures. These temperatures are in the range 2600–2800°K. Since exotic fuels such as cyanogen are generally ruled out on grounds of safety or economics, and since we are interested in conventional fuels such as coal, gas, or oil, it becomes necessary either to preheat the air strongly or to use air enriched with oxygen. Economics usually favors use of preheated air.

Some flame temperatures calculated for various fuel–oxidant mixtures are shown in Fig. 1. Owing to dissociation the flame temperature increases rather slowly, at the higher levels, with increasing temperature of air preheat. The actual carrying out of the calculations leading to the flame temperature is best done in two stages. One first calculates the equilibrium gas compositions at a set of assumed pressures and temperatures; then one makes an enthalpy balance and computes the amount of heat that must be added per kilogram mole of fuel to provide the increase in enthalpy for the change from reactants to equilibrium products. All enthalpies consist of a heat of formation part and a sensible heat part. Basic data are provided

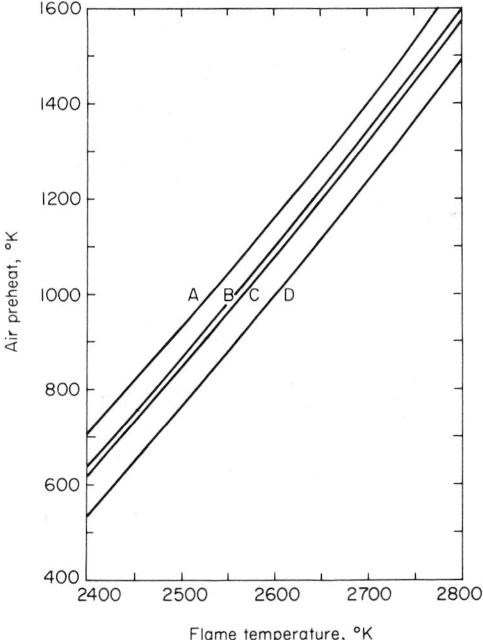

Fig. 1. Air preheat temperatures. Wet coal and moist air or char and moist air: 3% moisture in coal, 1% moisture in air, 5% heat loss. (A) Coal–air, $\phi = 1.05$, 5 atm. (B) Coal–air, $\phi = 0.95$, 5 atm. (C) Coal–air, $\phi = 0.95$, 6 atm. (D) Char–air, $\phi = 0.95$, 5 atm.

by sources such as the JANAF Tables (Stull and Prophet *et al.*, 1970). Calculation procedures are described by Penner (1957) and Way (1965), or in other standard thermochemical treatises. For the case of zero required heat addition we have the adiabatic flame temperature.

The calculation of equilibrium gas compositions is necessary not only as a step toward ascertaining the temperature following combustion (or degree of air preheat required) but also to identify the various properties that must be known to make calculations in the MHD generator. If we regard pressure and temperature as independent variables (for a given oxidant–fuel combination), the properties we are chiefly concerned with are the following: density, enthalpy, entropy, molecular weight, specific heats, electrical conductivity, and electron mobility. From enthalpy and entropy values a Mollier diagram may be constructed as shown in Fig. 2. Such diagrams are very useful in analysis of the process in the MHD generator duct.

In order to achieve the desired electrical conductivity at a reasonable temperature, one generally adds a "seeding" material such as potassium or

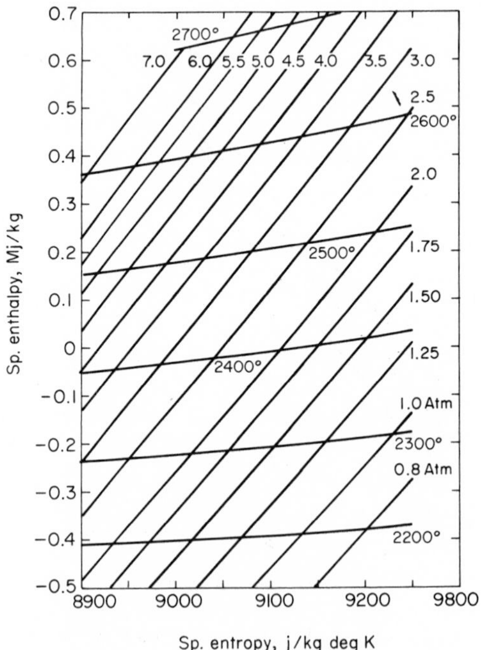

Fig. 2. Enthalpy–entropy diagram. Products of wet coal and moist air at $\phi = 0.95$. Seed 0.5% cesium by weight.

cesium. The low ionization potentials of these elements (4.34 eV for K, 3.89 eV for Cs) make it possible to achieve a degree of ionization, with corresponding release of free electrons, sufficient to give adequate conductivity at temperatures of 2200 to 2600°K. This aspect is treated by Way (1965), Frost (1961), and Rosa (1968). The flame temperatures shown in Fig. 1 are corrected to take account of the seeding material added. For the case of potassium, 0.5 to 1% by weight in the gas is generally sufficient; for the case of cesium, 0.25 to 0.5% by weight suffices. Use of additional seeding compound (beyond these proportions) is usually not advisable for several reasons: (1) cost of the seeding material, (2) the heat absorption caused by dissociation of the seeding compound (e.g., K_2SO_4), (3) the adverse effect of the alkali atoms themselves on the electron mobility. One must remember that the electrical conductivity depends on the product $n_e\mu_e e$ and although increase of seeding may lead to increased electron density n_e, it also leads to reduction of electron mobility μ_e because of the large cross section of alkali atoms. In the expression above, e is the electronic charge.

From the standpoint of securing good electrical conductivity it is desirable to use fuels of low hydrogen content. By the same token, dry air is superior to moist air and dry seeding compound is superior to seed solution. The reasons for this are that with hydrogen present there is a certain amount of formation of KOH (or CsOH) under the equilibrium conditions, and there is formation of OH. The former ties up alkali atoms that would otherwise be available for ionization, and the latter effect leads to electron attachments forming OH^- ions, thereby diminishing the number of free electrons.

Carbon, or CO, would be an ideal fuel for the MHD process. From the standpoint of MHD power generation coal is a superior fuel to gas or oil, and char or coke is superior to coal. If gasification processes are considered, a gas producer operated with air or oxygen would be preferred to one which requires steam.

In order to see more precisely the levels of temperature needed from the combustion chamber we may refer to Fig. 3. The conductivity required depends on the permissible length of the generator duct and on the cycle pressure ratio. Since MHD power systems usually operate on a Brayton cycle, we will need a pressure change of about 2.5 atm to realize an attractive cycle efficiency.* The acceptable length of the generator duct for a large power plant may be about 18 m. The conductivity appropriate under these conditions is about 6 mho/m at the inlet where the pressure is about

* This corresponds to about 5 : 1 cycle pressure ratio, since about 1.5 atm drop occurs in the nozzle, and backpressure is about 1 atm.

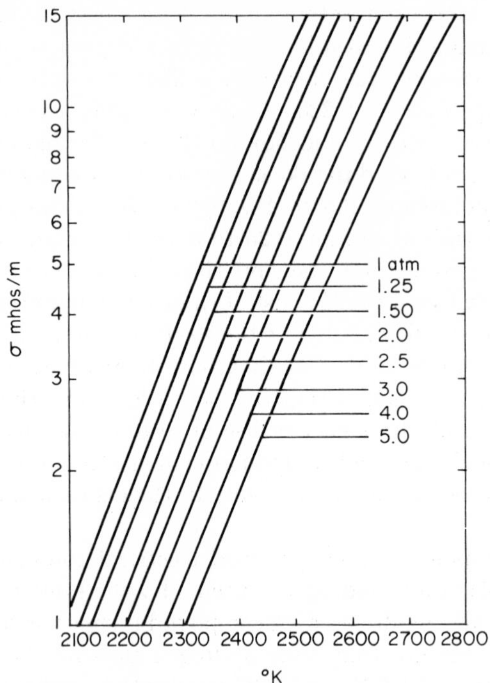

Fig. 3. Conductivity. Products of combustion of wet coal and moist air with 0.5 wt. % cesium seeding. $\phi = 0.95$.

3.5 atm. Figure 3 indicates that for coal and air products, at $\phi = 0.95$ the inlet gas temperature must be about 2545°K. To this must be added the temperature drop associated with expansion in the inlet nozzle (to give a velocity of 725 m/sec) which amounts to about 115°K. Hence we should have for the coal and air products leaving the combustion chamber a temperature of 2660°K at a pressure of about 5 atm. Preheat temperature for assumed 5% heat loss from the combustor (5% of fuel heating value) would be 1240°K, by Fig. 1.

It might be considered advantageous to operate the combustion chamber with less than stoichiometric air, say $\phi = 0.90$, where ϕ is oxygen equivalence ratio. This has a beneficial effect on electrical conductivity* as shown in Fig. 4. In the case of char, with 0.35 wt. % cesium seeding, and $\phi = 0.90$ an appropriate combustion chamber exit condition would be 2650°K, 4.5 atm, and the corresponding air preheat is 1100°K (Way, 1971b).

* Also, NO formation is greatly reduced.

The combustion system for an MHD power plant is characterized by the following requirements:

(a) exit gas temperatures 2600–2700°K, or in special cases 2800°K;
(b) ability to give a uniform outlet gas mixture with the seeding material well distributed, and uniform temperature;
(c) heat losses at a minimum;
(d) good durability of interior walls; and
(e) moderate pressure loss.

In the case of combustion systems for coal or char there are additional factors to consider. These relate to the presence of the ash in the fuel. It is a matter of the designer's choice whether to (a) seek a clean process with no ash or slag carryover, (b) seek a system giving a moderate level of slag carryover, say 5 or 10%, or (c) design for full discharge of ash constituents with the gases flowing through the MHD generator. Advantages of ash-free gases are that deposits on heat exchange surfaces are reduced, condensation

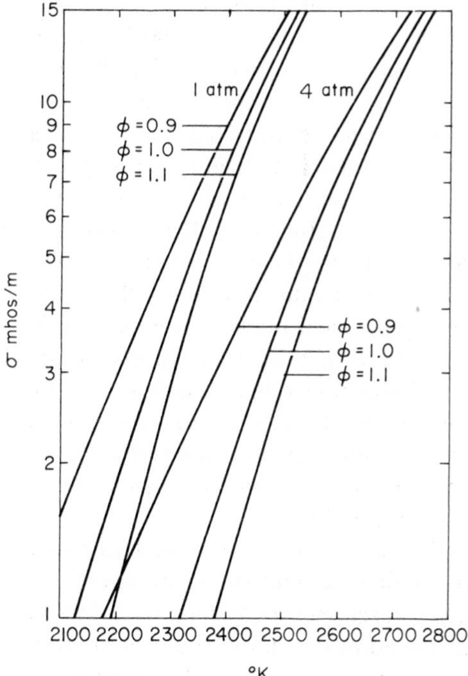

Fig. 4. Conductivity. Effect of oxygen equivalence ratio ϕ. Char and moist air with 0.35 wt. % cesium seed.

of slag on generator walls and electrodes is eliminated, and air preheaters of regenerative type may be used to give very high and thermodynamically desirable air preheat. To obtain such ash-free product gases, with coal or char as fuel, it is necessary first to form an ash-free fuel from the coal or char, or else contrive a combustion system that rejects all the ash. The rejection of 90% of the ash, option (b), is not nearly as complex or costly as (a) but still offers advantages of nonfouling of the generator duct and low ash loading through the heat exchange system and in the final stack gas cleaner.

The following portions of this discussion will deal in turn with combustors designed for experimental MHD setups, combustors for oil- or gas-burning MHD plants, and combustors for MHD plants burning coal or char.

III. COMBUSTION CHAMBERS OF EXPERIMENTAL MHD FACILITIES

One type of combustor that has been found to give exceptionally good service, even for temperatures as high as 2900°K with operation on raw oxygen, is the vortex burner with water-cooled walls. The vortex tends to retain the hotter gases near the core and the cooler layers of gas near the wall. Energy release can be made quite high, of the order of 400 MW/m^2 of cross section, per atmosphere. A diagram of a typical construction is shown in Fig. 5. Chambers of this type have been used successfully in several laboratories (Way and Hundstad, 1962; Ressick *et al.*, 1965).

It is not advisable to discharge the flow from the vortex chamber directly into the MHD generator. A plenum section should be provided to destroy the swirl and to accomplish good mixing, to secure uniform temperature. The plenum can be lined with refractory, to give nearly adiabatic wall conditions. Stabilized zirconium oxide or magnesium oxide can be used for this purpose. Ceramic lining alternatively to water-cooled walls could also be used for the vortex chamber proper. In any case, however, it might be advisable to apply water cooling to the walls as a precautionary measure against cracks and "see-throughs."

In the vortex combustion chamber, the oxidant is introduced tangentially through swirl openings around the upstream end. The liquid fuel is admitted by a burner nozzle centrally located in the upstream face. If gaseous fuel is used, the admission ports could be similarly located.

It is important to secure positive and reliable ignition so as not to load the system with combustible mixture. Positive ignition can be assured by use of an igniter torch, which is itself spark ignited; it may be used to

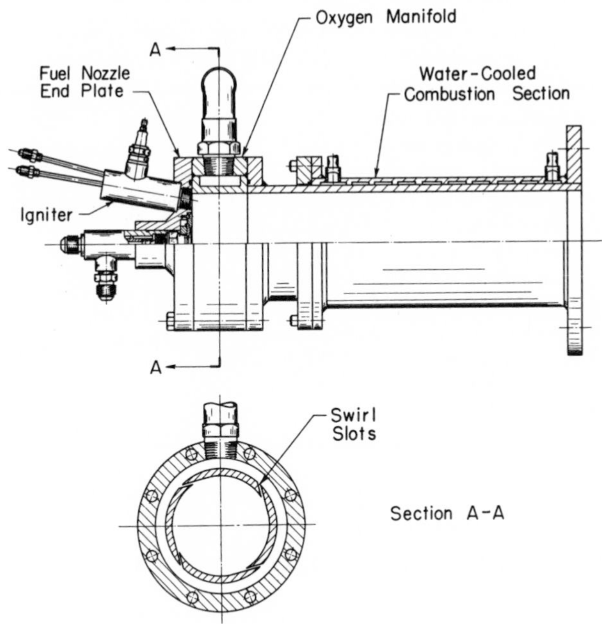

Fig. 5. Combustion chamber.

project a strong flame into the main burner space (a distance of several diameters). Igniter torches have been used which burn propane and oxygen, for example. A swirl torch can be run uncooled as the hottest portion of the flame remains in the center.

The operating sequence for starting the combustion system of the experimental MHD facility is given in Table I. The generator would normally be on open circuit setting during start-up of the combustor. To conserve fuel or oxygen, the magnet would normally have been energized before starting the combustor. As one proceeds now to load the generator, the pressure drop through the duct will increase because of the braking action of the Lorentz forces. This will, in the case of subsonic flow generators, reduce the flow rates and increase the combustion chamber pressure; therefore further readjustment of flows and pressures will be necessary as the generator is loaded. These adjustments may be very slight. Temperature level should be controlled by adjustment of preheat temperature rather than fuel rate adjustment since the ϕ value is an important test parameter.*

* ϕ is the oxidant equivalence ratio.

TABLE I

Sequence of Operations for Starting the Combustion System of the Experimental MHD Facility

1. Adjust all cooling water flows
2. Turn on spark in igniter torch
3. Turn on fuel and oxygen simultaneously to igniter
4. Check temperature or flame indicator to verify presence of igniter flame
5. Turn on burner oxidant to one-fourth load setting
6. Turn on burner fuel to one-fourth load setting
7. Verify burner ignition
8. Increase flows simultaneously to one-half load setting
9. Increase flows simultaneously to three-fourths load setting
10. Increase flows simultaneously to full load setting
11. Turn off igniter torch
12. Trim and adjust flows and pressure to test values

IV. COMBUSTION CHAMBERS FOR LIQUID AND GASEOUS FUELS FOR PROTOTYPE PLANTS

We deal here with combustion chambers such as those that may be used for pilot plants or first generation MHD power plants. Although experience with such plants is lacking, there is a body of technology regarding gas and oil combustion systems.

Gas-fired furnaces for high-output temperatures can be made with zirconium oxide liners (Kirillin and Scheindlin, 1968). Operation is found to be highly satisfactory.

A fairly reliable system is that in which one has a vortex flow, followed by a blending region to minimize swirl of gases entering the generator. It may be advisable to use opposed burners to eliminate swirls, as shown in Fig. 6.

The writer, because of favorable past experiences, would prefer to design the gas- or oil-fired combustor as a vortex burner. Walls may be of stabilized zirconia. Magnesia would withstand the temperature but there would be gradual wastage due to vaporization. This wastage is estimated, at 2300°K wall temperature, to be 6 cm in 1000 hr. With zirconia we would estimate material disappearance at the rate of 5 mm/yr.

The ceramic-lined walls should be of sufficient thickness to keep the heat loss to a fairly low value. The combustor heat loss through the walls must be absorbed by cooling water, and in central station plants it is desirable to keep such heat absorption at a minimum, in order not to affect adversely

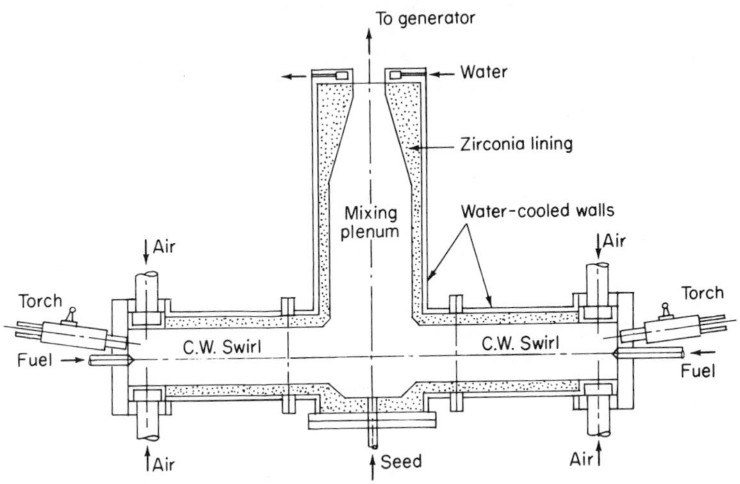

Fig. 6. Two-burner opposed-swirl combustor.

the regenerative feed heating system. A reasonable design goal is a wall heat loss of 2% of the thermal input for gas- or liquid-fueled systems.

The question arises as to the proper basis for specification of thermal loading. For combustion chambers of a type wherein the reactions tend to be area related, as in cyclone furnaces, it is appropriate to base the loading on the area of cross section and atmospheres of pressures, in other words, MW/m^2 atm. Use of this pressure relation helps assure that with change of pressure the pressure loss ratio will be constant, providing velocities are unchanged. Also, in burners having mixing limitations, the burning rate tends to be about proportional to the pressure.

In cyclonic combustion chambers for liquid or gaseous fuels the fuel is normally injected from a source at the center of the upstream end plate. With liquid fuels a swirl-type atomizer can be used. Some primary air admission around the burner is desirable. With gaseous fuels a multiple-port injector is appropriate but with swirling primary air to improve circumferential uniformity.

Alternatively to vortex combustors, straight-through plug-flow burners may be used. In such burners, the burning takes place throughout the volume and a volume basis for loading designation is appropriate. An example of a combustor of this type is the oil-burning chamber used in the British Marchwood experiments (Heywood and Womack, 1969). The ultimate goal was a 2000-MW thermal power plant, and this was to be preceded by a 60-MW(th) plant. To ensure a sound design for the 60-MW

TABLE II

PARAMETERS OF THE EXPERIMENTAL COMBUSTION RIG

Total thermal input	12 MW
Assumed residence time	50 msec
L/D ratio of combustor	5 : 1
Diameter of chamber	0.33 m
Length of chamber	1.53 m
Maximum operating pressure	9 atm
Air preheat	1700°K
Equivalence ratio (design)	1.025
Oil heating value (HHV)	42.9×10^6 J/kg
Design airflow	2.9 kg/sec

unit an experimental rig was first designed and operated having the parameters given in Table II. The thermal input here includes the air preheat energy. By the fuel oil alone the thermal input is about 9 MW. The chamber is of cylindrical form with water-cooled metal walls (11% Cr, 0.6 Ni, 0.8 Mo, 0.25 V). A single oil burner (Babcock and Wilcox Mark P Simplex) was used, with water jacket. Best results were obtained with an inverted "flowerpot"-type mixer with eight radial air admission holes, the ring of holes being about 0.1 m below the burner. When operated in a plug-flow mode, without the mixer, the combustion was incomplete and very smoky. Typical test conditions in the stirred mode were 2.64 kg/sec air, 0.195 kg/sec oil. Heat fluxes averaged about 1 MW/m^2 in the combustion chamber. (Calculated value with unit emissivity was 4.7 MW/m^2, which included 0.6 MW/m^2 convective portion). Flame temperature about 2700°K was measured at 1500°K preheat. A general sketch is shown in Fig. 7.

In the Russian U-02 test facility (Kirillin *et al.*, 1971), latest modification, experimental high-intensity combustors have been operated on natural gas at heat release rates up to 407 MW/m^3 atm, or 116 MW/m^2 atm. Losses due to incomplete combustion were 1.5 to 3%. This is apparently a vortex-type combustor, with zirconia lining. The unit has run over 1000 hr.

The Russian U-25 combustor (25 MW MHD) burns 4.75 kg/sec natural gas with a preheated mixture of 10.2 kg/sec O_2, 33.7 kg/sec air, and 1.35 kg/sec recycled seed solution. Burning is at 2.75 atm and the outlet temperature is 2870°K. In this case a ram-jet-type burner is used. Gas is injected by narrow longitudinal slots in the side wall upstream of an array of water-cooled tubes which serve as a radiation screen as well as flame anchors (Kirillin *et al.*, 1968).

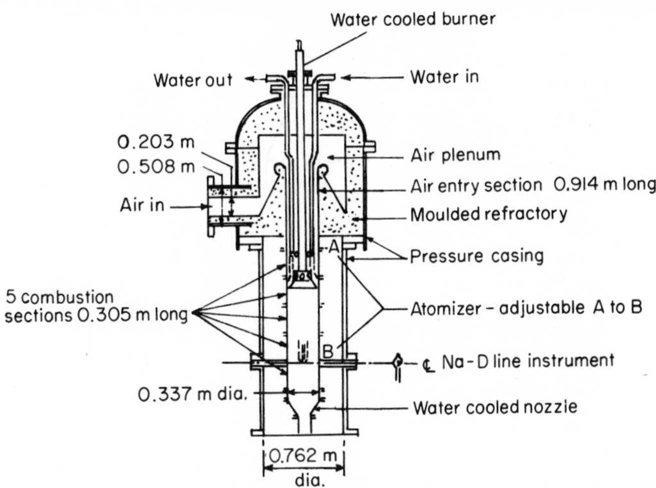

Fig. 7. Layout of liquid fuel MHD combustor. [From Heywood and Womack (1969), by courtesy of the Central Electricity Generating Board, U. K.]

V. SOLID FUEL COMBUSTION CHAMBERS

For combustion of coal in MHD systems one must first of all consider the question: Should the ash be retained in the combustion chamber and drained off as slag, or should it be sent through the MHD generator and caught in a cyclonic heat exchanger?

Combustion systems for coal which send the ash through the generator duct have been studied by the AVCO Corporation and by the CEGB group in England. The former used what is referred to as a ram-jet-type

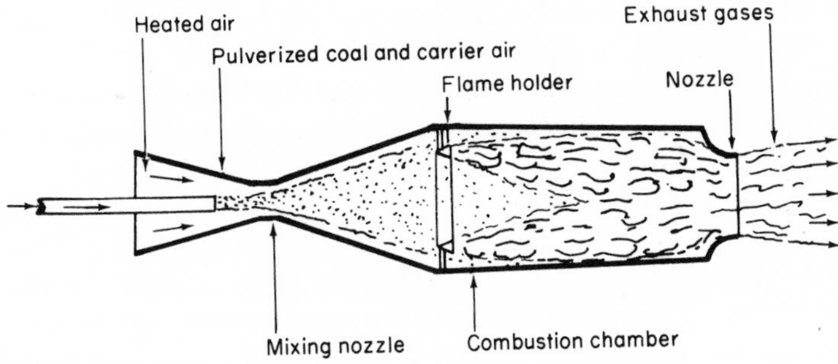

Fig. 8. Arrangement of ram-jet-type combustion chamber for pulverized coal. [See Hals (1967).]

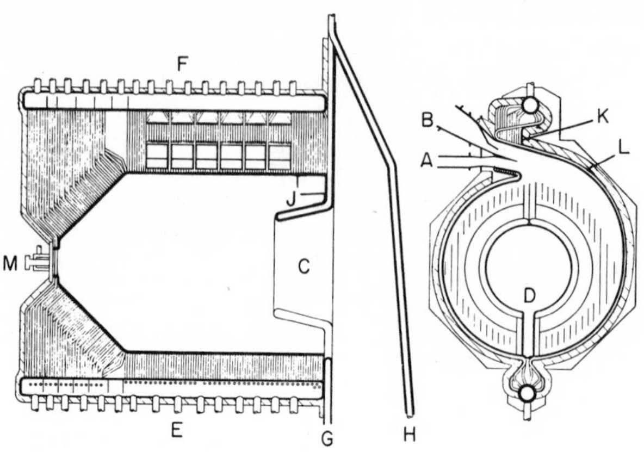

Fig. 9. Cyclone furnace of German Babcock and Wilcox design used at Wilton Works of Imperial Chemical Industries, England. Furnace diameter 10 ft, length 20 ft. Features are designated as follows: (A) primary air–coal inlet; (B) secondary air inlet and control dampers; (C) throat for exit gases; (D) slag tap slot, (E) supply tubes and bottom header; (F) risers and top header; (G) front wall tubes; (H) target wall and slag screen; (J) plastic chrome ore on studded tubes; (K) pressurized casing; (L) high-temperature insulation; (M) pressurized inlet for lighting up oil burners. [From Holmes-Smith *et al.* (1961). *J. Inst. Fuel* **34**, 307.]

combustor, as it involves flame holders to anchor the flame in a mixture of pulverized coal and air (Fig. 8). The investigators in England favor a plug-type burner similar in some respects to that of AVCO. In both cases satisfactory combustion of coal was obtained (Heywood and Womack, 1969; Hals, 1967).

If ash is sent directly through the generator, there may well be condensation of the ash constituents in the downstream section. There can also be condensation even in the upstream section on cooled walls. The behavior of the generator when coated with layers of slag (Rosa, 1971), which is electrically conducting at high temperatures, will most certainly be impaired. Therefore it seems appropriate to consider means of assuring flow in the generator free of condensed slag. This may be accomplished by use of a cyclone combustor.

Conventional cyclone furnaces (e.g., those of the Babcock and Wilcox Company) will discard 80 to 90% of the coal ash as liquid slag. Temperatures in the cyclone chamber will be 2000–2100°K. It is important that the mean temperature exceed by a margin of a few hundred degrees the value T_{250}, that is, the temperature at which the slag viscosity is 250 P. It is important to drain the slag away continually, to prevent excessive accumulation in the chamber, and to ensure removal of liquid iron. Liquid iron is most likely to appear in reducing atmospheres. Slight excess air is therefore desirable in the cyclone furnace usually about 10%. A sketch of a conventional cyclone furnace as used by the German Babcock and Wilcox Company is shown in Fig. 9. The coal as well as the air are brought in from tangential admission ports. (Actually, the coal injection is along a secant direction to prevent erosion by particle scouring of the inner surface.) Fineness of pulverization is approximately 50% minus 300 μm. Cyclone chambers, because of the high temperatures prevailing, are capable of burning a wide variety of coals, ranging from lignite to anthracite, and even coal-water slurries. They can also be operated on gas or oil. The ability to retain 80 to 90% of the ash is of considerable advantage in reducing fouling in the boiler, and facilitating final stack gas clean up.

Heat release in conventional cyclones, as reported by Seidl (1955, 1962), is such that, very nearly,

$$qD = 2.8 \times 10^6 \quad \text{Btu/ft}^2\,\text{hr} \qquad \text{or} \qquad 8.8 \quad \text{MW/m}^2,$$

where q is Btu/ft³ hr and D is chamber diameter in feet. For chambers of $L/D = 1.5$ this is equivalent to

$$Q/A = 4.2 \times 10^6 \quad \text{Btu/ft}^2\,\text{hr} \qquad \text{or} \qquad 13.2 \quad \text{MW/m}^2,$$

where Q is Btu/hr thermal rate and A is chamber cross section. The relations above are equivalent to the statement that $qQ^{1/2}$ is constant at the limiting thermal loading. Reference may be made to Fig. 10 from Seidl (1962).

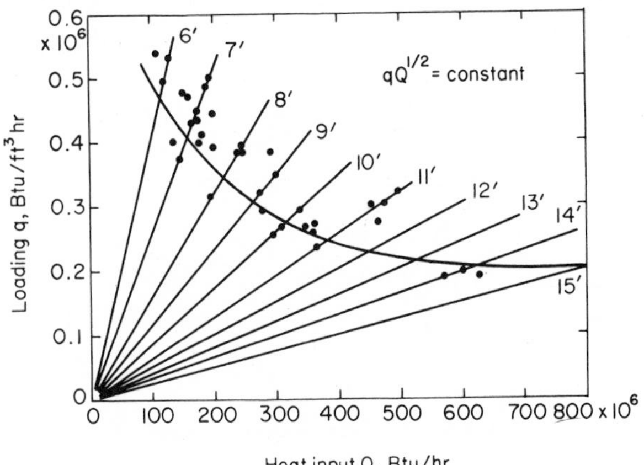

Fig. 10. Typical relation of heat release rate to total heat input for horizontal cyclone burners, from German experience. [From Seidl (1962).]

Flow patterns in cyclone chambers generally show an annular region of upstream swirling flow surrounding a core region flowing downstream. In the core region, near the outlet, there may also be a reversal of flow, since a low-pressure region exists at the center of the vortex. Typical flow patterns (Seidl, 1962) are shown in Fig. 11.

In application of the cyclone combustion chamber technique to the MHD power plant we must seek to develop outlet temperatures close to 2700°K. Such high temperatures are, of course, attainable by air preheating or oxygen enrichment. However, a problem would arise with excessive vaporization of the ash constituents. Therefore, one may have recourse to a two-stage cyclone combustion system (Way, 1967a,b).

An example of a two-stage cyclone chamber design may be described. Suppose the chamber is to burn a typical Pennsylvania bituminous coal of high heating value on a dry, ash-free basis of 36.31×10^6 J/kg. Assume there is 3% moisture in the coal and 7.9% ash. Assume also 26.5% volatile and 65.6% fixed carbon. The stoichiometric ratio of moist air (1% moisture) to dry ash-free coal is 12.00. Assume coal and air are reacted in the first stage at equivalence ratio $\phi = 1.6$ (60% excess air). In that case all the air may enter the first cyclone stage but only 65.6% of the coal, assuming we want an overall ϕ value of 1.05. (The ratio 1.05 : 1.60 is 0.656.) The furnace temperature in the first stage is then estimated to be about 2300°K, taking account of the heat loss to the walls which in this case may approach 10% of the thermal input to the first stage. This temperature is sufficiently low to prevent excessive vaporization (fuming) of the ash constituents, but there will be some entrainment of fine droplets of liquid slag. With good design (properly chosen area of outlet throat and swirl velocity) we may expect no more than 10% ash carryover to the second stage. For 1 kg of

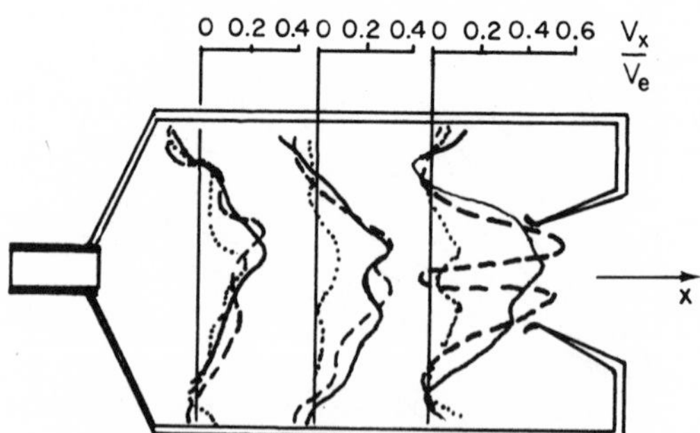

Fig. 11. Cyclone combustor. Typical distribution of axial velocities, where the solid curves denote coal; the dashed curves, diesel oil; the dotted curves, cold air. [From Seidl (1962.]

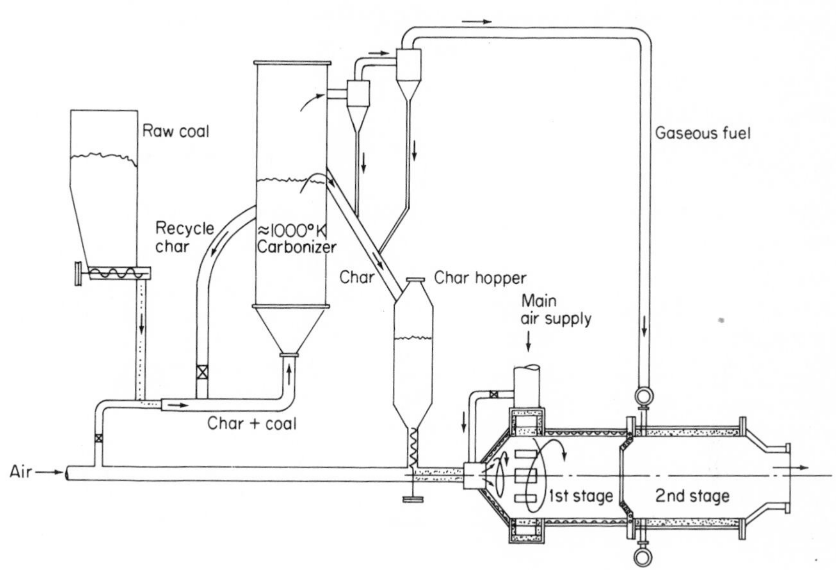

Fig. 12. Two-stage cyclone chamber with carbonizer.

dry coal fired we will have 0.079 kg ash. The 65.6% going to the first stage will carry 0.0518 kg ash, of which about 0.00518 kg will be carried over to the second stage. The second stage would be fired with the remaining 0.344 kg of the coal, containing 0.0272 kg ash. Total ash received in the second stage is then 0.0272 + 0.00518 or 0.0324 kg. Owing to the higher temperature in the second stage (about 2650°K) more ash is vaporized and probably no more than 50% of the ash can be retained (rejected as slag). On this basis, about 0.0162 kg of the ash goes on downstream to the generator. This is 20.5% of the original ash fired. This indicates the approximate ash retention capability of a two-stage chamber with coal firing in both stages.

If still lower ash carryover is desired from the two-stage combustor, an ash-free fuel can be burned in the second stage. A relatively simple way of accomplishing this objective is to carbonize the coal to char, while driving off the volatile fractions. The vapors are then burned in the second stage while the char is burned with excess air in the first stage. Total ash carryover can then be held to about 10%. The system would appear as shown in Fig. 12. The second stage would be designed with refractory walls, preferably stabilized zirconia, so that it would have very low heat loss.

An example will now be given for such a chamber with char fired to the

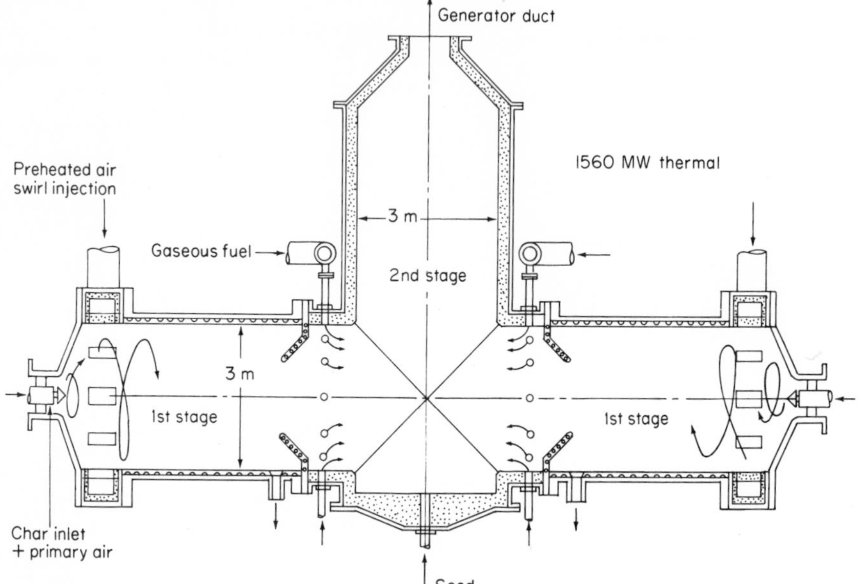

Fig. 13. Schematic of two-stage cyclone combustor.

first stage and pyrolysis gas to the second stage (see Fig. 13). We assume the same initial coal type as before. To make the example more realistic we shall base the design on 1000 MW thermal input to the first stage. The char produced by partial devolatilization of the coal will have a heating value of about 33.54×10^6 J/kg (ash-free basis), and the stoichiometric ratio of moist air to ash-free char is 11.35. We shall assume an overall oxygen equivalence ratio for the two-stage combustor of 1.05. Various numerical values will be as given in Table III for the example.

The gaseous fuel calculated here, coming from the pyrolysis unit, is slightly more than the initial volatile content of the coal. Actually, in the example, no account was taken of the air required to operate the carbonizer (see Fig. 12). The 14.57 kg represents the mass of that part of the coal fired which does not become char. It will be reacted with the air which goes to the carbonizer, so the total gas flow will be more than 14.57 kg, and the gas will contain some CO, CO_2, and N_2.

We would try to design the first stage for no more than 10% ash carry-over, and 90% retention.

If one takes account of the fact that the ash which passes downstream is ultimately recycled (with the recycled seed), the net amount of ash dis-

TABLE III

DATA FOR TWO-STAGE CYCLONE COMBUSTOR

Parameter	Unit	Value
Thermal input, first stage	MW	1000
Equivalence ratio, first stage	ϕ	1.60
Air preheat temperature	°K	1200
Heat loss, first stage	%	10
Temperature in first stage	°K	2285
Assumed loading, first stage	MW/m^2 atm	15
Working pressure	atm	5
Cross-section area, first stage	m^2	13.3
Chamber diameter if two are used	m	2.92
Char heating value	MJ/kg (ash free)	33.54
Char feed rate, first stage	kg/sec (ash free)	29.81
Ash supplied with char	kg/sec	3.69
Air rate, first stage at $\phi = 1.60$	kg/sec	541.5
Total coal supply for $\phi = 1.05$	kg/sec (dry, ash free)	42.98
Ash fired with coal	kg/sec	3.69
Water fired with coal	kg/sec	1.40
Gas fired to second stage, including H_2O	kg/sec	14.57
Gas fuel to second stage, without H_2O	kg/sec	13.17
Coal heating value, dry, ash free	MJ/kg	36.31
Thermal input from coal supplied	MW	1560
Thermal input to second stage	MW	560
Assumed loading, second stage	MW/m^2 atm	15
Area required, second stage	m^2	7.47
Diameter of second stage (one chamber)	m	3.08
Coal as fired	kg/sec	48.07
Ash carried downstream	kg/sec	0.37

charge downstream can be calculated from the relation

$$0.1(3.69 + x) = x \qquad \text{or} \qquad x = 0.41 \quad \text{kg/sec.}$$

The 0.41 kg ash is recycled, combines with the incoming 3.69 kg to make 4.10 kg and the 10% of this carried over becomes the 0.41 kg. Note that all the ash is then rejected as slag, namely 3.69 kg. An electrostatic precipitation of 99.5 to 99.8% effectiveness would be used to capture the seed and ash from the exhaust stream. Note that no elaborate means is needed to separate seed and the recycled ash. Most of the seed will end as Cs_2SO_4 and can be dissolved in water for further processing to remove sulfur.

The ash carryover will depend on entrainment, vaporization, and volatilization. A distinction is made in the latter two terms to signify the difference between ordinary vaporization and formation of gaseous ash products as a result of chemical decomposition. Thus, SiO_2 may evolve in the form SiO in the gas phase. It is most probable that, owing to the short residence time in the chamber, equilibrium will not be approached and we would simply have vaporization.

Vapor pressure curves for several species are shown in Fig. 14 (Ferguson and Wilkins, 1963). Table IV gives the ash composition of several coals that may be of interest (Selvig and Gibson, 1956).

The relatively high vapor pressure of silica, and its plentiful supply in the ash, would indicate that the vaporization process in the first stage might consist largely of formation of silica vapor. The alkali constituents and the sulfur would probably also enter the gas phase. Insofar as liquid droplets of slag are entrained, these would be converted totally to vapor, if they are carried over to the second stage.

The amount of ash carryover is very difficult to calculate and such calculations might not be reliable. The quantity vaporized from the slag layer on the wall depends on the exposure time as well as temperature conditions. We would not expect equilibrium values of vapor pressure to be reached.

The particle (droplet) entrainment will depend on a balance of forces. A particle will be driven inward by the inward radial component of velocity as the vortex of flow contracts while approaching the outlet nozzle, but it

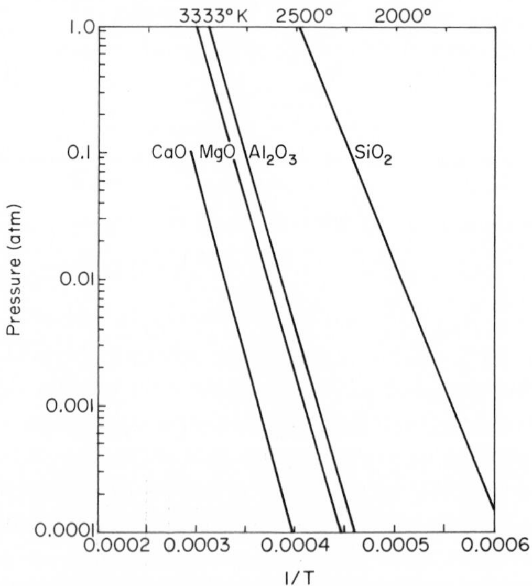

Fig. 14. Vapor pressures of ash constituents.

TABLE IV

ASH COMPOSITIONS[a]

Coal	Lower Kittaning Mine #72	Pittsburgh Banning #2	Clearfield B Cunard #1	Illinois #6 Old Ben #10
Ash %	6.8	8.7	8.5	8.6
SiO_2	42.4	56.1	43.6	46.2
Al_2O_3	33.9	27.8	29.3	22.9
Fe_2O_3	18.8	9.9	19.9	7.7
CaO	1.4	1.0	1.4	10.1
MgO	0.4	0.6	0.6	1.6
TiO_2	1.0	1.6	1.3	1.0
K_2O	0.9	0.8	1.1	0.8
Na_2O	0.5	1.0	0.3	0.7
SO_3	0.9	0.7	2.0	8.9
P_2O_5	0.10	0.32	0.14	—

[a] See Selvig and Gibson (1956).

will also be subjected to the radially outward inertia force mu^2/r. The inward aerodynamic force may be expressed in terms of the conventional terminal settling velocity v_t, in a gravitational field, as mgv/v_t; that is, when $v = v_t$ we know the drag force is mg. Equating mgv/v_t to mu^2/r gives $v_t = grv/u^2$. If r is the radius of the outlet nozzle, and u and v are known from the cyclone flow pattern, the terminal velocity v_t can be calculated, and a conventional calculation then yields the diameter of the particle which would just barely be carried through. Larger particles would be separated and retained. For example, with $r = 0.5$ m, $v = 20$ m/sec, $u = 100$ m/sec we find $v_t = 9.81 \times 0.5 \times 20/100^2 = 0.0098$ m/sec, and the critical particle diameter would be about 28 μm. To promote less entrainment carryover we should reduce v, increase u, and reduce r. Since normally, in the vortex, u increases as r decreases, the reduction of outlet throat helps to ensure low particulate carryover. This will, however, increase the cyclone pressure drop.

Increase of operating pressure in the cyclone combustion chamber will tend to reduce the ash carryover. This is because any constituents which tend to vaporize could not surpass a partial pressure corresponding to the local temperature, whereas the total pressure may be increased to 5 or 8 atm.

The overall cyclone pressure drop depends on the inlet jet dynamic head, as well as on the outlet configuration. It was found (Roberts, 1964) from

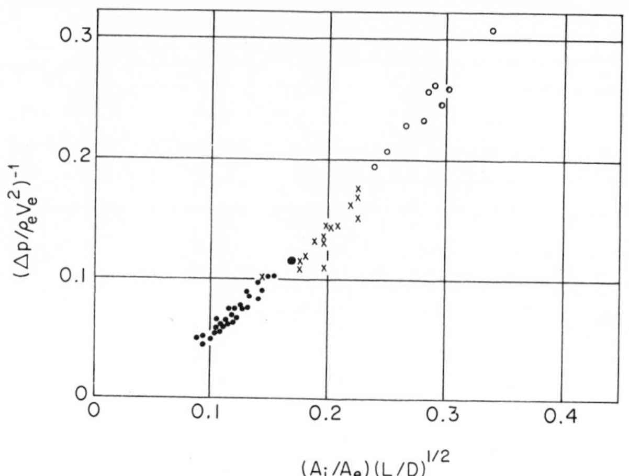

Fig. 15. Pressure drop correlation for cyclone furnaces using Yakubov formulation. Experimental points for Imperial Chemical Industries units at Wilton and Witton. A_i is air inlet opening area; A_e is outlet throat area; L/D is length-to-diameter ratio of chamber; Δp is total pressure drop; $\frac{1}{2}\rho_e V_e^2$ is dynamic head at outlet. [From Holmes-Smith *et al.* (1961). *J. Inst. Fuel* **34,** 307.]

British cyclone experience that $\Delta p/(\rho_e V_e^2)$ correlates with the expression $(A_i/A_e)\ (L/D)^{1/2}$ as shown in Fig. 15.* In the two-stage chamber example treated above the pressure drop can be estimated. From the loading criterion of 75 MW/m² the relation of char feed rate to sectional area A_0 is

$$m_c = 2.235A_0 \quad \text{kg/sec.}$$

The total flow in the first stage is 19.28 times the ash-free char rate, so

$$\rho_e V_e A_e = 19.28 \times 2.235A_0 = 43.1A_0.$$

If we assume $d/D = 0.67$, then $A_e/A_0 = 0.449$. Consequently,

$$\rho_e V_e = 96.0 \quad \text{kg/m}^2\,\text{sec.}$$

Outlet density ρ_e, at 5 atm, 2200°K, is 0.82 kg/m³, so $V_e = 118.5$ m/sec. We then find $\rho_e V_e^2 = 11510$ N/m². With $L/D = 1$ and $A_i/A_e = 0.4$ we would have approximately $\Delta p/\rho_e V_e^2 = 2.5$, and in that case $\Delta p = 28800$ N/m². With total pressure 5 atm, or 506,000 N/m², this is a fractional pressure drop of 5.7%. The ratio of airflow to total gas flow in the first stage is 0.942; since the density ratio of the incoming preheated air to the out-

* A_i/A_e is ratio of air inlet area to outlet throat area. The formulation shown is due to Yakubov (1957).

going gas is about 2.0, the air injection velocity will be given by

$$V_i = 0.942V_eA_e/2.0A_i \qquad \text{or} \qquad V_i/V_e = 0.471A_e/A_i = 1.177.$$

With $V_e = 118$ we have $V_i = 140$ m/sec. The swirl velocity at the radius of the outlet nozzle would be 208 m/sec. Pressure drop could be reduced in this design by selecting larger d/D.

The heat transferred to the wall depends on a balance between heat received (mainly by radiation) and heat transferred through the slag layers by conduction. There will be some heat consumed also in the partial vaporization of the slag. A thicker slag layer tends to reduce the heat transported. It was shown theoretically by Reid and Cohen (1944) that for given slag flow properties (e.g., viscosity) the heat transferred through the walls tends to vary as the (slag flow rate)$^{-1/3}$. In the excellent review paper by Roberts (1964) results of several other investigators are reported. For example, Marshak tested vertical cyclones and found a decrease by 40% of heat flux when ash quantity was increased 110%. Percentage heat loss was about 5% for the lower slag flow. Other investigators have also generally observed 4 to 5% heat loss.

An interesting MHD coal combustor experiment has recently been reported. Tager *et al.* (1971) report tests with a vertical cyclone burning pulverized coal with an outlet temperature of 2300°C. A sketch is reproduced in Fig. 16. This chamber, running at 1 atm, had heat release of 20 MW/m^2 and it is estimated it would have a capacity of 100 MW/m^2 at 5 atm. It retains 80% of the coal ash at 1 atm. Increase of pressure should increase ash retention capability.

We should mention, finally, MHD systems using coal as fuel, which transform the coal into a clean fuel gas. In principle we could consider, again, the two-stage cyclone with the first stage operating as a gasifier. Thus the first stage of the two-stage system could be operated at equivalence ratio of about $\phi = 0.6$. It would be an air-blown gas producer rather than one requiring steam. (We wish to keep hydrogen input to a minimum.) A gas consisting of CO, CO_2, and N_2, with slight H_2, would be generated. This gas would then be burned in the second-stage combustion chamber. Such arrangements have been considered from time to time by various investigators (Hoover *et al.* 1966; Hoy and Watt, 1970).

The problem with the cyclone, air-blown gas producer is that a certain amount of fuel or ash particles will be aerodynamically entrained. Even with close attention to all design details probably at least 5% of the ash would be carried over. It can thus not be classed as a *clean gas* producer, though it may still be of interest as a two-stage system giving reasonably good ash retention. The use of excess fuel in the first stage presents problems

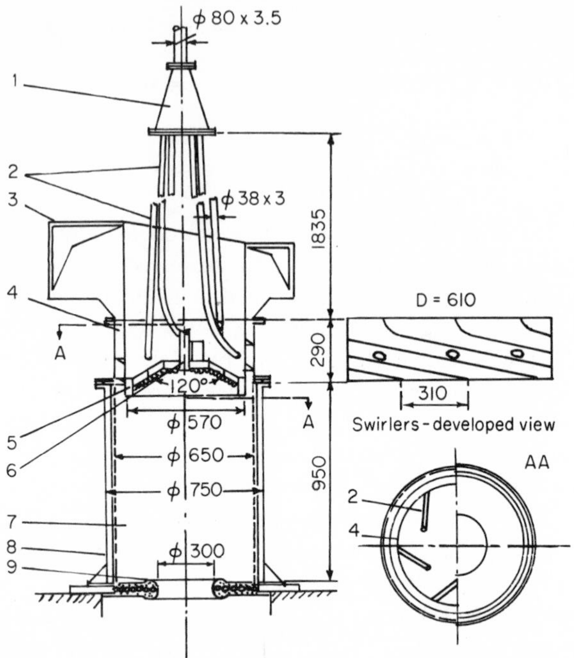

Fig. 16. Single-stage cyclone coal-burning chamber for MHD generator application as developed at Krzizanovsky Power Institute. Features are designated as follows: (1) coal distributor; (2) injection tubes; (3) air manifold; (4) air swirl vanes; (5) roof structure; (6) igniter torch; (7) cyclone chamber; (8) cooling jacket; (9) outlet. Dimensions are in millimeters.

of liquid iron generation and production of H_2S, which might be corrosive. There may also be problems of carbon loss in the slag. These disadvantages are offset by the relatively simple procedure of merely injecting air in the second stage.

If a slagging fixed-bed gasifier is used, producing a CO-rich gas, there will be problems of prevention of caking and carryover of particles. Caking might be prevented by appropriate pretreatment of the coal. Particulate carryover can be retarded by maintenance of low velocities, use of sized coal or char, and use of swirl in the outlet section. The higher pressures present in the MHD system are beneficial in keeping the size of the gasifier units in reasonable bounds.

The distinctive features of the coal gasifier for the MHD plant are (a) it should preferably operate with air rather than steam, (b) it should use highly preheated air, (c) it should be of slagging type, (d) sensible heat should be conserved, and (e) it should operate at pressures of 4 to 8 atm.

It is not absolutely essential that steam be avoided. The gains from having a clean fuel are so considerable that the presence of some hydrogen could be accepted.

With clean gas supplied to the MHD plant the exhaust gas stream will be free from ash, and it will be possible to operate regenerative air preheaters to very high temperatures. Such heaters, featuring pure MgO matrix material, have been demonstrated to 1900°K. Resulting power plant station efficiencies would be over 60% (Way, 1969, 1971a).

In the work in England in connection with the Marchwood project (Heywood and Womack, 1969) both gasifying and nongasifying coal combustion systems were considered. Schematic diagrams are shown in Fig. 17. The short time scale imposed in the British effort required concentration on a single-stage system, although it was felt that ultimately a two-stage arrangement might be preferable. It was felt, rightly or wrongly, that an axial flow combustor, with turbulent plug flow, would yield higher volumetric heat release in large units, since cyclone burners tend to follow the energy per unit area loading criterion. Therefore an axial flow combustor

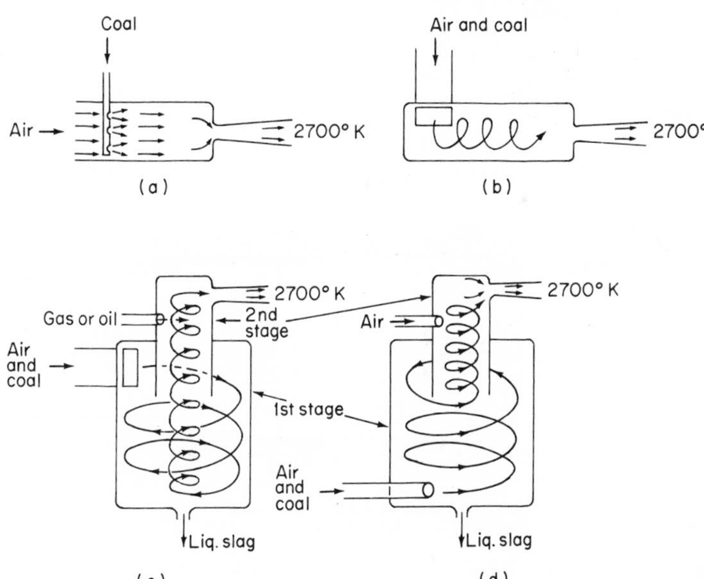

Fig. 17. Schematic representation of type of coal combustors for MHD considered in Marchwood experiments. (a) Axial flow, with fully mixed plug flow. (b) Cyclone single-stage chamber with combustion largely at walls. (c) Two-stage cyclone system with clean fuel combustion in second stage. (d) Two-stage cyclone combustion with coal gasification in first stage. [From Heywood and Womack (1969).]

of the general configuration of Fig. 18 was used in experiments to establish the design for the 60-MW(th) power plant. The approaching airstream, however, was given a moderate swirl. The coal injector is a small cup that contains a high-speed vortex of coal and air mixture, which emerges from

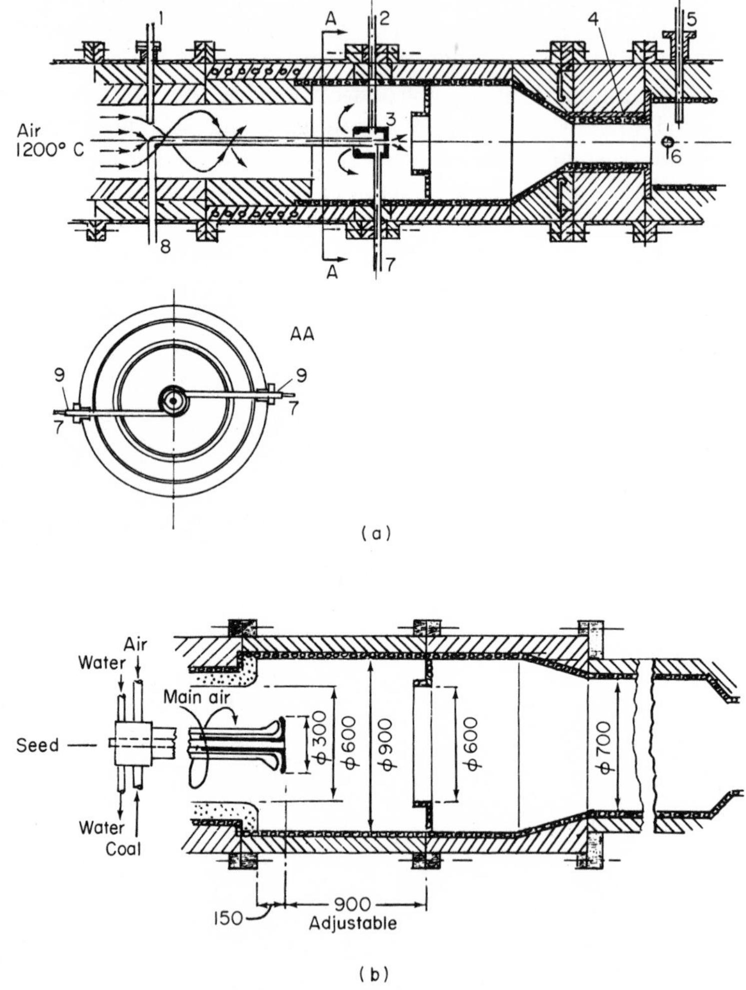

Fig. 18. Plug-type coal burners for British Marchwood project. (a) Pilot-scale 45-mm combustor. Features are designated as follows: (1) gas sampling probe; (2) disperser air inlet; (3) coal disperser; (4) conductivity cell; (5) gas sampler; (6) Na-line reversal; (7) coal inlet; (8) seed inlet; (9) cooling water. (b) Combustor for proposed 60-MW thermal unit. [From Heywood and Womack (1969).]

the upstream face in counterflow to the main airstream. Seed (dry) is injected axially. The baffle device downstream of the injector was found necessary for good mixing. Some data on typical operating conditions for this experimental combustor, in comparison with conditions for the 60-MW plant, are shown in Table V.

TABLE V

MARCHWOOD PROJECT IN ENGLAND. TYPICAL CONDITIONS AND PERFORMANCE OF PILOT-SCALE COMBUSTOR AND PROPOSED 60-MW COMBUSTOR[a]

Parameters	60-MW unit	Pilot unit
Total air + oxygen (kg/sec)	19.2	0.88
Fraction of oxidant for coal dispersion (%)	3	5
Oxygen concentration in oxidant (wt. %)	29	33
Oxidant preheat (°K)	1470	1470
Oxidant equivalence ratio (ϕ)	1.05	1.04
Coal size index	90%–75 μm	90%–75 μm
Mass flow (kg/sec)	2.6	0.13
Thermal input (MW)	73	3.8
Ash content (%)	10	10
Ash silica ratio (%)	80	81
Seed	Dry K_2SO_4	Dry K_2SO_4
Concentration in gas (mole % °K)	0.5	1.0
Combustor pressure (atm)	5	5
Gas residence time (msec)	45	48
Gas velocity (m/sec)	45	21
Volume (m^3)	1.4	0.07
Wall area (m^2)	7.9	1.00
Surface area of baffle (m^2)	0.7	0.15
Combustion intensity (MW/m^3)	52	52
Performance[b]		
Percentage of carbon reacted	97.5	98
Heat transfer to walls (MW/m^2)	0.660	0.440
Heat transfer to baffle (MW/m^2)	0.940	0.580
Losses—% thermal input		
Cooling combustor	10	18
Seed evaporation	3.5	6
Heat to slag	3.5	3
Outlet temperature (°K)	2690	2640

[a] See Heywood and Womack (1969).

[b] *Note:* Since the heat invested in the seed and slag components remains in the gas stream, it is not necessary to regard these heat quantities as "losses," as long as they are taken into consideration in calculating flame temperature.

VI. STATUS OF THE MHD COMBUSTION PROBLEM

For liquid or gaseous fuels combustion chambers can be designed and built for MHD systems either by following the principles of vortex burner design, or using a straight-through flow with a mixing device at the upstream end. With vortex burners heat releases of up to 400 MW/m^2 atm can be obtained in laboratory-scale combustors and over 100 MW/m^2 atm in combustors where pressure loss is a more important consideration.

For coal-fired combustion chambers, experimental work both in the U.S. and the United Kingdom with ram-jet or plug-flow-type combustors has shown that coal can be burned successfully. With this type of burner the ash is all carried through the MHD generator. To avoid the ash carryover cyclone-type chambers which retain a major part of the slag have been investigated. The single-stage chamber at the Krzyzanovsky Power Institute has been successful in retaining 80% of the ash with a single stage. Two-stage cyclones with the first stage serving as a gasifier have been proposed, as have also two-stage systems with the first stage running with excess air. The latter appears to be the most promising because it is free from many of the problems that beset the fuel-rich gasifier-type first stage. Experimental programs with two-stage cyclone combustors have been virtually nonexistent, and it is hoped that this situation will be rectified in the near future. This type of combustor would appear to offer the best prospects for early realization of coal firing in an MHD generator system.

Another topic of great importance for MHD is the perfection of suitable gasification units for production of a clean gas from coal or char. Of particular interest is a gas producer which reacts hot exhaust products with coal or char to give a fuel gas of high CO concentration. This is the method referred to as chemical regeneration. It makes possible an appreciable advance in cycle efficiency and reduction of air preheat temperatures. The problems to be solved relate to isolation of seed from ash or slag, and achievement of compact size for the gasifier.

References

Cohen, P., and Reid, W. T. (1944). Furnace Performance Factors. *Trans. ASME* (*Suppl.*) **66,** 83.

Ferguson, H. F., and Wilkins, D. M. (1963). Some Data Relevant to the Volatilization of Ash Components from Slags at High Temperatures. *BCURA Mon. Bull.* **27,** No. 10.

Frost, L. S. (1961). Conductivity of Seeded Atmospheric Pressure Plasmas. *J. Appl. Phys.* **32,** 2029.

Hals, F. (1967). Magnetohydrodynamic Power Generation. Paper 67-PWR-12. *ASME Joint Power Conf., Detroit, September 1967.*

Heywood, J. B., and Womack, G. J. (1969). "Open Cycle MHD Power Generation." Pergamon, Oxford.

Holmes-Smith, E. H., Lister, J., and Liddell, P. V. (1961). New High Pressure Cyclone Boiler at Wilton. *J. Inst. Fuel* **34,** 307.

Hoover, D. Q., Somers, E. V., Tsu, T. C., Way, S., and Young, W. E. (1966). Feasibility Study of Coal Burning MHD Generation. Appendix 8, Vol. II. Final Rep., Office of Coal Res. Contract 14-01-0001-476, (Westinghouse Res. Lab.).

Hoy, H. R., and Watt, J. D. (1970). Effect of Inorganic Matter on Combustion. *North Amer. Fuel Technol. Conf. Ottawa, Canada, May–June, 1970.*

Kirillin, V. A., and Scheindlin, A. E. (1968). "Magnetohydrodynamic Method of Electrical Power Generation." Energia, Moscow (in Russian).

Kirillin, V. A., Neporozhniy, P. S., and Scheindlin, A. E. (1968). A 25000 kW Pilot Power Plant. *Symp. Eng. Aspects MHD, 9th, Tullahoma, Tennessee April 1968.*

Kirillin, V. A., *et al.* (1971). Investigations at U.02 MHD Plant—Some Results. *Proc. Int. Conf. MHD Elec. Power Generation, 5th, Munich.*

Penner, S. (1957). "Chemistry Problems in Jet Propulsion." Pergamon, Oxford.

Ressick, K., Eustis, R. H., and Kruger, C. H. (1965). Design and Performance of the Stanford Combustion MHD Generator. *Symp. Eng. Aspects MHD, 6th, Pittsburgh, Pennsylvania.*

Roberts, A. G. (1964). The Cyclone Furnace. *BCURA Mon. Bull.* **28,** No. 1.

Rosa, R. (1968). "Magnetohydrodynamic Energy Conversion." McGraw-Hill, New York.

Rosa, R. (1971). Design Considerations for Coal Fired MHD Generator Ducts. *Int. Conf. MHD Power Generation, 5th, Munich.*

Seidl, H. (1955). The Development and Practice of Cyclone Firing in Germany. *Proc. Joint Conf. Combust. ASME-I.Mech.E. M.I.T., Cambridge, Massachusetts, June, 1955.*

Seidl, H. (1962). Eleventh Coal Science Lecture. *BCURA Gazette* No. 46, Inst. of Civil Eng., London, October 1962.

Selvig, W. A., and Gibson, F. H. (1956). Analysis of Ash from United States Coals. Bull. 567, U. S. Bur. of Mines.

Stull, D. R., and Prophet, H., *et al.* (1970). "JANAF Thermochemical Tables," 2nd ed. U. S. Govt. Printing Office, Washington, D. C.

Tager, S. A., *et al.* (1971). The Development and Investigation of a High Temperature Combustor to Be Used for a Solid Fuel MHD Generator. *Int. Conf. MHD Elec. Power Generation, 5th, Munich April 1971* (Krzizanovsky Power Inst., Moscow).

Way, S. (1965). Chemical Aspects of MHD Power Generation. *A.I.Ch.E.–I.Chem. E. Symp. Mater. Associated Direct Energy Conversion, London, June 1965.*

Way, S. (1967a). Problems of the Coal Fired MHD Power Plant. *Proc. Biennial Gas Dynamics Symp., 6th, Northwestern Univ., August 1965.* Northwestern Univ. Press, Evanston, Illinois.

Way, S. (1967b). U.S. Patent 3358624, December 1967.

Way, S. (1969). Char Burning MHD Systems. *ASME Winter Annual Meeting, November 1969,* Paper WA-69-Ener-13.

Way, S. (1971a). MHD Power Generation and Coal Gasification. "Supplementary Volume of Encyclopedia of Chemical Technology (Kirk–Othmer)." Wiley, New York.

Way, S. (1971b). MHD Power Plant for Early Realization. *Int. Conf. Magnetohydrodynamic Elec. Power Generation, 5th, Munich April 1971.*

Way, S., and Hundstad, R. L. (1962). Direct Generation of Power from a Combustion Gas Stream. *Int. Symp. Combust., 8th, 1960,* p. 241. Williams & Wilkins, Baltimore, Maryland.

Yakubov, G. V. (1957). *Izv. Akad. Nauk. Kaz. SSR Energy Ser.* **12,** No. 1, 105.

XII

Temperature Measurements and Gas Analysis in Flames and Plasmas Using Spectroscopic Methods*

GUNTER J. PENZIAS

NORCON INSTRUMENTS, INC.
SOUTH NORWALK, CONNECTICUT

I. INTRODUCTION

The methods described in this chapter deal with spectral radiation measurements made over a narrow wavelength interval; a prism or grating spectrometer is needed to isolate the wavelength interval of interest. This is in contrast to discussions in previous chapters where gas radiation is treated in terms of total radiation, that is, spectral radiation integrated

* This work was done while the author was a Senior Research Scientist with the Warner & Swasey Company, Control Instrument Division, Flushing, New York.

over a broad wavelength interval to encompass all the bands of the radiating gas molecule.

The discrete character of gas radiation as opposed to continuum radiation in flames is shown in Fig. 1, Chapter VIII. Infrared spectra typical of nonluminous hydrocarbon flames are shown in Fig. 6. Each band of radiation is attributable to a particular combustion gas species such as CO_2 and H_2O. In some portions of the spectrum these bands overlap.

Since the spectral transmittance of a gas is related to the number of molecules along the optical path it is possible to determine combustion gas concentrations *in situ* from spectral measurements. These techniques are described in Section II. Gas temperatures may also be determined from spectroscopic measurements as described in Section III.

II. SPECTROSCOPIC METHODS OF GAS ANALYSIS

A. Background

Knowledge of species concentrations as functions of space and time in a flame would help to establish the physical and chemical processes which occur in combustion. It is very desirable to determine these concentrations without withdrawing samples for analysis; only in this way can we ensure that a representative analysis is obtained, and that the measuring process does not disturb the phenomena being observed. In order to develop techniques of combustion gas analysis *in situ*, we have studied the quantitative relations between infrared spectral transmittances of combustion gases and concentrations of the respective infrared-absorbing species (Penzias and Tourin, 1962a; Babrov and Tourin, 1963; Penzias and Maclay, 1963; Penzias, 1966). These techniques were primarily intended for application in supersonic combustion research, shock tube studies, or other high-speed flow or short-duration phenomena.

The general principles of infrared gas analysis are well known (Harrison *et al.*, 1948). If a beam of infrared radiation is sent through an infrared-absorbing gas, the gas will absorb some of this radiation whenever the frequency of the radiation equals a characteristic vibration frequency of the gas molecule. At such a frequency, the beam emerging from the gas will be weaker than the incident beam, by the amount of energy absorbed. The energy absorbed is, in turn, dependent on the number of absorbing molecules in the gas path traversed by the infrared beam. The spectral transmittance of the gas τ is given by

$$\tau = I/I_0, \tag{1}$$

where I_0 is the strength of an incident beam of infrared radiation of narrow spectral width centered at ν, a characteristic vibration frequency of the gas, and I is the strength of the transmitted beam emerging from the gas. The quantity $\alpha_\lambda = 1 - (I/I_0)$ is the spectral absorptance of the gas.

In the usual procedure for infrared gas analysis, the infrared-absorbing gas is identified (qualitative analysis) by varying the frequency of the infrared beam and observing the characteristic frequencies at which absorption occurs. Quantitative analysis is commonly performed by comparing the absorptance of the unknown sample at a characteristic wavelength to the absorptance of a series of known samples until a match is obtained. Both unknown and known samples are maintained at the same pressure, temperature, and path length. This procedure is standard for cases where the gas sample of interest can be placed in a gas cell of known length, and where there are no temperature or pressure effects.

For *in situ* analysis absorptance data as a function of composition, geometrical path length, temperature, and pressure at all anticipated experimental conditions are required. To use a purely empirical approach of setting up calibration curves for infrared analysis *in situ* would require an impracticably large number of experimental measurements for all possible temperatures, pressures, path lengths, and mixture ratios. Therefore, we sought to reduce the amount of experimental data required by making use of the theory of molecular spectra to establish quantitative relationships. It was found that the simple Beér–Lambert absorption law usually does not hold for gases (Tourin and Babrov, 1962) and more sophisticated relationships are required. Mathematical models (Goody, 1964) of infrared absorption bands, which correlate the measured transmittance with the gas parameters, were investigated. The band model which best correlated the experimental data (Babrov and Tourin, 1963; Penzias and Maclay, 1963) was the statistical or random band model (Goody, 1964) with random line spacing, constant intensities, and constant widths.

Spectroscopic band model parameters were determined for CO_2 and water vapor over a wide range of temperatures. The method for using the statistical model for gas analysis was tested by the determinations of CO_2 concentrations in shock-heated gases from transmittance measurements.

From the results of this work (Penzias and Maclay, 1965; Penzias, 1966) we have demonstrated how to use the statistical model to determine CO_2 and H_2O concentrations *in situ* in combustion gases.

B. Theory

Detailed mathematical derivations of the statistical model based upon spectroscopic considerations have been given by Goody (1964), Plass

(1958), and Oppenheim (1963). The resulting equations pertinent to the present work are given below. The statistical model relates the transmittance τ to absorbing path length l, absorber pressure P_a, and nonabsorbing broadening gas pressure P_b, using three spectroscopic parameters, as follows:

$$\ln \tau^{-1} = 2\pi P_a[(\gamma_a^0/d) + m(\gamma_b^0/d)]f(x), \tag{2}$$

where $m = P_b/P_a$ and $f(x) = xe^{-x}[J_0(ix) - iJ_1(ix)]$. The terms $J_0(ix)$ and $J_1(ix)$ are Bessel functions. The argument x is defined as

$$x = (S^0/d)l/2\pi[(\gamma_a^0/d) + m(\gamma_b^0/d)], \tag{3}$$

where S^0/d is the line strength parameter, γ_a^0/d and γ_b^0/d are the half-width parameters for self-broadening and for foreign gas broadening, respectively, per unit pressure and averaged over the spectral interval. Tabulated values of $f(x)$ are available (Kaplan and Eggers, 1956).

The statistical model parameters S^0/d, γ_a^0/d, and γ_b^0/d in Eqs. (2) and (3) are determined from transmittance measurements made at a specified wavelength and temperature. Equation (2) can be rewritten as

$$(\ln \tau^{-1})/P_a = C(S^0/d, \gamma_a^0/d, \gamma_b^0/d, m, l), \tag{4}$$

where C, constant with pressure, is the slope of a straight line that results from plotting $\ln \tau^{-1}$ versus P_a for a given mixture ratio and cell length. If the plot of $\ln \tau^{-1}$ versus P_a is not a straight line which intersects the origin, then the statistical model is not applicable to the transmittance data.

Since there are three unknowns, transmittance measurements at three mixture ratios and/or cell lengths must be made and the values of C determined in order to determine the unknown parameters.

From Eqs. (2) and (4) we see that the equations to be solved have the following form:

$$C_i = 2\pi[(\gamma_a^0/d) + m_i(\gamma_b^0/d)]f\{(S^0/d)l_i/[2\pi(\gamma_a^0/d) + m_i(\gamma_b^0/d)]\}, \tag{5}$$

where $i = 1, 2, 3$ correspond to three measurements made at three different mixture ratios and cell lengths. An iterative method for the solution of these equations has been described by Babrov (1960) and also by Krakow *et al.* (1966).

In principle, one could obtain the band model parameters from only three measurements. The accuracy of the values obtained, however, depends on the sensitivity of the particular experiments conducted. In practice the following experimental conditions are used to obtain maximum accuracy:

(1) pure absorber and long path length—high x ($x > 5$);

(2) large pressure of broadener relative to the pressure of absorber, short path length—low x ($x < 0.1$);

(3) moderate pressure of broadener relative to pressure of absorber, long path length—high x.

The accuracy of S^0/d depends on measurements (2) and (3). The greater the ratio of high x to low x, the greater the accuracy of the value determined for S^0/d. The ratio of path length in the two experiments should be as great as possible. This difference in path length is desirable because one cannot achieve optimum accuracy by manipulation of mixture ratio alone.

The value of $\gamma_a{}^0/d$ is determined from the self-broadening experiment (1) above, the experiment with pure absorber.

The accuracy of $\gamma_b{}^0/d$ is influenced by the mixture ratio used in experiment (3), the high x measurement. To avoid large errors in the determination of $\gamma_b{}^0/d$ the mixture ratio used should be chosen so that the total half-width is due to a significant extent to the foreign gas broadening.

Since $\ln \tau^{-1}$ varies linearly with pressure at a constant mixture ratio, the value of C in Eq. (4) can be determined accurately from the slope of the best straight line drawn through a graph of the results of a series of measurements made at different absorber pressures but with the path length and mixture ratio as specified above, in order to obtain maximum accuracy. Clearly it is undesirable to measure absorptances near unity or zero for input data to a statistical band model fit because the error in the transmittance becomes extremely large.

The discussion above pertains to transmittance measurements made at the same wavelength and temperature. To obtain the temperature variation of transmittance as a function of path length, pressure, and concentration, it is necessary to perform a series of measurements as described above at various temperatures and obtain the band model parameters. The parameters can then be graphed as a function of temperature, for interpolation. The techniques described above have been used to obtain band model parameters for CO_2 and H_2O. The experimental approach to obtaining the necessary transmittance data is strongly influenced by the accuracy requirements given above.

C. Experimental Apparatus

To provide experimental data for determining the statistical model parameters for CO_2 and water vapor, transmittances for CO_2 and CO_2–N_2

mixtures and for water vapor and water vapor–nitrogen mixtures were measured at temperatures between 1273 and 600°K in electrically heated infrared gas cells for various cell lengths and mixture ratios. CO_2 transmittance data, at temperatures above 1273°K, were obtained from shock tube measurements.

1. Instrumentation for Gas Cell and Combustion Gas Measurements

A general experimental setup for measuring infrared transmittances of hot gases, including details of the optical system, is shown schematically in Fig. 1. The optical system can be used either with gas cells heated in an electric tube furnace or with flames and combustion gases. As shown, the source and receiver are disposed opposite one another on either side of the sample of interest which is located at a common focal point of the source and receiver. If the flame or combustion gas stream is very large, both source and receiver are focused at infinity. The instrument functions as follows: infrared radiation from the globar G (an electrically heated silicon carbide rod providing an arbitrary, but constant intensity) in the source unit passes through chopper Ch, a rotating sector disk operating at a fixed frequency. The resulting pulsed infrared beam is reflected from plane mirror M_1 to spherical mirror M_2. Then M_2 sends the pulsed beam through the hot gases to spherical mirror M_3, which collects the steady radiation beam from the hot gases as well as the pulsed beam from M_2. Radiation collected by M_3 is reflected by plane mirror M_4 to plane mirror M_5. A second chopper

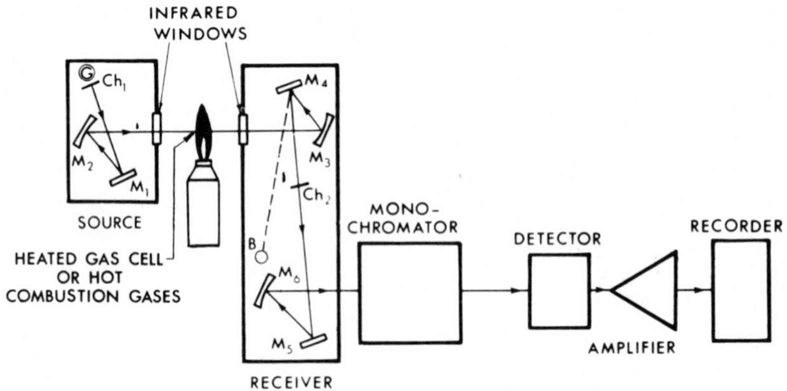

Fig. 1. Schematic diagram of spectroradiometer pyrometer for measuring transmittances of hot gases and temperatures of flames and plasmas. [From Penzias and Tourin (1962). *Combust. Flame* **6**, 149.]

Ch_2 is located at a focal point of M_3 between M_4 and M_5. Radiation relayed by M_5 is collected by spherical mirror M_6 and focused on the entrance slit of the monochromator. A prism or grating monochromator may be used; its function is to select a narrow wavelength band from the incoming radiation and transmit that band to the infrared detector. The detector converts the radiant energy it receives to an electrical signal, which is amplified and indicated on the recorder.

To measure the transmittance of the hot gases, the chopped radiation from the globar is first measured without any hot gases in the optical path (I_0). The measurement is then repeated with the hot gases in the optical beam. In this case, the radiation reaching the detector consists of the chopped globar radiation diminished due to the absorptance of the gas and the steady radiation emitted by the hot gases. Only the chopped signal from the globar is amplified by the ac amplifier which is tuned to the chopper frequency. Therefore, only the radiation transmitted by the hot gases (I) is recorded. The spectral gas transmittance τ is determined from the two measurements above as per Eq. (1).

The reason for using the two-chopper optical system, illustrated in Fig. 1, is to facilitate the measurement of gas emission as well as absorption in order that the gas temperature may be determined using the infrared emission–absorption method (Tourin, 1962). This method is discussed in detail in Section III. To measure radiation from the hot gas, a shutter is interposed in front of the globar G, blocking the radiation from the source unit. Chopper Ch_2 then modulates the radiation from the hot gases, thus producing an ac signal for the detector and electronic system. For calibration purposes, mirror M_4 can be rotated so that radiation from the blackbody standard illuminates the measuring system.

Specially constructed quartz gas cells of varying lengths, heated electrically in a tubular furnace, were used to obtain transmittance data up to 1273°K. The gas composition was carefully controlled (Babrov and Tourin, 1963) to provide the desired mixture ratios required for accurate determination of the band model parameters.

2. Instrumentation for Shock-Heated Gas Measurements

Infrared transmittances of shock-heated gases were measured with a high-speed spectroradiometer (Penzias *et al.*, 1966, p. 225). The elements of the high-speed spectroradiometer are essentially the same as those shown in Fig. 1. However, the chopper frequency in the source unit and the electronics differ because of the short test times (of the order of 1 msec) available. The high-speed spectroradiometer, consisting of a source and a receiver unit, is shown at the shock tube viewing section in Fig. 2. The

Fig. 2. High-speed spectroradiometer for measuring emission and transmittance of shock-heated gases and detonations, positioned at the shock tube viewing section. [From Penzias, G. J., Dolin, S. A., and Kruegle, H. A. (1966). *Appl. Optics* **5,** 225. Copyright 1966, the Optical Society of America.]

source unit generates a beam of radiant energy modulated at 90 kHz, directed through the viewing windows to the receiver unit.

The modulated radiation beam from the source and the unmodulated radiant emission of the shock-heated gases are collected simultaneously by the receiver unit, which contains a grating monochromator set at a particular wavelength where the gas emits and absorbs. The radiant energy within the infrared beam passed by the monochromator is focused on an indium antimonide infrared detector. The detector converts the radiation signal to an electrical signal which is amplified and displayed on a dual-beam oscilloscope. Since the source emission is modulated and the emission of the shock-heated gases is unmodulated, it is possible to discriminate between them and to separate them electrically, thereby yielding simultaneous outputs indicative of gas emission and gas absorption.

A typical oscillogram of shock-heated CO_2–N_2 mixture obtained with the high-speed spectroradiometer is shown in Fig. 3. As can be seen in Fig. 3, the dual-beam oscilloscope used to record the data is triggered 75 μsec before the arrival of the shock wave at the viewing section. This provides a measure of the initial modulated source signal (upper trace in Fig. 3). When the shock wave passes the windows, the hot gas radiation is detected, resulting in the signal on the lower trace in Fig. 3. Simultaneously, the modulated signal on the upper trace is diminished as the heated gases absorb some of the source energy. With the arrival of the contact surface, the gases are cooled, and the signal on the lower trace gradually decreases to the zero emission level, while the upper, modulated signal returns to its initial value.

The transmittance is obtained by measuring the peak-to-peak amplitude I_0 of the modulated source signal before the arrival of the wave and the peak-to-peak amplitude I of the decreased modulated signal during the steady hot flow portion.

A single-pulse shock tube system, shown in Fig. 2, was employed. The shock tube low-pressure section was a 5.72-cm-i.d. stainless steel tube,

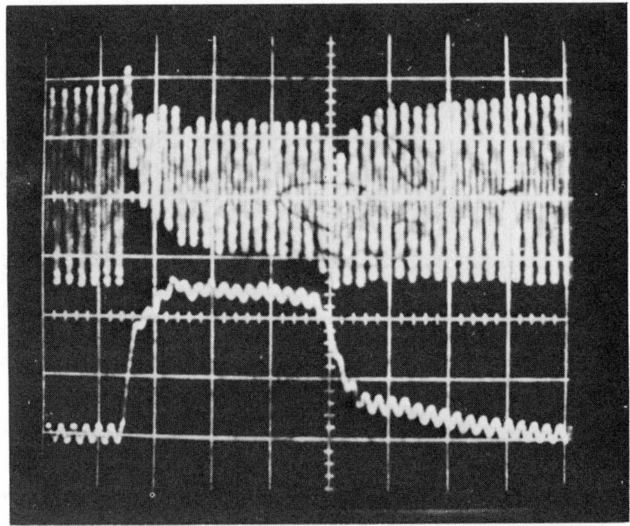

Fig. 3. Typical oscillogram of shock-heated CO_2–N_2 mixture obtained with high-speed spectroradiometer. Upper modulated trace shows the transmitted intensity, from which the spectral absorptance may be determined. Lower trace shows the sum of the transmitted and emitted intensities. Sweep rate 50 μsec/cm (laboratory time). [From Penzias, G. J., Dolin, S. A., and Kruegle, H. A. (1966). *Appl. Optics* **5**, 225. Copyright 1966, the Optical Society of America.]

4.57-m long. The driver section was an 11.42-cm-i.d. stainless steel tube, 1.22-m long. The driver and driven sections were connected by a converging transition section containing a 5.08-cm valve to permit rapid evacuation of the shock tube. Aluminum or Mylar diaphragms, held at the transition section, were ruptured by increasing the driver gas pressure. The shock tube viewing section, which contained two sapphire windows mounted in line, was located 4.12 m from the diaphragm.

Platinum thin-film resistance probes (Gaydon and Hurle, 1962), mounted flush along the shock tube, were used to measure the shock velocity. The probe outputs were displayed on an oscilloscope modified for raster sweep operation. The raster sweep was timed using a time-mark generator with pulses every 10 μsec. The shock velocity measurements were accurate to within ±0.5%. The signal generated by one of the thin-film resistance probes was also used to trigger the dual-beam oscilloscope which recorded the emission and absorption measurements.

D. Results and Applications

1. CO_2 Statistical Model Parameters

The statistical model parameters, to relate spectral transmittance to CO_2 concentration, were obtained at 4.40 μm. This wavelength was selected because there is no interference from atmospheric CO_2, the transmittance is lowest (highest absorptance) at this wavelength, and there is least interference from CO (4.6 μm fundamental band) and H_2O (6 μm band) which may be present in combustion gases.

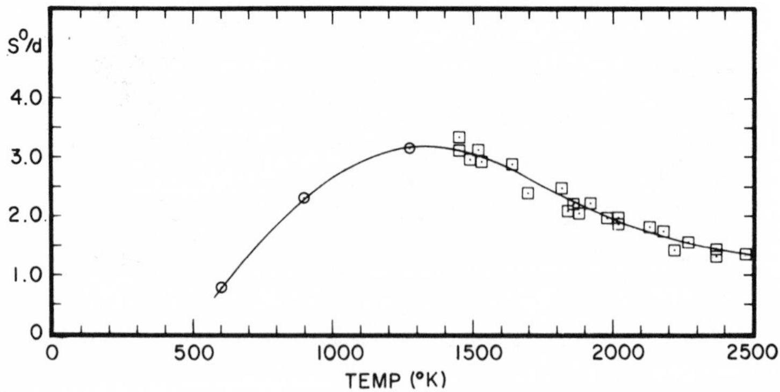

Fig. 4. Strength parameter ($S°/d$) of CO_2 at 4.40 μm versus temperature. ⊙ Furnace data; ⊡ shock tube data.

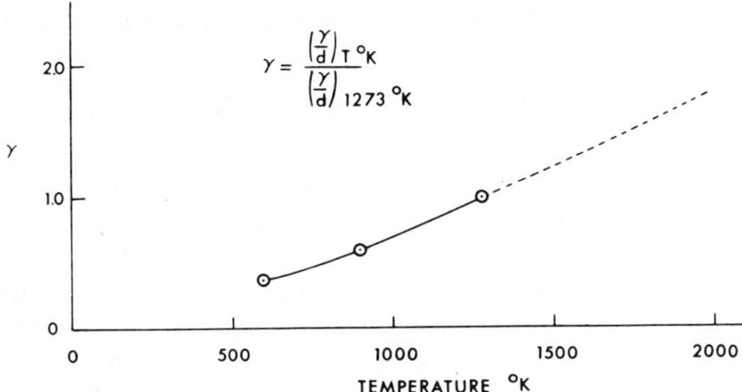

Fig. 5. Ratio of total half-width parameter (γ°/d) at temperature T to half-width at 1273°K for CO_2–N_2 at 4.40 μm versus temperature. ⊙ Furnace data. [Penzias and Maclay (1965). *Int. Symp. Combust., 10th,* p. 191. Combust. Inst., Pittsburgh, Pennsylvania.]

The band model parameters obtained over the temperature range from 600 to 2500°K are shown in Figs. 4 and 5. The strength parameter (S^0/d) of CO_2, plotted as a function of temperature, is shown in Fig. 4. The half-width parameter temperature dependence is shown in Fig. 5, plotted as the ratio of the half-width parameter at T to its value at 1273°K. In Fig. 5, the half-width parameters for self-broadening and foreign gas broadening have been combined and are represented as γ^0/d. The values of $(\gamma_a^0/d)CO_2$ and $(\gamma_b^0/d)N_2$ at 1273°K are 1.67 and 0.63 atm^{-1}, respectively. For convenience, the half-width parameter temperature variation was presented as shown in Fig. 5 to simplify using the data for concentration determinations.

Note that the half-width parameters increase with temperature, whereas, in general, one expects the actual half-width of a single line to decrease with increasing temperature. This apparently anomalous behavior of the CO_2 half-width parameter is due to the increasing number of hot bands and the resulting decrease in the mean line spacing. The half-width of the individual lines may be decreasing with increasing temperature but the mean line spacing is decreasing more rapidly.

The dashed portion of the curve in Fig. 5 above 1273°K represents an extrapolation of the data. It was not possible to determine the half-width parameters above 1273°K precisely, because the data obtained from the shock-tube measurements were made at low values of x [Eq. (3)]. The strength parameters, however, can be determined from low x measurements. For low values of $x, f(x) \cong x$ and Eq. (2) reduces to

$$\ln \tau^{-1} = P_a(S^0/d)l. \tag{6}$$

For values of $x < 0.1$ the error in determining S^0/d using Eq. (6) is less than 5% and increases to 10% for $x < 0.3$.

2. Application of the Statistical Model to CO_2 Determination in Shock-Heated Gases

The statistical model relationship for determining CO_2 concentrations *in situ* from infrared transmittance measurements was tested using the shock tube system. The transmittance of shock-heated CO_2 at 4.40 μm for several CO_2–N_2 mixtures was measured using the high-speed spectroradiometer. The CO_2–N_2 mixtures were prepared by the method of partial pressures in a mixing chamber and were allowed to stand for 24 hr before use. The shock tube was evacuated to a pressure of a few microns before filling, and the shock tube system had an overall leak rate of less than 1 μm Hg/min. Initial pressures in the low-pressure section were measured with a mercury manometer using a double-reduction technique (Babrov, 1960) which permitted the initial pressure to be measured with an experimental error of less than 0.25%. The gas temperatures and pressures were determined from the measured shock velocity using the method suggested by Gaydon and Hurle (1962), assuming a chemically frozen and vibrationally equilibrated state behind the incident shock.

To determine the concentration of CO_2 in the shock-heated gas, the procedure used was as follows:

1. The shock velocity and infrared transmittance at 4.40 μm were measured.
2. From the shock velocity, the gas temperature and pressure were determined. (Gas temperature can also be determined from the infrared emission and absorption as described in Section III.)
3. Using the gas temperature determined in step 2, the strength and half-width parameters were obtained from Figs. 4 and 5.
4. The geometrical path length (in centimeters) l corresponds to the distance between the windows in the shock-tube observation section and was known.
5. The values of the strength and half-width parameters and geometrical path length were substituted into Eqs. (2) and (3), which were solved for the absorber pressure P_a. Note that $m = (P_t - P_a)/P_a$, where P_t is the pressure determined in step 2.

The results of the shock-tube measurements are shown in Table I. The agreement between the actual concentration and the concentration determined from the measured transmittance using the statistical model relationship is better than 8%. This is within the error of the strength and half-width temperature values in Figs. 4 and 5.

TABLE I

COMPARISON OF EXPERIMENTAL AND CALCULATED CONCENTRATIONS OF SHOCK-HEATED CO_2, DETERMINED *in situ*, USING THE STATISTICAL MODEL AT 4.40 μm[a]

T (°K)	Pressure (atm)	l (cm)	τ	CO_2 partial pressure (atm)	
				Experimental (initial mixture)	Calculated (from statistical model)
2390	0.847	6.56	0.639	0.0423	0.0398
2325	0.781	6.56	0.317	0.112	0.105
2210	0.922	6.56	0.227	0.141	0.131
2090	0.805	6.56	0.590	0.0403	0.0398

[a] Penzias and Maclay (1965). *Int. Symp. Combust., 10th,* Combust. Inst., Pittsburgh, Pennsylvania.

3. WATER VAPOR STATISTICAL MODEL PARAMETERS

To determine water vapor band model parameters, regions where the water vapor absorptance is strong (i.e., transmittance is low) must be sought. This is complicated by the presence of water vapor in the atmosphere which must be eliminated from the optical path. This interference can be minimized by either evacuating or flushing the system with an inert gas such as nitrogen. For practical application in combustion studies, flushing the optical path with dry nitrogen is preferred.

When the optical path was flushed with dry nitrogen, interference from atmospheric H_2O was eliminated in two wavelength regions, at 2.51 and 2.85 μm, where the hot water vapor absorptances were relatively strong. The transmittance is lower at 2.85 μm than at 2.51 μm, which reduces the experimental error for transmittance measurements at low water vapor concentrations. However, at 2.85 μm, CO_2 also has an absorption band. Therefore, when CO_2 is present in the combustion gas stream, measurements must be made at 2.51 μm, where CO_2 does not interfere. Because of interest in hydrogen–air combustion, it was felt that the additional effort in obtaining data at two wavelengths is justified.

The statistical model parameters for water vapor have been determined at 2.51 and 2.85 μm at 1273 and 637°K using gas cells. The parameters are listed in Table II.

The limited amount of data available for water vapor is due in great part to the experimental difficulties involved in getting accurate data at elevated temperatures. Additionally, measurement of water vapor transmittances

TABLE II

H_2O STATISTICAL MODEL PARAMETERS

λ (μm)	$S°/d$	$\gamma_a°/d$	$\gamma_b°/d$	T (°K)
2.51	0.175	0.486	0.085	1273
2.51	0.184	0.706	—	637
2.85	0.311	0.617	0.12	1273
2.85	0.362	1.088	—	637

are strongly dependent on the spectral resolving power of the instrument used to obtain the data. (This is not the case for CO_2.) The narrower the spectral slit width, the higher the measured absorptance. The data in Table II were obtained from spectra measured with an instrument having a resolution of approximately 2 cm^{-1}, which is a practical upper limit to the resolution that can readily be achieved with standard grating instrumentation. When these data are to be applied to determining water vapor concentrations *in situ*, the same spectral resolution must be used.

Because of the above-mentioned difficulties, other approaches such as the use of integrated absorption (Krakow *et al.*, 1966) and high-resolution measurements to determine individual line strengths and widths (Babrov and Casden, 1968; Babrov and Healy, 1969; Krakow and Healy, 1969) appear to offer a more fruitful means for developing *in situ* gas analysis for water vapor from infrared transmittance measurements.

III. SPECTROSCOPIC METHODS OF TEMPERATURE MEASUREMENT

A. General

To apply the methods discussed above for determining species concentration, a knowledge of the gas temperature is required. For the same reasons that sampling is not possible, it may not be possible to determine gas temperatures by conventional immersion methods. Additionally, at the higher temperatures of combustion flames and plasmas, the use of temperature probes becomes impracticable. A natural alternative is to make use of the radiation emitted by the hot gases as a means for determining gas temperatures. Radiation pyrometry allows temperature determination without being in contact with the hot body. The temperature measuring range is unlimited and since no probe is used, the measurement is independent of gas velocity.

The idea of determining temperatures of hot gases from their radiation is not new. An excellent monograph by Tourin (1966a) covers this subject in detail. It is my intention to describe briefly some of the spectroscopic methods of temperature measurement used in this laboratory and illustrate their application to flames, shocks, detonations, and plasmas. For greater detail, the interested reader is referred to the aforementioned monograph by Tourin (1966a).

Most methods of pyrometry can be grouped into two classes—radiometric and spectrometric.

Radiometric methods involve only thermodynamics, without reference to the mechanisms of radiation. The principal radiometric method employed in this laboratory for measuring flame, shock, and detonation temperatures is the infrared emission–absorption method (Tourin, 1962) described in Section III,B.

Spectrometric methods entail explicit use of the quantum theory of optical spectroscopy. These methods have been applied extensively to plasmas and are described in Section III,C.

B. Flames, Shocks, and Detonations

In the infrared emission–absorption method, the spectral radiance and spectral absorptance of the hot gas are measured in a selected narrow infrared band. Gas temperature is then determined by means of Kirchhoff's law,

$$N_\lambda(T)/\alpha_\lambda(T) = N_\lambda^{\mathrm{b}}(T), \tag{7}$$

and the Planck radiation law,

$$N_\lambda^{\mathrm{b}}(T) = (C_1\lambda^{-5}/\pi)[\exp(c_2/\lambda T) - 1]^{-1}, \tag{8}$$

where $N_\lambda(T)$ (W/cm^2 sr μm), $\alpha_\lambda(T)$, and $N_\lambda^{\mathrm{b}}(T)$ are, respectively, the spectral radiance of the gas at a wavelength λ, the spectral absorptance of the gas at the same wavelength, and the spectral radiance of a blackbody at that wavelength; T is the absolute temperature; c_1 and c_2 are constants. Measurements of $N_\lambda(T)$ and $\alpha_\lambda(T)$ determine $N_\lambda^{\mathrm{b}}(T)$; T is then easily obtained from Eq. (8).

The essential elements of a spectroradiometer pyrometer for measuring gas temperatures are shown schematically in Fig. 1. The functioning of the instrument was described in Section II,C. To measure temperature by the infrared emission–absorption method, the emission and absorption measurements must be made through identical optical paths (Babrov, 1961). This requires that the optical paths be carefully matched throughout the optical train, from the radiating source in the source unit through to the detector, so that the optical throughput is constant. This requirement is satisfied by the optical system shown in Fig. 1.

An emission and absorption spectrum typical of nonluminous hydrocarbon–oxygen flames is shown in Fig. 6. These data are from a 12-cm-diameter methanol–oxygen flame measured 14 cm from the base of the flame. From the emittance and emissivity data of Fig. 6 the flame temperature can be determined by means of Eqs. (7) and (8), noting that for a gas the spectral radiance is equal to spectral emittance divided by π and that emissivity is identical to absorptance. Figure 7, curve A, shows flame temperature, plotted against wavelength, obtained from the data of Fig. 6. A similar plot for a point 23 cm from the base of the flame is shown in Fig. 7, curve B.

Figure 7 has a number of interesting features. The indicated temperature is different at different wavelengths, corresponding to the nonuniformity of the temperature distributions across the flame. Higher temperatures are obtained from wavelengths where the emissivity is low than from wavelengths where the emissivity is high. This can be seen by comparing the

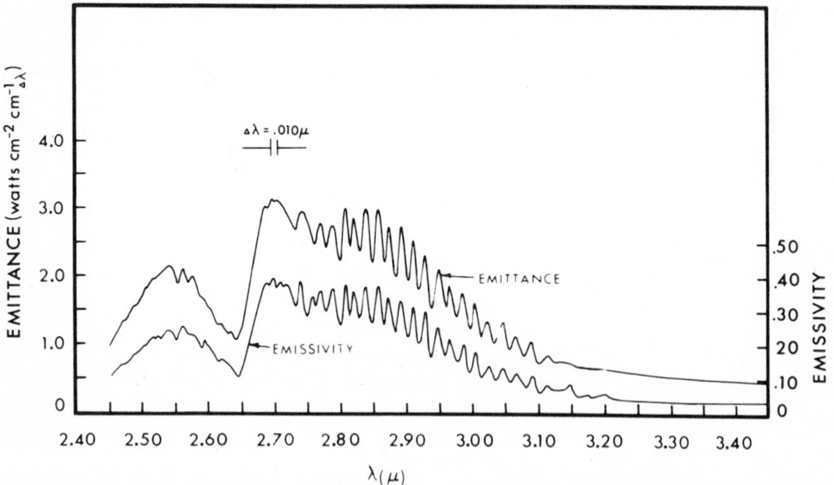

Fig. 6. Typical spectral emission and emissivity plots of nonluminous hydrocarbon flame. These curves were obtained from measurements of emission and transmittance of a 12-cm-diameter methanol–oxygen flame, 14 cm from the base of the flame. [From "Temperature, Its Measurement and Control" by R. H. Tourin. Copyright (C) 1962 by Litton Educational Publishing, Inc., by permission of Reinhold Publishing Corporation.]

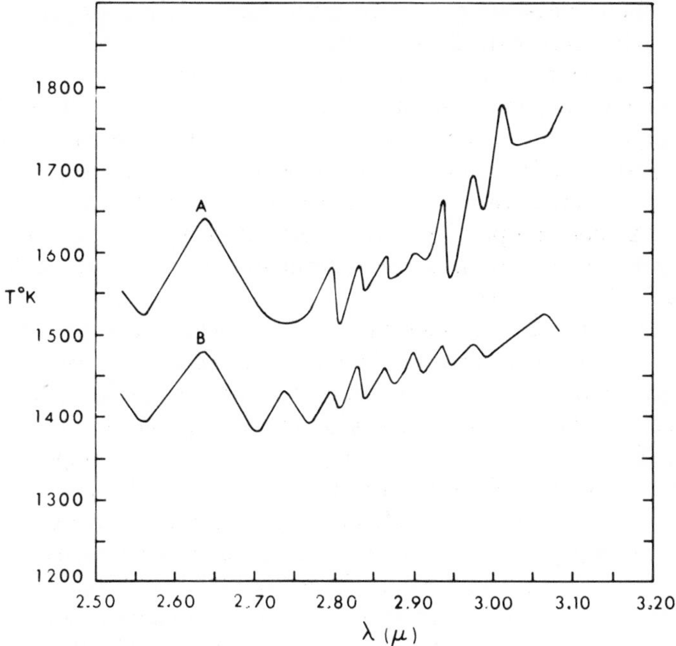

Fig. 7. Flame temperature determined by infrared emission–absorption method. Curve A is a plot of methanol–oxygen flame temperature versus wavelength from the data of Fig. 6. For convenience, alternate spectral peaks have been omitted. Curve B shows a corresponding plot for the same flame, measured 9 cm further downstream than for curve A. [From "Temperature, Its Measurement and Control" by R. H. Tourin. Copyright (C) 1962 by Litton Educational Publishing, Inc., by permission of Reinhold Publishing Corporation.]

emissivity plot in Fig. 6 with the data in curve A, Fig. 7. The reason for this phenomenon is clear. When the emissivity is high, radiation emitted from the flame interior suffers considerable reabsorption; hence, in effect we are primarily measuring radiation from the cooler outer boundary of the flame and the indicated temperature is relatively lower. On the other hand, at wavelengths where the emissivity is low there is less reabsorption and the pyrometer receives radiation from the hotter inner core of the flame, resulting in higher indicated temperatures.

Comparison of the two curves in Fig. 7 brings out the effect of different temperature gradients. The data for curve A were obtained in the flame reaction zone where relatively steep gradients existed due to the mixing pattern. The steep rise in temperature at the long-wavelength end corresponds to the temperature in the core of the flame. Contrasting this to curve B where measurements were made further downstream and substantial mixing occurred, the temperature gradient is much less and the corresponding temperature wavelength plot is nearly level.

The infrared emission–absorption method, when applied to inhomogeneous flames, results in a weighted average of the temperatures across the flame. A qualitative indication of the temperature variation can be derived from plots of temperature versus wavelength similar to those of Fig. 7. However, this does not provide a strict point-by-point determination of the temperature profile. Methods for determining temperature profiles are discussed by Tourin (1966a).

Shock and detonation temperatures have been measured with the high-speed spectroradiometer pictured in Fig. 2 (Penzias *et al.*, 1966, p. 225). The spectral radiance and absorptance are obtained from oscillographic records such as the one shown in Fig. 3. The absorptance is obtained from the upper trace in Fig. 3 by measuring the peak-to-peak amplitude I_0 of the modulated source before the arrival of the wave and the peak-to-peak amplitude I of the decreased modulated signal during the steady hot flow portion. The ratio I/I_0 equals the gas transmittance. One minus the transmittance equals the absorptance $\alpha_\lambda(T)$ of Eq. (7). The increase in voltage in the lower trace of Fig. 3 corresponds to the gas radiation, which is converted to spectral radiance, $N_\lambda(T)$, by means of a blackbody calibration. The gas temperature is then readily determined from Eq. (8).

Gas temperatures in shock-heated CO_2–N_2 mixtures were determined by simultaneously measuring the infrared spectral radiance and absorptance of hot CO_2 at 4.40 μm. The temperature measurements were made in the incident shock and were compared with values calculated from the measured shock velocities, using the method suggested by Gaydon and Hurle (1962) for a chemically frozen and vibrationally equilibrated state behind the incident shock. The results are shown in Table III. The error in the calculated temperature due to the uncertainty of the velocity measurement is 15°K.

In order for the measured temperature to have significance, equilibrium between the vibrational, rotational, and translational degrees of freedom must be established. The measurements were made 50 to 100 μsec after the arrival of the shock wave, to allow sufficient time for vibrational relaxation and the establishment of thermal equilibrium. The maximum difference between measured and calculated temperature was 100°K. This is within the experimental error of 4%.

Detonation temperatures were determined for hydrogen–oxygen, propane–oxygen, and propane–oxygen–nitrogen mixtures, at 1 atm initial pressure, from spectral radiance and absorptance measurements at various

TABLE III

INCIDENT SHOCK TEMPERATURES IN DEGREES KELVIN[a]

Measured[b]	Calculated[c]	Measured[b]	Calculated[c]
1485	1490	2122	2091
1510	1520	2118	2207
1690	1700	2248	2325
1840	1860	2284	2390
2010	1980	2480	2480

[a] From Penzias, G. J., Dolin, S. A., and Kruegle, H. A. (1966). *Appl. Optics* **5**, 228. Copyright 1966, the Optical Society of America.

[b] From infrared emission and absorption of hot CO_2 at 4.40 μm.

[c] From shock velocity by the method of Gaydon and Hurle (1962).

wavelengths in the 2.5-μm H_2O band and the 4.3-μm CO_2 band. The detonations were produced by filling the low-pressure sections of the shock tube with the premixed combustible gases, and igniting with a spark.

The measured temperatures and velocities were compared with theoretical values computed on the basis of the Chapman–Jouget theory (Emmons, 1958), as shown in Table IV. The differences between theoretical and experimental temperatures are within the experimental error of 4%.

In the above-mentioned applications of the infrared emission–absorption method, the gas absorptance in Eq. (7) was obtained from measurements of gas transmittance. For flames containing solid particles, such as pulverized coal and "sooty" flames,

$$\alpha_\lambda(T) \neq 1 - (I/I_0), \tag{9}$$

because the solid particles scatter radiation from the source unit and interfere with the I measurement.

The infrared emission–absorption method could still be applied if the effect of scattering on the transmittance measurement could be determined, or alternatively, if the emissivity of the solid particles could be calculated from other information, the absorptance measurement could be eliminated. Formulas for estimating the emissivity of solid particles in industrial flames are given in Chapter VIII.

Another approach is to apply the "two-color" method. From Eq. (7) the spectral radiances and absorptances at two wavelengths λ_1 and λ_2 are related

TABLE IV

COMPARISON OF EXPERIMENTAL RESULTS WITH THEORY FOR SELF-SUSTAINING DETONATIONS[a]

Mixture	Velocity (m/sec)		Temperature (°K)	
	D_{th}	D_{exp}	T_{th}	T_{exp}
$2H_2 + O_2$	2826[b]	2780	3646[b]	3670[e]
$H_2 + O_2$	2306[b]	2200	3423[b]	3340[e]
$0.4H_2 + 0.6O_2$	2082[b]	2130	3172[b]	3246[e]
$C_3H_8 + 5O_2$	2362[c]	2345	3830[d]	3700[f]
$C_3H_8 + 5O_2 + 3N_2$	2177[c]	2210	3580[d]	3490[f]

[a] From Penzias, G. J., Dolin, S. A., and Kruegle, H. A. (1966). *Appl. Optics* **5**, 230. Copyright 1966, the Optical Society of America.

[b] Bollinger and Edse (1961).

[c] Manson *et al.* (1963).

[d] Rouge (1964) (private communication from N. Manson, 1964).

[e] H_2O—2.50 μm.

[f] CO_2—4.50 μm.

as follows:

$$[N_{\lambda_1}(T)/\alpha_{\lambda_1}(T)]/[N_{\lambda_2}(T)/\alpha_{\lambda_2}(T)] = N_{\lambda_1}{}^{b}(T)/N_{\lambda_2}{}^{b}(T). \quad (10)$$

If we assume $\alpha_{\lambda_1}(T) = \alpha_{\lambda_2}(T)$, then Eq. (10) reduces to

$$N_{\lambda_1}(T)/N_{\lambda_2}(T) = R_b(\lambda_1\lambda_2, T), \quad (11)$$

where

$$R_b(\lambda_1, \lambda_2, T) = N_{\lambda_1}{}^{b}(T)/N_{\lambda_2}{}^{b}(T) \quad (12)$$

is a known function of temperature. The flame temperature can readily be determined from Eq. (11) by measuring $N_{\lambda_1}(T)$ and $N_{\lambda_2}(T)$. The advantage of this method is that only emission is measured.

The two-color method should be employed with caution, recognizing the assumption made. In practice, only sooty flames or flames containing solid particles are sufficiently gray for this method to be applicable.

C. Plasmas

1. GENERAL PLASMA CHARACTERISTICS

Plasma may be defined as gas which is at least partially ionized. Plasmas are usually hotter than combustion gases and are electrically conductive.

Plasmas in the range up to approximately 8000°K are composed mainly of atoms and diatomic molecules, plus a few percent of ions (atomic and molecular) and electrons. Few polyatomic molecules are found in plasmas because they are dissociated by the high temperature. As one goes to higher temperatures, dissociation and ionization become progressively greater, until eventually the plasma consists of atomic nuclei and free electrons. Obviously the study of plasma covers an enormous range of phenomena. Relatively cool plasmas produced by plasma jets and electric arcs, in the temperature range 5000–20,000°K, are involved in engineering applications such as metal cutting, flame spraying, hypersonic wind tunnels, and simulation of space vehicle reentry. Experiments with exploding wires involve hotter plasmas, up to perhaps 50,000°K. Still higher temperatures, to millions of degrees, are the domain of plasma physics, in which the plasma is usually confined by magnetic field and in which the main practical objective is the release of energy by controlled nuclear fusion.

The plasma jet is a stream of gas, issuing from a nozzle, which has passed through an electric arc and been heated by the arc, but which does not carry appreciable electrical current. The term "arc" is usually reserved for the plasma region between electrodes, in which significant amounts of electric current are carried. In plasma jets and electric arcs, temperatures are usually below 20,000°K; the plasma consists of diatomic molecules, atoms, ions, and electrons.

The characteristics of the spectra of plasmas determine the appropriate spectroscopic methods of the temperature measurement to be used. Plasma jets and arcs typically exhibit strong emission lines of gas atoms and ions, some diatomic molecular bands, and some continuum. Absorption is usually very weak in the visible and near ultraviolet regions, and therefore the plasma is said to be optically thin. However, absorption in the infrared region may be quite strong.

Figure 8 shows emission and absorption spectra of an argon plasma jet in the near infrared (Tourin, 1966b). The lowermost spectrum in Fig. 8 is the emission spectrum. At the top of Fig. 8 is the continuous spectrum of a tungsten strip lamp used as a source and measured with the plasma jet off (I_0 measurement). Just below the strip lamp continuum is the spectrum of the strip lamp transmitted through the plasma, showing the characteristic absorption by the plasma. The spectral lines are due to neutral argon. Absorption is quite strong in the near infrared region, decreasing toward the red. The presence of measurable absorption enables us to determine the absorptance at various infrared wavelengths, and thereby to measure the plasma temperature by the emission–absorption method [Eqs. (7) and (8)].

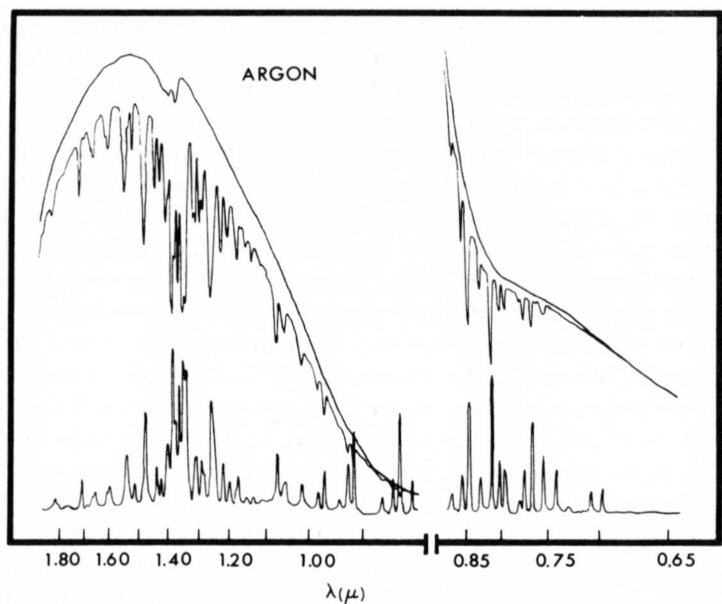

Fig. 8. Emission and absorption spectra of an argon plasma jet. [From Tourin (1966b). *Appl. Spectrosc.* **20,** 2.]

In spectral regions where absorption is strong, the plasma is said to be optically thick. At the same time the plasma may be optically thin, that is, no measurable absorption in the visible region. Figure 9 shows the emission spectrum of a nitrogen plasma jet in the visible wavelength region near 4000 Å at two distances above the nozzle (Tourin, 1966b). The spectra are due to atomic nitrogen and the nitrogen molecular ion N_2^+. The optically thin spectra are useful for the spectroscopic temperature measurement methods described below, which have been widely used to measure plasma and arc temperatures.

2. The "One-Line" Method

The expression for the integrated radiance of a spectral line not affected by absorption is

$$N_\lambda^{nm}(T) = \int_{\text{line}} N_\lambda(T)\, d\lambda = 1.582 \times 10^{-20} (g_n A_{nm}/Q\lambda_{nm}) n_0 l \exp(-E_n/kT) \tag{13}$$

in watts per unit area per unit solid angle, where the indices n and m refer to the upper and lower energy states, respectively; A_{nm} is Einstein's coefficient of spontaneous emission (transition probability), λ_{nm} is the wavelength of the line in microns, l is the path length in centimeters, n_0 is the number density of neutral atoms, g_n is the statistical weight of the upper state of the transition, Q is the partition function, E_n is the energy of the upper state, k is Boltzmann's constant, and T is the absolute temperature.

The one-line method is useful to measure the temperature of a hot plasma containing appreciable numbers of excited atoms, provided that the transition probability A_{nm} is known for one or more lines of the gas. The other quantities in Eq. (13) are readily determined, with the exception of the number density of neutral atoms n_0 and the partition function Q, which are a function of temperature. For the noble gases, the ideal gas law can be used to calculate n_0 at temperatures where ionization is negligible. At temperatures higher than about 10,000°K, the ideal gas law is too inaccurate, and n_0 must be determined from the Saha equation (Saha and Srivastra, 1958).

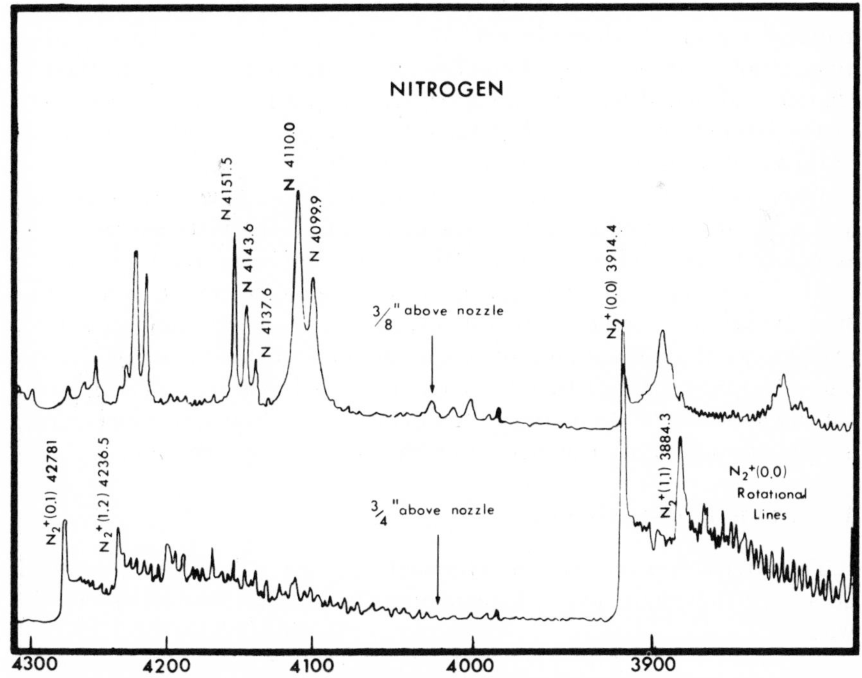

Fig. 9. Spectra of a nitrogen plasma jet at different points on the plasma jet axis. [From Tourin (1966b). *Appl. Spectrosc.* **20**, 2.]

The line radiance N_λ^{nm}, given by Eq. (13), increases with increasing temperature until a maximum is reached at a temperature which depends on the particular atomic line. The maximum occurs when the effect of increasing temperature on the excitation of radiant emission is counterbalanced by the reduction in radiating species concentration, due to ionization. For example, this temperature is 15,500°K for the 4158-Å argon line. If the temperature at the axis of the plasma is higher than that for which line emission is a maximum, the peaking function method (Fowler and Milne, 1924) is preferable to the one-line method, because it requires only a measurement of emission strength relative to the maximum, rather than absolute measurement.

3. Two-Line Ratio Method

It follows from Eq. (13) that for two atomic emission lines

$$\ln(N_1/N_2) = \ln[g_1A_1\lambda_2/g_2A_2\lambda_1] - (E_1 - E_2)/kT, \tag{14}$$

where the subscripts 1 and 2 refer to the two spectral lines. In this case, one need not know the number density or path length, both of which are required for the one-line method. Moreover, one can use relative rather than absolute transition probabilities. A disadvantage of this method is that the calculated temperature is critically dependent upon the accuracy of the ratio of line intensities, if the difference in the upper energy states of the two lines is small (say $E_1 - E_2 < 0.4$ eV).

When both lines are from the neutral atom, the two-line ratio method yields an atomic excitation temperature, that is, the temperature corresponding to the equilibrium population of neutral atom electronic energy levels. If both lines are from an ion, we obtain an ionic excitation temperature. If excitation and ionization should share a common equilibrium, then one can use the ratio of an atomic line radiance to an ionic line radiance. The advantage of this is that the ion and atomic levels are generally much further apart (by 10 to 50 times) than two levels of the same atom or ion, so that the sensitivity is much better for the atom–ion line ratio.

4. Atomic Boltzmann Plot

The extension of the two-line ratio method to many atomic lines is called an atomic Boltzmann plot. Changing units in Eq. (13) and solving for T, we get

$$T = -625E/[\log(N\lambda/gA) - C]. \tag{15}$$

When $\log(N\lambda/gA)$ is plotted against $E \times 10^3$ cm^{-1}, the temperature is given by the reciprocal of the slope.

5. Molecular Boltzmann Plot

In the cooler parts of a plasma jet, molecular emission bands are observed in the visible and ultraviolet regions. A Boltzmann plot can be constructed for a series of lines in one of these bands. For example, for the R branch of the molecule N_2^+, the equation used is (Tourin, 1966a)

$$\ln(N/2J') = \ln(C\nu^4/Q_r) - [BJ'(J'+1)hc/kT], \tag{16}$$

where N is the radiance of a line in a diatomic molecular vibration–rotation band, J' is the rotational quantum number, C is a constant incorporating the transition probability, ν is the wavenumber in reciprocal centimeters, B is the rotational constant for the upper state, and Q_r is the rotational partition function.

The quantity $\ln(N/2J')$ is plotted against $(B/k)J'(J'+1)hc$ and the temperature is given by the reciprocal of the slope.

6. Application

Plasma spectra are obtained with instrumentation similar to that shown schematically in Fig. 1. Usually a grating monochromator is employed in order to provide sufficient resolution to measure accurately the line spectra occurring in plasmas as shown in Figs. 8 and 9. For measurements in the visible, multiplier phototubes are used as detectors because they offer good sensitivity and response.

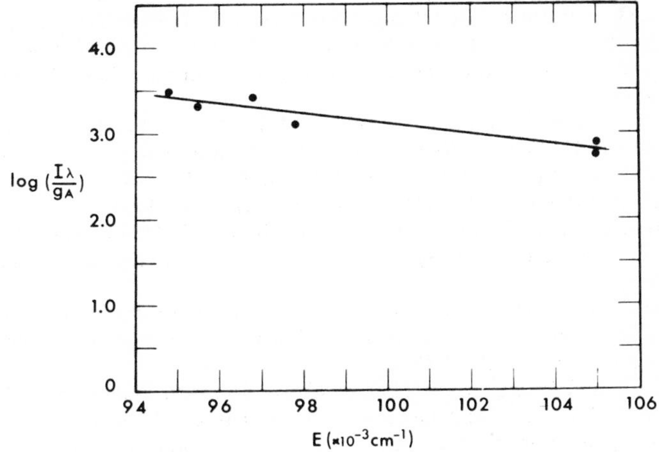

Fig. 10. Atomic Boltzmann plot for a nitrogen plasma jet, measured 1 cm above the nozzle. $T = 625E/[\log(I\lambda/qA) - C] = 10750°\mathrm{K}$. [From Ryan *et al.* (1963).]

Temperatures in an argon plasma jet at 1 atm pressure have been measured using the single-line method. The measured radiances of five spectral lines between 4300 and 4164 Å resulted in an average temperature of 11,210 ± 160°K (Tourin, 1966a).

To determine temperatures in a nitrogen plasma jet at 1 atm, atomic and molecular Boltzmann plots have been used. An atomic Boltzmann plot is shown in Fig. 10. To check the temperature obtained from the atomic Boltzmann plot given in Fig. 10, the nitrogen plasma was seeded with sufficient argon in order to obtain argon lines of sufficient strength so that the single-line method may be used. The resulting one-line temperature for argon was 10,930°K, in good agreement with the atomic Boltzmann plot.

References

Babrov, H. J. (1960). Dissertation, Univ. of Pittsburgh [also available as AFOSR TR59-207 (1959)].

Babrov, H. J. (1961). *J. Opt. Soc. Amer.* **51,** 171.

Babrov, H. J., and Casden, F. (1968). *J. Opt. Soc. Amer.* **58,** 179.

Babrov, H. J., and Healy, A. R. (1969). *J. Opt. Soc. Amer.* **59,** 779.

Babrov, H. J., and Tourin, R. H. (1963). *J. Quant. Spectrosc. Radiat. Transfer* **3,** 15.

Bollinger, L. E., and Edse, R. (1961). *Amer. Rocket. Soc.* **31,** 251.

Emmons, H. W. (ed.) (1958). "High Speed Aerodynamics and Jet Propulsion," Vol. III, Fundamentals of Gas Dynamics, pp. 574–686. Princeton Univ. Press, Princeton, New Jersey.

Fowler, R. H., and Milne, E. A. (1923). *Mon. Not. Roy. Astron. Soc.* **83,** 403.

Fowler, R. H., and Milne, E. A. (1924). *Mon. Not. Roy. Astron. Soc.* **84,** 499.

Gaydon, A. G., and Hurle, I. R. (1962). "The Shock Tube in High-Temperature Chemical Physics," pp. 37–39, 113. Van Nostrand-Reinhold, Princeton, New Jersey.

Goody, R. M. (1964). "Atmospheric Radiation. I. Theoretical Basis." Oxford Univ. Press, London and New York.

Harrison, G. R., Lord, R. C., and Loofbourow, J. R. (1948). "Practical Spectroscopy," Chaps. 14 and 17. Prentice-Hall, Englewood Cliffs, New Jersey.

Kaplan, L. D., and Eggers, D. F. (1956). *J. Chem. Phys.* **25,** 876.

Krakow, B., and Healy, A. R. (1969). *J. Opt. Soc. Amer.* **59,** 1490.

Krakow, B., Babrov, H. J., Maclay, G. J., and Shabott, A. L. (1966). *Appl. Opt.* **5,** 1791.

Manson, N., *et al.* (1963). *Int. Symp. Combust., 9th,* p. 461, Fig. 2. Academic Press, New York.

Oppenheim, U. P. (1963). *Int. Symp. Combust. 9th,* p. 96. Academic Press, New York.

Penzias, G. J. (1966). NASA CR-54914, Final Rep. on Contract NAS 3-6282. NASA Lewis Res. Center, Cleveland, Ohio.

Penzias, G. J., and Maclay, G. J., (1963). NASA CR-54002, Final Rep. on Phase III, Contract NAS 3-1542, NASA Lewis Res. Center, Cleveland, Ohio.

Penzias, G. J., and Maclay, G. J. (1965). *Int. Symp. Combust. 10th,* pp. 189. Combust. Inst., Pittsburgh, Pennsylvania.

Penzias, G. J., and Tourin, R. H. (1962). *Combust. Flame* **6,** 147, 149.

Penzias, G. J., Dolin, S. A., and Kruegle, H. A. (1966). *Appl. Opt.* **5,** 225.

Plass, G. N. (1958). *J. Opt. Soc. Amer.* **48,** 690.

Rouge, M. (1964). Private communication from N. Manson.

Ryan, L. R., Babrov, H. J., and Tourin, R. H. (1963). Infrared Spectra and Temperatures of Plasma Jets. Aeronaut. Res. Labs. Rep. No. ARL63-35, February 1963, Fig. 13, p. 25. Wright-Patterson AFB, Ohio.

Saha, M. N., and Srivastra, B. N. (1958). "Treatise on Heat." Hafner, New York.

Tourin, R. H. (1962). *In* "Temperature, Its Measurement and Control in Science and Industry" (C. M. Herzfeld, ed.), Vol. 3, Part 2, p. 455. Van Nostrand-Reinhold, Princeton, New Jersey.

Tourin, R. H. (1966a). "Spectroscopic Gas Temperature Measurement: Pyrometry of Hot Gases and Plasmas." American Elsevier, New York.

Tourin, R. H. (1966b). *Appl. Spectrosc.* **20,** 3.

Tourin, R. H., and Babrov, H. J. (1962). *J. Chem. Phys.* **37,** 581.

XIII

Furnace Analysis: A Comparative Study

ROBERT H. ESSENHIGH

FUEL SCIENCE SECTION
DEPARTMENT OF MATERIAL SCIENCES
THE PENNSYLVANIA STATE UNIVERSITY
UNIVERSITY PARK, PENNSYLVANIA

ARVIND C. THEKDI*

RESEARCH AND DEVELOPMENT DEPARTMENT
SURFACE COMBUSTION DIVISION
MIDLAND-ROSS CORPORATION
TOLEDO, OHIO

G. MALHOUITRE

GROUPE AERODYNAMIQUE ET PHYSIQUE DES ÉCOULEMENTS,
ELECTRICITÉ DE FRANCE
DIRECTION DES ÉTUDES ET RECHERCHES
CHATOU (NEAR PARIS), FRANCE

YIH-WAN TSAI*

MELTING AND FORMING DIVISION
PITTSBURGH PLATE GLASS INDUSTRIES, INC.
CREIGHTON, PENNSYLVANIA

* Formerly: Combustion Laboratory, The Pennsylvania State University, University Park, Pennsylvania.

Equations developed previously (Essenhigh and Tsai, 1969, 1970) to predict overall furnace performance have been tested for general validity using about 35 data sets generated by about 20 different furnaces and engines. Most of the data were obtained from previously published reports in the literature. Furnace types included shell and water tube boilers, glass tanks, blast furnaces, billet heating furnaces, petroleum heaters, steam turbines, diesel engines, and special experimental furnaces. The performance equations tested predicted the functional form of firing rate, efficiency, and heat utilization factor (HUF), with output. Test of validity depended on the last, where theoretically, the HUF declined linearly with output. For all systems examined, either the linear decline was observed or else the decline was so insensitive that it was obscured by experimental scatter and the HUF was approximately constant. The equations show that this latter condition is obtained as a first approximation for the condition that the range of output is small compared with a theoretical maximum for that system. The performance of all systems examined is therefore described by one single principal equation relating firing rate to output involving only two variables and three "constants," with differences between systems appearing only as differences in the values of the constants. The efficiency and HUF equations are then derived directly from the firing rate equation through their respective definitions. By reasonable extension it is argued that these three equations may well describe general overall thermal performance of virtually all process and power devices, to within reasonable margins of error.

I. INTRODUCTION

Furnace analysis is a general theory of furnace behavior and overall thermal performance (Thring and Reber, 1945; Thring, 1962; Hottel and

Sarofim, 1967) that has been modified recently (Essenhigh and Tsai, 1969, 1970) to bring together in a few simple equations the integrated effects of reaction kinetics, aerodynamics, and, above all, heat transfer in any continuous or periodic furnace. Extension to batch furnaces and power systems is also potentially possible. The equations obtained (Essenhigh and Tsai, 1969, 1970) provide explicit relationships for the variation of firing rate, thermal efficiency, and heat utilization factor (Thring and Reber, 1945), with output. The equations also contain "constants" which vary with furnace type, shape, and size, that are implicit functions of the combustion science components: kinetics, aerodynamics, and so on. The heat utilization factor is particularly important as a parameter representing the intrinsic efficiency of a furnace that is determined primarily by the internal heat transfer from the flame and furnace fabric to the load.

Since development of the equations (Essenhigh and Tsai, 1969, 1970) was on a predominantly phenomenological basis the results should be independent of mechanisms. The equations should therefore apply generally, whatever the function of the furnace, providing only that it is operating continuously, or approximately so (e.g., periodic furnaces operating through a campaign). As yet, however, the equations have only been tested on data from two small, special-purpose furnaces (MacLellan, 1965; Thekdi, 1971; Thekdi *et al.*, 1970); so it is our objective in this chapter to show that the theoretical generality can also be substantiated experimentally by additional tests using data obtained from previously published reports in the literature (Armstrong, 1927; Evans and Bailley, 1928; Sarjant, 1937, 1950; Seddon, 1944; Moxon *et al.*, 1944; Cressy and Lyle, 1956; Kruszewski, 1957; Anon., 1958; Lester, 1963; DeBaufre, 1931; Wilson *et al.*, 1932; Blizard, 1944; Lobo and Evans, 1940; Kirov and Ritchie, 1950; Wheater and Howard, 1950; Hemenway and Wheater, 1950; Corey and Cohen, 1950; Trinks, 1951; Hammond and Sarjant, 1953; Orning *et al.*, 1958; Richardson, 1962; Amer. Gas Ass., 1968; Pattison, 1968; Doolittle, 1941; Warren and Knowlton, 1941; Datschefski, 1967; Behringer, 1970).

Since the equations are fundamental to engineering understanding of furnace operation and performance, their value will be self-evident to those interested in practical applications of combustion theory to industrial systems; at the same time the equations are something of an endpoint for those interested only in the more basic aspects of combustion science. Since the equations, therefore, help to bridge the difficult gap between combustion science and combustion engineering it is clear that their value in this "bridging" role is a direct function of their generality of application. This, therefore, provides further self-evident incentive for this comparative study.

II. THEORY (SUMMARY)

Furnace analysis originated as an empirical study. Initiation may be attributed to Hudson (1890) who used an equation subsequently modified by Orrok (1926) to predict boiler performance. Later contributions came from studies of other systems including forge furnaces, melting and reheating furnaces, blast furnaces, and glass tanks. In a number of these (Armstrong, 1927; Evans and Bailley, 1928; Sarjant, 1937, 1950; Seddon, 1944; Moxon *et al.*, 1944; Cressy and Lyle, 1956; Kruszewski, 1957; Anon., 1958; Lester, 1963) the variation of firing rate with output was reported as rising linearly. The thermal efficiency rose correspondingly to an asymptotic upper limit. Further investigations (DeBaufre, 1931; Wilson *et al.*, 1932; Blizard, 1944; Lobo and Evans, 1940; Kirov and Ritchie, 1950; Wheater and Howard, 1950; Corey and Cohen, 1950; Trinks, 1951; Hammond and Sarjant, 1953; Orning *et al.*, 1958; Richardson, 1962; Amer. Gas Ass., 1968; Pattison, 1968; Doolittle, 1941; Warren and Knowlton, 1941; Datschefski, 1967; Behringer, 1970), however, over a wider relative range of outputs, modified this simple picture. At higher outputs the firing curves were found to be nonlinear, becoming concave upward. The corresponding efficiency curves were then found to peak at some optimum output and to decline thereafter. Uniquely for glass tanks, however, linear firing curves continued to be reported (Seddon, 1944; Moxon *et al.*, 1944; Cressy and Lyle, 1956; Kruszewski, 1957; Lester, 1963), this behavior being apparently anomalous compared with other systems although it is now known to be due to the effects of heat recovery.

Theoretical furnace analysis was developed in due course to explain all this experimental behavior (Thring and Reber, 1945; Thring, 1962; Hottel and Sarofim, 1967). There is no space here for a detailed history of the development though a few highlights need to be mentioned. Initially, attention was largely directed toward the problem of internal heat transfer, and work on these lines through the 1930s (largely dominated by Wohlenberg and Hottel, in classic investigations) ultimately led to emergence of a few tentative expressions for efficiency of a few groups of furnaces [see, e.g., McAdams and Hottel (1942)] but with something better than empiricism as a basis. The equations developed were still partly empirical with correlative terms incorporated as the result of the initial theoretical approach that made the equations rather more informative than the purely empirical curves and relationships existing up to that time. The predominantly mechanistic approach to analysis, however, tended to fragment the subject since this opened up the prospect of having to divide the total set of all possible furnaces into a limited but still large number of subsets, each representing one particular mechanistic model and each with its associated specific analysis.

Up to 1945 this fragmentation would appear to have been accepted as inevitable, but in that year Thring and Reber (1945) suggested the possibility of developing a theory with total generality. They cannot be said to have succeeded since they too incorporated mechanistic aspects into their analysis so the results should apply only to their plug-flow model considered. Their approach, however, can be seen to be predominantly phenomenological with mechanistic components embedded in it. In comparison, mechanistic aspects predominated in Hottel's prior and later investigations carried out with a number of different associates [see Hottel and Sarofim (1967)]. Since the total theoretical generality is only possible in most instances where the method of analysis is phenomenological, Thring and Reber's approach seemed to have far greater potential for development than any other and, as explained elsewhere (Essenhigh and Tsai, 1969), it was on this basis that we set out to develop the necessary generality by removing the purely mechanistic components from their analysis. Such an approach does, of course, exclude knowledge of internal behavior in this first instance, but a general approach still seemed to us to be worthwhile, in spite of its evident limitations, for about the same reasons that Ohm's law is still valuable in spite of development of modern theories of electrical conduction. Our equations might, in fact, be thought of as "Ohm's law for furnaces" with Hottel's type of mechanistic analysis subsequently used for prediction of constants appearing in the phenomenological equations. The two approaches therefore complement each other.

The phenomenological equations then obtained (Essenhigh and Tsai, 1969, 1970) for firing rate H_f, efficiency η, and heat utilization factor α as functions of thermal output H_s are as follows:

$$H_f = H_f^0 + H_s/(1 - H_s/H_s^m)\alpha^0, \tag{1}$$

$$\eta = H_s/H_f = \frac{\{\alpha^0[1 - (H_s/H_s^m)](H_s/H_s^m)\}}{\{\alpha^0(H_f^0/H_s^m) + [1 - \alpha^0(H_f^0/H_s^m)](H_s/H_s^m)\}}, \tag{2}$$

$$\alpha = H_s/(H_f - H_f^0) = \alpha^0(1 - H_s/H_s^m), \tag{3}$$

where H is enthalpy, with subscripts f for the fuel and s for the load. The superscript zero on H_f is the value of firing rate (or holding heat) at zero output and depends on wall losses. The superscript m on H_s is the enthalpy carried out of the furnace by the load at a theoretical maximum output, this value being obtained by methods described below. The ratio H_s/H_s^m is therefore a normalized output with values potentially ranging from zero to unity. Finally, α^0 is a parameter measuring the intrinsic efficiency of a furnace, and depends on internal heat transfer to the load. It is obtained as the limiting value of the heat utilization factor α at zero output [see Eq. (3)], where α is defined by the two left-hand side terms in Eq. (3).

The essential simplicity of the three equations is underscored by their incorporation of only two variables, H_f and H_s and three "constants," H_f^0, H_s^m, and α^0. The equations as written are given in thermal flow units, for example, Btu/hr, kcal/hr, or MW, but conversion to operating data as used typically in actual plant [lb, ft^3, gal, kg, etc., per hour (or day, etc.)] only requires division of H_f by the calorific value of the fuel h_f, and H_s by the specific enthalpy of the load h_s: h_f is, of course, a constant by definition; h_s is not necessarily so, but this is generally the case on a given furnace for operational engineering reasons [see Essenhigh and Tsai (1969)].

In deriving the equations, the important one is Eq. (1) for the firing curve. The other two depend on the definitions of η and α, respectively, and are easily obtained from the definitions by elimination of H_f and $H_f - H_f^0$, respectively, using Eq. (1). To derive the firing equation we adopted Thring and Reber's approach by considering a general heat balance across a thermodynamic surface enclosing any furnace in general, and comparing the heat balances thus obtained at two different levels of output. If H_w and H_g are the enthalpy losses through the wall and in the stack gases, we have, in general,

$$H_f = H_w + H_g + H_s. \qquad (4)$$

At zero load a holding heat H_f^0 is still required to balance wall losses and maintain the furnace at operating temperature, so we have

$$H_f^0 = H_w^0 + H_g^0. \qquad (5)$$

Thring and Reber then assumed, as we also do here, that the wall loss is substantially independent of output. Their argument was that wall losses are dominated by load temperature and this latter, for engineering reasons, is maintained fairly constant. Consequently, when $H_w \simeq H_w^0$, subtraction of Eq. (5) from Eq. (4) eliminates the wall loss. However, if the wall loss does rise with output, the next approximation correction required is easily introduced and has no effect on the forms of the three equations (1), (2), or (3), but only on α^0 (Essenhigh and Tsai, 1970).

We now have to introduce some argument to allow us to handle H_g. This represents our point of departure from Thring and Reber's analysis. The first step is simple. If we write

$$H_g = h_g F = (h_g/h_f) H_f, \qquad (6)$$

which as a definition merely states that a specific gas-exit enthalpy h_g

can be defined, per unit of fuel input, that from simple considerations of stoichiometry is clearly proportional to the true specific gas-exit enthalpy per unit mass of gas leaving. The manipulation indicated by Eq. (6) enables us to factorize out H_f in the heat balance expression. Next, to allow for the real possibility that the gas-exit temperature (and enthalpy) will rise with output—as argued elsewhere (Essenhigh and Tsai, 1970) in general terms that it must—we can allow h_g to be expressed as a series expansion of H_s/H_s^m, the normalized output. Finally, therefore, we introduce the essentially heuristic assumption (see also Section V) for limiting the series expansion, that we only need to consider seriously the first two terms of the expansion. So, if we write

$$h_g \simeq h_g^0[1 + b(H_s/H_s^m)], \tag{7}$$

this may be sufficient for our purposes. The justification for the assumption depends on the results. If the derived equations fail, the limitation is too restrictive; but if not, it is acceptable. As we see below, the general agreement obtained is satisfactory.

III. PRINCIPAL PERFORMANCE CURVES

The principal performance curves are those for H_f, η, and α [testing Eqs. (1)–(3)]. Supplementary behavior also exists that is concerned with such additional factors as excess air effects, heat recovery, and so forth, and these are considered in Section IV. In what follows in this section, our objective is to show that the three performance equations do indeed describe behavior of a wide range of furnaces. In establishing this, the crucial test is to show that Eq. (3) for the HUF α is satisfied, with α declining linearly with output. To establish this we first need the holding heat, obtainable by extrapolation of the firing curve.

A. The Firing Curve

Reports in the literature describe the firing curve as being either linear (Armstrong, 1927; Evans and Bailley, 1928; Sarjant, 1937, 1950; Seddon, 1944; Moxon *et al.*, 1944; Cressy and Lyle, 1956; Kruszewski, 1957; Anon., 1958; Lester, 1963) or concave upward (DeBaufre, 1931; Wilson *et al.*, 1932; Blizard, 1944; Lobo and Evans, 1940; Kirov and Ritchie, 1950; Wheater and Howard, 1950; Corey and Cohen, 1950; Trinks, 1951; Hammond and Sarjant, 1953; Orning *et al.*, 1958; Richardson, 1962; Amer. Gas Ass., 1968; Pattison, 1968; Doolittle, 1941; Warren and

Knowlton, 1941; Datschefski, 1967; Behringer, 1970) as the firing rate H_f increases with output H_s. From Eq. (1) it will be seen that at small outputs (say for $H_s/H_s^m < 0.1$), the curve approximates to a straight line. Small output therefore seems to be adequate reason for most of the reports of linearity. For some others, notably glass tanks, we now have reason to believe that the firing curves are not quite straight lines as still commonly reported (Seddon, 1944; Moxon *et al.*, 1944; Cressy and Lyle, 1956; Kruszewski, 1957; Lester, 1963), but the degree of scatter in relation to the output range reported is mostly high, and definite curvature of the lines was until recently very difficult or impossible to establish. The small curvature is also partly due, it would seem, to the consequences of heat recovery (see Section IV,E).

For other furnaces, the lines are concave upward and Fig. 1 is typical of the behavior reported in qualitative agreement with Eq. (1), including now some recently published data on glass tanks (Datschefski, 1967;

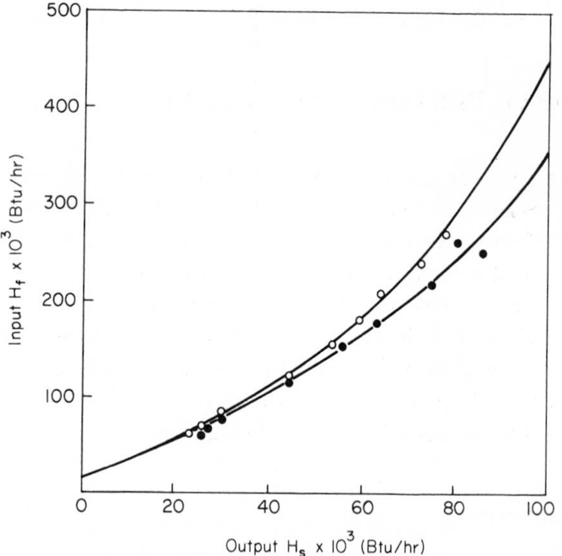

Fig. 1. Firing curve for a continuous furnace—variation of thermal input (firing rate) with output for an experimental furnace, illustrating typical curvature: ○ non-luminous flame; ● luminous flame. Low-temperature load of high absorptivity. [Data from Thekdi (1971). Similar behavior has been reported by the Amer. Gas Ass. (1968), Behringer (1970), Blizard (1944), Corey and Cohen (1950), Datschefski (1967), De-Baufre (1931), Doolittle (1941), Hammond and Sarjant (1953), Hemenway and Wheater (1950), Kirov and Ritchie (1950), Lobo and Evans (1940), Orning *et al.* (1958), Pattison (1968), Richardson (1962), Trinks (1951), Warren and Knowlton (1941), Wheater and Howard (1950), and Wilson *et al.* (1932).]

Behringer, 1970). These particular data in Fig. 1 are for the Pennsylvania State University (PSU) experimental furnace (Thekdi, 1971; Thekdi *et al.*, 1970). They also illustrate a typical problem involved in determining H_f^0. The lowest output recorded is frequently at an appreciable level above zero output so that extrapolation to zero output always entails uncertainties. For the PSU furnace, determination of H_f^0 was aided by additional measurements of wall loss, but in general these are very rarely available.

We have so far now assembled about 35 reports of data providing firing curves of which two-thirds show the concave-upward curvature, and one-third are reported as linear.

B. The Heat Utilization Factor α

When H_f^0 has been determined by extrapolation of the firing curve to zero output, α can be calculated, and Eq. (3) can be tested. For those furnaces where the firing curve is reported as a straight line, α never differs more than marginally from a constant. For such furnaces the output range is insufficient to test Eq. (3), except to the extent that $\alpha \simeq \alpha^0$ is a good approximation at small outputs.

For furnaces with a wider range of outputs, however, the upper line of Fig. 2 (solid squares) illustrates typical behavior. These data are for a steel

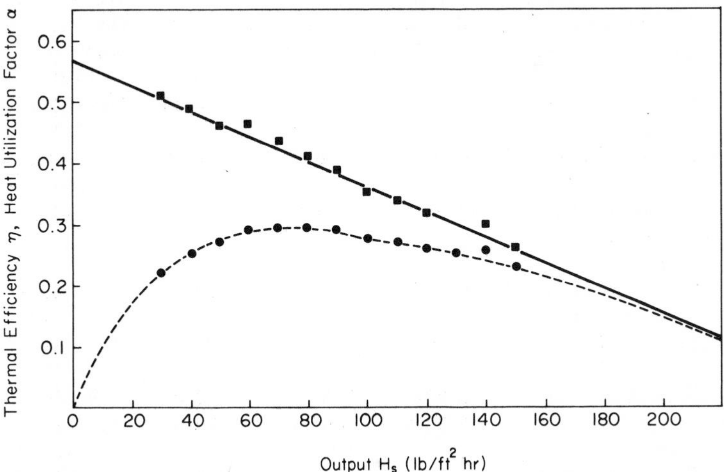

Fig. 2. Performance curves for a continuous furnace. Variation of: ■ heat utilization factor α; and ● thermal efficiency η with output for a steel plant (split flame) furnace (billet heating). Multiply by 350 to obtain H_s in Btu/ft² hr. [Data from Trinks (1951). Similar behavior for α and η has been reported by the Amer. Gas Ass. (1968) and others (see list in Fig. 1).]

TABLE I

VALUES OF HEAT UTILIZATION FACTOR α AND LIMIT OUTPUT H_s^m FOR DIFFERENT FURNACES[a]

Number	Type of furnace	α^o (dimensionless)	H_s^m Btu/hr $\times 10^6$	H_s^m MW	Reference
1	Glass tank	0.413	∞	∞	Seddon (1944)
2	Glass tank	0.453	∞	∞	Moxon *et al.* (1944)
3	Glass tank	0.564	∞	∞	Kruszewski (1957)
4	Glass tank	0.610	∞	∞	Kruszewski (1957)
5	Glass tank	0.417	∞	∞	Lester (1963)
6	Glass tank	0.542	∞	∞	Amer. Gas Ass. (1968)
7	Glass tank	0.6	6.0	1.76	Datschefski (1967)
8	Glass tank	0.51	5.0	1.47	Datschefski (1967)
9	Annealing furnace	(4.35[b])	—	—	Armstrong (1927)
10	Billet reheating furnace	(20.00[b])	—	—	Armstrong (1927)
11	Single zone billet furnace	0.67	0.625	0.18	Pattison (1968)
12	Split flame furnace	0.57	0.95	0.28	Trinks (1951)
13	Top fired furnace	0.57	0.625	0.18	Trinks (1951)
14	Petroleum heaters	0.765	12.1	3.55	Wilson *et al.* (1932)
15	Diesel engine	0.7262	0.13	0.04	Doolittle (1941)
16	Lancashire boiler	0.945	0.28	0.08	Kirov and Ritchie (1950)
17	Lancashire boiler	0.925	0.475	0.14	Kirov and Ritchie (1950)
18	Lancashire boiler	0.935	0.275	0.08	Kirov and Ritchie (1950)
19	Lancashire boiler (mechanically fired)	0.9148	62.0	18.2	Richardson (1962)
20	Lancashire boiler (hand fired)	0.965	42.0	12.3	Richardson (1962)
21	Pulverized coal-fired boiler	0.963	74.7	21.3	Richardson (1962)
22	Water tube boiler	0.694	1300.0	382.0	DeBaufre (1931)
23	Power plant steam generator	0.7076	780.0	229.0	Wheater and Howard (1950)
24	Power plant steam generator	0.74	1470.0	431.0	Corey and Cohen (1950)
25	Lancashire boiler experimental furnaces	0.91	0.02	0.006	Thring and Reber (1945)
26	Water-cooled coils	0.400	∞	∞	MacLellan (1965)

TABLE I (Continued)

Number	Type of furnace	α^0 (dimensionless)	H_s^m Btu/hr × 10^6	H_s^m MW	Reference
27	Experimental furnace shining load (nonluminous flame)	0.42	0.21	0.06	Thekdi (1971)
28	Experimental furnace shining load (luminous flame)	0.39	0.28	0.08	Thekdi (1971)
29	Experimental furnace black load (nonluminous flame)	0.5455	0.18	0.05	Thekdi (1971)
30	Experimental furnace black load (luminous flame)	0.5435	0.22	0.06	Thekdi (1971)
31	Experimental furnace high-temperature (900°C) load	0.212	0.29	0.09	Thekdi (1971)

[a] Parameter values in Eq. (3).
[b] α is dimensional (cwt. of steel/cwt. of fuel).

furnace (Trinks, 1951), and the graph shows both α and η on the same plot. Within reasonable limits of error, α is declining linearly with output, as required by Eq. (3). Similar linear declines have been obtained with data from about a dozen other furnaces (DeBaufre, 1931; Wilson *et al.*, 1932; Blizard, 1944; Lobo and Evans, 1940; Kirov and Ritchie, 1950; Wheater and Howard, 1950; Hemenway and Wheater, 1950; Corey and Cohen, 1950; Trinks, 1951; Hammond and Sarjant, 1953; Orning *et al.*, 1958; Richardson, 1962; Amer. Gas Ass., 1968; Pattison, 1968; Doolittle, 1941; Datschefski, 1967; Behringer, 1970) including boilers, some glass tanks, steel billet furnaces, and petroleum heaters after correction for excess air (see Section IV,B). From this type of plot we obtain by extrapolation the two other constants, α^0 and H_s^m, required for a complete description of the three performance curves. In all, a total of about 20 data sets have been found showing this linear decline. In all other reports so far obtained α was effectively constant.

Table I lists about 20 values of α^0 and H_s^m [see Eq. (3)] for a number of different furnace types.

C. Efficiency Curves

The lower curve of Fig. 2 (solid circles) is typical of the peaking efficiency curve obtained from furnaces having nonlinear firing curves (and linearly declining heat utilization factors). Plots of this shape were first predicted theoretically by Thring and Reber (1945). Characteristics of these curves are their inverted, asymmetric U-shape, and the asymptotic approach of η to α as H_s goes to its limit H_s^m. A further important characteristic not illustrated here but discussed previously (Essenhigh and Tsai, 1969) concerns the effect of varying the wall loss H_w by increasing the insulation. If the wall loss is decreased, the α line remains unchanged but the efficiency is increased at all outputs, with greatest effect at the lower outputs with the peak efficiency, in consequence, moving to lower output.

The peaking efficiency plot of Fig. 2 has been found to be characteristic for 20 out of 35 furnaces whose data have been examined. Those not agreeing with this form had linear firing curves.

D. Comparison of Theories

It is convenient at this point to compare several different theories using the efficiency plot as the basis for comparison (Fig. 3). The plots are

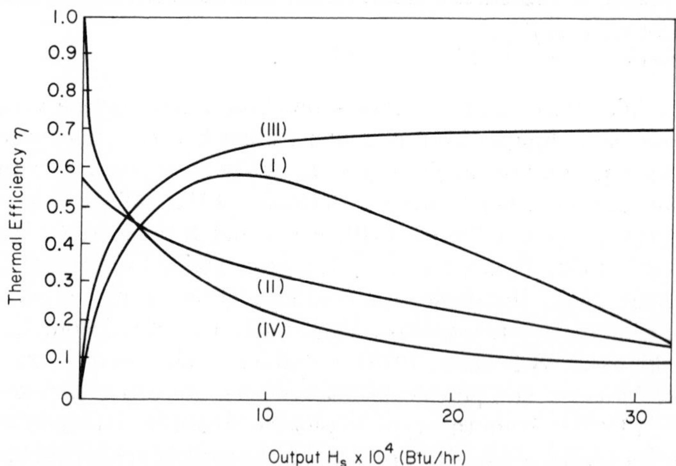

Fig. 3. Comparison of theories predicting variation of thermal efficiency with output for continuous furnaces. (i) Actual performance, predicted by Thring and Reber (1945) and Essenhigh and Tsai (1969, 1970); (ii) Hottel and Sarofim (1967); (iii) SGT and HAPF equations (Seddon, 1944; Moxon *et al.*, 1944); (iv) Hudson–Orrok (Hudson, 1890; Orrok, 1926) equation.

illustrated on a comparative basis. Line i is typical of the actual plots discussed in Section III,C and described quantitatively by Eq. (2), or by Thring and Reber's equations for certain systems. Line ii illustrates predictions from Hottel's "firing density" equation (Hottel and Sarofim, 1967; Hottel, 1961), apparently developed during the 1950s, and a counterpart, using his approach, to Eq. (1). The two theories agree qualitatively at high outputs but differ markedly at low outputs. The discrepancy is due to different assumptions made in the different analyses concerning wall loss. This is assumed to be zero to negligible in the Hottel analysis, so his equations are either applicable only to low loss systems such as boilers or they represent instead the heat utilization factor or something close to it. Where wall loss is important as in steel plant furnaces, glass tanks, and the like, the early empirical equations based on linear firing curves predicted almost the complete opposite in trend, as illustrated by line iii. This represents the typical behavior that would be obtained from the Society of Glass Technology or Hartford Average Practice Formula equations for glass tanks (Seddon, 1944; Lester, 1963), with efficiency rising asymptotically to an upper limit. Finally, for completeness, the Orrok–Hudson (Hudson, 1890; Orrok, 1926) prediction has been included (line iv). This also shows continued rising efficiency with decreasing output for about the same reasons as for the Hottel curves.

IV. SUPPLEMENTARY RELATIONS

The data outlined in the previous section reasonably support the contention that the three performance equations [(1)–(3)] do generally seem to describe behavior of a number of furnaces. In this section we are concerned with modifications, supplementary relations, and extensions to cases not yet considered.

A. Specific Gas-Exit Enthalpy

One supplementary equation of significance that has already been examined indirectly is Eq. (7), representing the variation of specific gas-exit enthalpy with output. The significance of this relation is that it is also responsible for the linear decay of α with H_s predicted by Eq. (3) and examined in Section III,B. Otherwise, if h_g is constant, α is constant; if it rises nonlinearly, α falls nonlinearly.

Direct investigation of Eq. (7) is therefore desirable since α may show a variation with H_s that is reasonably linear at one level of approximation but not at a finer level of approximation, so it then becomes difficult to determine whether the variations are real or just due to errors of measurement. It is therefore of significance that, where the gas exit temperature was recorded, and h_g could be calculated, h_g was found to rise linearly with output (DeBaufre, 1931; Kirov and Ritchie, 1950; Wheater and Howard, 1950; Hemenway and Wheater, 1950; Orning *et al.*, 1958; Richardson, 1962) in support of the conclusion of Section III,B. An almost linear rise of h_g with output has also been obtained in recent calculations (Tsai and Essenhigh, 1969) from a mathematical model of simple systems using the radiation equation of transfer. The model also predicted H_f varying with H_s substantially in line with expectations. The physical reason for the increase (Essenhigh and Tsai, 1969, 1970) is essentially that each unit mass of combustion products spends less time in the furnace as the firing rate and volume throughput increase. There is therefore less time for each unit mass to lose heat so, for a given rate of loss, its net loss is reduced as the firing rate increases and its emergent temperature must rise also. Thring and Reber (1945) described this as the heat transfer "bottleneck" effect.

B. Excess Air

In the previous section some mention has already been made of the influence of excess air. This, of course, is well known among combustion engineers as a prime factor needing control. It contributes additional "ballast" passing through the furnace, to be heated up for no purpose, and providing only an extra thermal load. This initially cools the furnace, and additional fuel is then required to return the furnace to its operating temperature.

Correcting for this is quite an involved procedure (Thekdi, 1971). In the first instance, for small increments in excess air (with firing rate increased to match this), it is possible to assume with some justification that, at constant output, all gas and material temperatures are substantially unchanged. As the excess air increases, however, the gas velocity through the furnace also increases. The same heat transfer "bottleneck" effect just described (Section IV,A) due to reduced residence time, again comes into play, and the gas exit temperature rises. This has been investigated directly in a few instances (Kirov and Ritchie, 1950; Orning *et al.*, 1958; Richardson, 1962), mostly on boilers, and in all cases the temperature was found to rise linearly. Using this information, excess air corrections can be made and, after correction, firing data that are otherwise highly scattered reduce to a common firing curve obeying Eq. (1) (Wilson *et al.*, 1932; Blizard, 1944; Lobo and Evans, 1940; Kirov and Ritchie, 1950; Wheater and Howard,

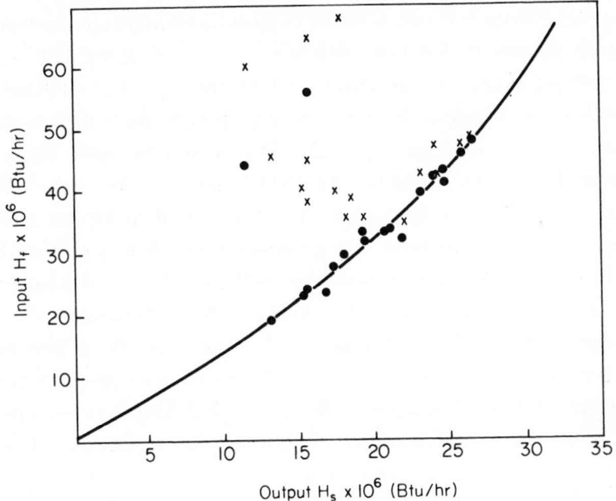

Fig. 4. Effect of excess air on firing (performance) curve for a continuous furnace. × Original data before corrections for excess air; ● data corrected for excess air. [Data from Lobo and Evans (1940). Similar behavior has been reported by Blizard (1944), Corey and Cohen (1950), Hammond and Sarjant (1953), Hemenway and Wheater (1950), Kirov and Ritchie (1950), Lobo and Evans (1940), Orning *et al.* (1958), Richardson (1962), Wheater and Howard (1950), and Wilson *et al.* (1932).]

1950; Hemenway and Wheater, 1950; Corey and Cohen, 1950; Hammond and Sarjant, 1953; Orning *et al.*, 1958; Richardson, 1962) (see Section III,B). A particularly good example of the importance of the correction is provided by some data on firing of petroleum heaters (tube stills) (Wilson *et al.*, 1932) where the variation of firing rate with output without the excess air correction showed a very badly scattered plot with a poor but negative correlation; this was changed to a positive correlation when the excess air correction was made. This is illustrated in Fig. 4 with crosses for the points before making the excess air correction, and solid circles after the correction.

Calculating the detailed correction at high excess air is an involved procedure. At low excess air, however, an approximate correction is given by

$$H_f = H_{fs}(1 + E), \tag{8}$$

where H_{fs} is the effective thermal input rate required at stoichiometric firing, and E is the fractional excess air.

C. Wall Loss

Further corrections may also have to be made for variable wall loss. There are two possible ways in which the wall loss can vary systematically and influence the firing curves, both of which have already been mentioned. This is a brief summary for completeness. (1) Changing the wall loss by changing the insulation has no direct effect on the heat utilization factor α. It affects the operational efficiency η, which rises as wall loss drops, with the greatest effect at low output. (2) The wall loss may also vary with output. Where data exist it has been found to increase linearly (MacLellan, 1965; Thekdi, 1971). As a first approximation in developing the theory it was assumed to be constant on the grounds that this is generally close to the truth. However, it is in any event usually small, so quite large increases are, overall, relatively unimportant. Thus a 10% increase in a 10% wall loss corresponds only to 1% change of the total input. If the wall loss is important, for example, in glass tanks, it can be assumed to rise linearly with output to provide a simple correction, and this has no effect on the functional form of the performance equations (Essenhigh and Tsai, 1969, 1970).

D. Heat Exchangers

Furnace analysis is not necessarily restricted to direct fired devices. Since the prime basis for the analysis is energy flow across a thermodynamic surface, the equations should also apply to heat exchangers and heat recovery units (e.g., waste heat boilers) so long as the basic assumptions of effectively constant wall loss and either a constant or linear rise in gas exit temperature with output are satisfied. The mechanistic theory of heat exchangers is, of course, already developed, so applying the phenomenological theory outlined in this chapter to heat exchangers by themselves is something of a redundant operation. It serves a useful purpose, however, when a heat exchanger is part of a furnace system and the influence of heat recovery on a furnace is to be included in the analysis (see Section IV,E).

It is inappropriate to develop in detail here the comparison between the furnace analysis and the conventional approach to heat exchangers, but two results will suffice for our purposes here. The first is that the wall loss is generally so small that H_f^0 may be taken as effectively zero. The second is that, in conventional theory of parallel or counterflow exchangers or their effective equivalent [see, e.g., Rohsenow and Choi (1961)], equations developed to predict efficiency include exponential decay terms for exchanger length. The characteristic decay distance is a function of throughput velocity, initially falling with increasing velocity and then rising again. This means that a range of velocities exists through which the characteristic decay distance is small enough for the exponential terms to be unimportant, and the exchanger efficiency is substantially constant. (Ultimately the velocities will become high enough for the exponential term to become significant, and efficiency declines, this being another example of the "bottleneck" effect.) To a first approximation, many heat exchangers do have reasonably constant efficiency over a useful range of flow rates. This is equivalent in Eq. (1) to saying that H_s^m is proportionately very large, and the first approximation, with H_f^0 taken as zero, adequately describes behavior. Direct test of these conclusions using data obtained from heat exchangers on operating furnace systems is difficult because the necessary data are not generally recorded. Data are available, however, for a superheater attached to a boiler, and for this the enthalpy output in the steam is very closely proportional to the enthalpy input carried by the hot gases leaving the combustion chamber, with all lines passing substantially through the origin.

E. Heat Recovery

The effect of a heat exchanger added to a furnace can now be included. For a full output range where the exchanger efficiency is assumed to obey Eq. (2), the equations developed are too complex to discuss here, but for the simplification discussed in Section IV,C that the efficiency is effectively constant, which would appear to apply in practice over a wide output range, the exchanger effect is easily calculated. By assuming that the fraction of heat recovered from the stack gas as air preheat can be exactly balanced on a Btu basis by a reduced firing rate H_f', we obtain

$$H_f' - H_f^{0\prime} = [H_s/\alpha^0(1 - H_s/H_s^m)][1 - \eta_h(1 - \alpha^0) - \alpha^0\eta_h(H_s/H_s^m)], \tag{9}$$

where the primed values refer to the thermal input rate with heat recovery, and η_h is the (constant) efficiency of the heat exchanger. This is an interesting equation since it shows that even at constant heat exchanger efficiency, the existence of the heat exchanger serves to offset the heat transfer "bottleneck" effect. If η_h were unity, the heat recovery would exactly compensate and the whole of the right-hand side of Eq. (9) would reduce to H_s. Naturally this limit is never reached, but even with α^0 and η_h less than unity there is partial compensation. This evidently is primarily responsible for the almost linear behavior of the firing curves for glass tanks. (The high wall loss also obscures the curvature until high outputs are reached.)

F. Power Systems

Finally, we may consider extension to power systems. Here again an energy balance across a thermodynamic surface should be a valid basis for generalizations; and it becomes again a matter of experiment to determine the most probable or applicable form of the variation of h_g with output. To determine this, data from a number of power systems have been examined. One set of data for steam turbines (Warren and Knowlton, 1941) shows an almost linear "firing" curve with relatively small variation of α with output. The great variations, leading to peaking efficiency curves, are to be found in the earlier data (1920s); later data show the general effects of improved design, with efficiency reaching a maximum and not apparently declining, the decline in α accordingly being barely perceptible.

The equations evidently also apply to reciprocating systems. Figure 5, which shows the three performance curves for a diesel engine, is a typical example (Doolittle, 1941). The α line is a little contorted, but within a rational level of error it is well approximated by a straight line.

It is also of interest that an early approach (Artsay, 1929) to prediction of boiler performance leading to an equation of form similar to that of Hudson and Orrok was based on a pelton wheel analogy. It is now clear that the analogy was reasonably successful because all the systems we have been discussing are evidently analogous.

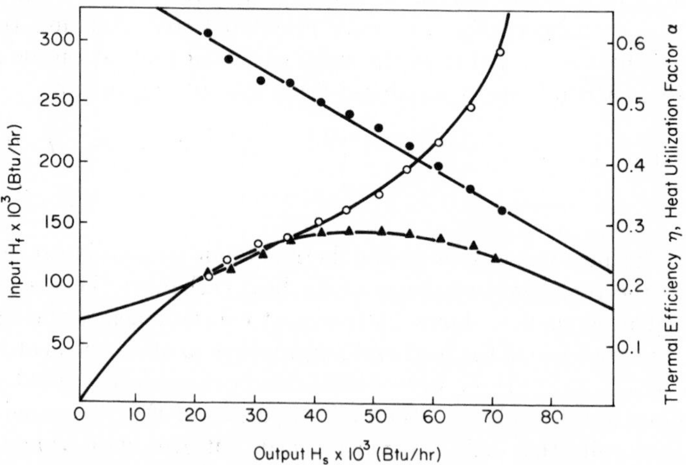

Fig. 5. Performance curve for a diesel engine illustrating obedience to performance equations (1)–(3) (see text) for a power generator. ○ Firing curve; ▲ efficiency; ● heat utilization factor. [Data from Doolittle (1941). Similar behavior has been reported by Warren and Knowlton (1941).]

V. DISCUSSION AND CONCLUSIONS

What we have now established by the data examination just discussed is that about 35 sets of performance data, obtained from about 20 furnaces, heat exchangers, and engines, can be described by the three performance equations [(1)–(3)], or their first approximations. With this approximation proviso included, there are then no exceptions. These data sets do, of course, only represent a small subset of all possible sets of data from all possible furnaces, but those in the subset were in effect chosen randomly, so the zero exceptions of obedience to the performance equations indicate a good probability that these same equations will apply universally. The agreement is, of course, only good to within a definable margin of error. Within that margin there is some scatter, in some cases appreciable, but overall the agreement seems to be satisfactory to about a first-order of magnitude. Some of the scatter is undoubtedly random error. If this can be reduced by improved measurement, then we should expect to see second- and third-order deviations from the common equations due to real differences in different systems. Otherwise it does seem that we can reasonably say that, to a first order of magnitude, all heating and power systems appear to behave alike, with differences only showing up as differences in values of the three constants of the equations.

If the equations are examined, the point at which we would expect to find system differences emerging when measurements are more accurate is in the variation of specific stack gas enthalpy h_g with output. It is indeed remarkable that a linear variation should be such a good approximation; it was certainly not to have been expected *a priori*. It is now clear that this is the parameter requiring most careful attention in any future measurements. For the time being, however, the linear rise can be accepted as reasonable operational behavior which introduces a major simplifying factor into the performance equations.

The variations of h_g with output do, of course, represent an element of empiricism in the analysis which is undesirable, although it is something that is also present in all other operational performance equations. Thring and Reber (1945), for example, had to introduce a "combustion delay" factor β to make allowance for the difference in theoretical and real flame temperatures. Likewise, Hottel (Hottel and Sarofim, 1967) introduced a Δ factor to represent a temperature difference between the average flame temperature in a (stirred) reactor and the exit stack gas temperature. Both represent essentially heuristic assumptions and, as is usual in such cases, evaluation of terms containing the β or Δ factors requires considerable "clairvoyance" (to use Hottel's description) in selection of appropriate values for either factor. It was, in fact, one of our arguments in developing the approach outlined in this chapter that if heuristic assumptions have to be introduced, they might as well be as general as possible, and so chosen as to reduce as far as possible the element of clairvoyance required for determining parametric values in any equations.

Our approach to this matter of parameter values is, in any event, a little different from that of some other investigators. The ultimate objective of all theories is to be able to predict all parameter values from more basic (mechanistic) considerations; but, in the first instance, merely tabulating values of the principal constants (H_f^0, α^0, H_s^m) obtained experimentally serves two functions. Tables of such values listed under furnace type, operational range, and so on, can be valuable for operational purposes in existing plant, or for assistance in design of new plant (just as tables of electrical resistance for different materials summarize Ohm's law behavior for future use and prediction). There is also some chance that such tables can also provide indications of further regular sources of behavior responsible for observed variations. Examples already developed previously (Thekdi, 1971; Thekdi *et al.*, 1970), for instance, include the following: α^0 is clearly a prime function of load emissivity and flame–load temperature difference, with α^0 rising as emissivity and temperature difference increase. The factor α^0 for a total system also increases with increased thermal recovery (see Section IV,E), a not unexpected result, but the equations can now be developed to quantify the effect; α^0 also decreases with increasing excess air. In contrast, α^0 was found to be relatively insensitive to flame emissivity, a factor traditionally assumed to be overriding in comparison with the other factors mentioned.

The effects of some of these factors are illustrated in Fig. 6 which shows the variation of thermal efficiency with output in a laboratory furnace (Thekdi, 1971), with different flame emissivities, load emissivities, and load temperatures. Clearly, output itself is the prime factor responsible for variations in efficiency. (We can also show indirectly that wall loss is important at low outputs.) Otherwise we see that changing the temperature difference between the flame and the load has a major effect on efficiency, with efficiency and H_s^m dropping sharply as load temperature rises. The next most important effect illustrated is the influence of load emissivity, with efficiency again dropping substantially with change from a black plate (emissivity about 0.9) to a shiny plate (emissivity about 0.4). Finally, flame emissivity has little more than marginal influence.

This leads us, in conclusion, to define the next stage of the problem, which is the matter of attempting prediction of the operational constants of the performance equations, H_f^0, α^0, and H_s^m. We now have to start considering

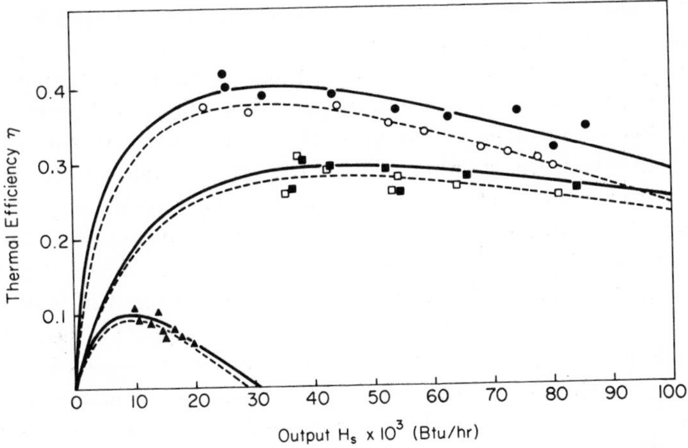

Fig. 6. Influence of flame and load characteristics on efficiency of a model furnace illustrating relative effects of changing flame emissivity, load reflectivity (and absorptivity), and flame–load temperature difference. Solid curves denote luminous flame; dashed curves, nonluminous flame. ○,● Black plate (80°C); □,■ shiny plate (80°C); △,▲ high-temperature load (900°C) at $\eta_L = 0.9$. [Data from Thekdi (1971).]

internal behavior inside the thermodynamic enclosure, which implies embarking on a course that should converge with Hottel's mechanistic methods of analysis and results. The initial steps in this have already been taken (Tsai and Essenhigh, 1969). The essentially inductive approach described in this chapter, however, is aimed at splitting the problem up into more manageable parts. The three operational constants represent physical behavior of the furnace at limiting conditions. For example, the values of H_f^0 and α^0 represent behavior at zero load, which usefully simplifies the problem of calculation. If we know the inside wall temperatures, wall thickness, materials, shapes, and so on, the wall loss can be calculated directly by conventional methods. If a static load is included in the furnace which must emerge, during furnace operation, at a known predetermined temperature T_s, this is the temperature at which the load must be held, at zero output operation. If we accept the reported conclusion (Thekdi, 1971; Thekdi *et al.*, 1970) that hot-wall furnace operation is relatively insensitive to flame emissivity, we can assume initially that the flame is transparent, so the load is sitting in a radiant enclosure with temperatures such that the load is maintained at its output temperature T_s. The inside wall temperatures will clearly be fairly close to T_s, depending on emissivities, view factors, and similar parameters. (This clearly cannot be used for cold-wall systems such as boilers.)

We can, likewise, build up additional programs to calculate gas temperatures, gas-exit enthalpies, and so on, based on appropriate assumptions of flame behavior, radiancy, and so forth, these being determinable in principle by calculation from appropriate aerodynamic and reaction kinetic models. This is, of course, the converse of the more conventional deductive approach in which the starting point is the sets of equations representing kinetic, aerodynamic, heat transfer, and associated behavior, which are then solved simultaneously in association with appropriate boundary conditions—an operation of extreme complexity, as Hottel (1969) has pointed out recently, due in part to the iteration required in handling the radiation equations. In spite of this extreme complexity, however, the ultimate integrated effects of all the interacting factors must still reduce in reasonable obedience to the phenomenological performance equations [(1)–(3)] if our conclusion about the generality of application is reasonably sound. We have here, then, the basis of the statement made in the opening sentence of this chapter.

The simplicity of the end result, however, does raise the question of the real need for excessively complex intermediate models and calculations, except where detailed information on the variation of thermal flux intensity on a load is needed. Extension of the work described herein is therefore planned initially to have a similar basis; for example, in examining the thermal interaction between flame, furnace, and load, the combustion cavity can also be treated phenomenologically. The necessary future work is being programmed, however, to include in due course the effects of aerodynamics, reaction kinetics, and so on, so that eventually fundamental information can be fed into a general mathematical model or set of models of combustion chambers and furnaces. In this way the influence of fundamentals on operational performance of practical devices can be assessed, which is after all a not unimportant reason for studying combustion processes in the first instance.

Acknowledgment

The work reported in this chapter was prepared at the Combustion Laboratory, Fuel Science Section, Department of Material Sciences, The Pennsylvania State University, University Park, Pennsylvania 16802, under Research Contract for Columbia Gas System Service Corporation and submitted as Technical Report No. FS/CGSSC-12/69-4. It was presented at the Symposium on "Control of Heat Exchange and Temperature During the Melting and Forming of Glass" which was organized by the Union Scientifique du Verre and held at Stresa, Italy, September 8–11, 1970.

REFERENCES

Amer. Gas Ass. (1968). Evaluation of a Production Glass-Melting Furnace. Project Status Rep. for Fourth Quarter, Inst. of Gas Tech. Chicago, Illinois. AGA, Washington, D.C.

Anonymous (1958). "Efficient Use of Fuel," 2nd ed., pp. 533–534, 487–488, 470–471. H M Stationary Office, London.

Armstrong, H. C. (1927). Characteristics of Furnace Curves as an Aid to Fuel Control. *Engineer* **144,** 445.

Artsay, N. (1929). Analysis of Heat Absorption in Boilers and Superheaters. *Trans. ASME* **22,** 247.

Behringer, H. (1970). Private communication. Brockway Glass Co., Brockway, Pennsylvania.

Blizard, J. (1944). Absorption of Heat by Walls of a Furnace. *Trans. ASME* **66,** 79.

Corey, R. C., and Cohen, P. (1950). Furnace Heat Absorption in Paddy's Run Pulverized-Coal-Fired Steam Generator. *Trans. ASME* **72,** 925.

Cressy, M. S., and Lyle, A. K. (1956). What's New in Container Plant Design. *Ceram. Ind.* **73,** 98, 102.

Datschefski, G. (1967). Natural Gas in a Cross-Fired Glass Melting Furnace. *Glass Technol.* **8,** 148.

DeBaufre, W. L. (1931). Heat Absorption in Water-Cooled Furnaces. *Trans. ASME* **53,** 253.

Doolittle, J. S. (1941). Effect of Variations in Atmospheric Conditions on Diesel-Engine Performance. *Trans. ASME* **63,** 91.

Essenhigh, R. H., and Tsai, Y. W. (1969). Furnace Analysis Applied to Operation and Design of Glass-Melting Tanks. *Glass Ind.* **50,** 278, 333.

Essenhigh, R. H., and Tsai, Y. W. (1970). Furnace Analysis Applied to Glass Tanks at High Output: The Heat Transfer 'Bottleneck' Effect. *Glass Ind.* **51,** 68, 108.

Evans, E. C., and Bailley, E. J. (1928). Blast Furnace Data and Their Correlation. *J. Iron Steel Inst.* **117,** 53.

Hammond, E., and Sarjant, R. J. (1953). Heat Release in Coal-Fired Combustion Chambers. *J. Inst. Fuel* **26,** 364.

Hemenway, H. H., and Wheater, R. I. (1950). Furnace Heat Absorption in Paddy's Run Pulverized-Coal-Fired Steam Generator, Part II. *Trans. ASME* **72,** 37.

Hottel, H. C. (1961). Radiative Transfer in Combustion Chambers. *J. Inst. Fuel* **34,** 220.

Hottel, H. C. (1969). A Proposed New Statement of Objectives of the Int. Flame Res. Foundation. Internal memorandum to the Amer. Flame Res. Committee.

Hottel, H. C., and Sarofim, A. F. (1967). "Radiative Transfer," Chapter 14. McGraw-Hill, New York.

Hudson, J. G. (1890). Heat Transmission in Boilers. *Engineer* **70,** 449, 483, 523.

Kirov, N. Y., and Ritchie, E. G. (1950). The Effects of Excess Air and Output on the Performance of a Lancashire Boiler. *J. Inst. Fuel* **23,** 203.

Kruszewski, S. (1957). Heat Transfer from Flames of Fuels Used in the Glass Industry. *One-day Symp. Flames Ind., Inst. Fuel Proc.* B5. Inst. of Fuel, London.

Lester, W. R. (1963). Average Practice Fuel Formula for Glass Melting Furnaces. *Glass Ind.* **44,** 260.

Lobo, W. E., and Evans, J. E. (1940). Heat Transfer in the Radiant Section of Petroleum Heaters, *Trans. ASME* **35,** 743.

McAdams, W. H., and Hottel, H. C. (1942). "Heat Transmission," 2nd ed., pp. 77–84. McGraw-Hill, New York.

MacLellan, D. E. (1965). Thermal Efficiency of Industrial Furnaces: A Study of the Effect of Firing Rate and Output. M.S. Thesis, Pennsylvania State Univ., University Park, Pennsylvania.

Moxon, L., Winus, F., and Dudding, B. P. (1944). A Test of the Thermal Performance Formulas Based on a Glass-Melting Tank Furnace Fired by Coke-Oven Gas. *J. Soc. Glass Tech.* **28,** 53.

Orning, A. A., Weintraub, M., Schwartz, C. H., Mihok, E. A., McCann, C. R., and Harrold, W. C. (1958). An Investigation of the Variation in Heat Absorption in a Pulverized-Coal-Fired Slag-Tap Steam Boiler at Blaine Island, Charleston, West Virginia. *Trans. ASME* **80,** 1239.

Orrok, G. A. (1926). Radiation in Boiler Furnaces. *Trans. ASME* **48**(3), 218.

Pattison, J. R. (1968). Continuous Reheating Furnaces in Steel Industry. *J. Inst. Fuel* **41,** 345.

Richardson, D. A. (1962). Factors Affecting the Efficiency of a Lancashire Boiler. *J. Inst. Fuel* **35,** 290.

Rohsenow, W. M., and Choi, H. V. (1961). "Heat Mass and Momentum Transfer." Prentice-Hall, Englewood Cliffs, New Jersey.

Sarjant, R. J. (1937). Fuel Economy in Melting and Reheating Furnaces. *J. Inst. Fuel* **10,** 335.

Sarjant, R. J. (1950). Furnace Design and Practice, *Proc. Inst. Mech. Eng.* **162,** 78.

Seddon, E. (1944). The Assessment of the Thermal Performance of Tank Furnaces for Melting Glass. *J. Soc. Glass Tech.* **28,** 33.

Thekdi, A. C. (1971). Studies in Furnace Analysis: Influence of Flame Emissivity on Furnace Performance. Ph.D. Thesis, Pennsylvania State Univ., University Park, Pennsylvania.

Thekdi, A. C., Tsai, Y. W., and Essenhigh, R. H. (1970). Application of Furnace Analysis to Determine Effect of Flame Luminosity on Glass Tank Performance. Paper presented at the 68th Annu. Convention of the Can. Ceram. Soc., Ottawa.

Thring, M. W. (1962). "The Science of Flames and Furnaces," 2nd ed., Chapter 7. Wiley, New York.

Thring, M. W., and Reber, J. W. (1945). The Effect of Output on the Thermal Efficiency of Heating Appliances. *J. Inst. Fuel* **18,** 12.

Trinks, W. (1951). "Industrial Furnaces," Vol. I, p. 210. Wiley, New York.

Tsai, Y. W., and Essenhigh, R. H. (1969). Calculated Effects of Radiative and Convective Heat Transfer Parameters on Glass Tank Performance. Tech. Rep. No. FS/CGSSC-11/69-3 to Columbia Gas Syst. Serv. Corp., Columbus, Ohio.

Warren, G. B., and Knowlton, P. H. (1941). Relative Engine Efficiencies Realizable from Large Modern Steam-Turbine-Generator Units. *Trans. ASME* **63,** 125.

Wheater, R. I., and Howard, M. H. (1950). Furnace Heat Absorption in Paddy's Run Pulverized-Coal-Fired Steam Generator, Part I. *Trans. ASME* **72,** 893.

Wilson, D. W., Lobo, W. E., and Hottel, H. C. (1932). Heat Transmission in Radiant Sections of Tube Shells. *Ind. Eng. Chem.* **24,** 486.

XIV

An Introduction to Stirred Reactor Theory Applied to Design of Combustion Chambers

ROBERT H. ESSENHIGH

FUEL SCIENCE SECTION
DEPARTMENT OF MATERIAL SCIENCES
THE PENNSYLVANIA STATE UNIVERSITY
UNIVERSITY PARK, PENNSYLVANIA

The art of constructing scientific theories lies less in knowing what to put in than in what to leave out (Attr: J. C. Maxwell).

In this chapter, the basic equations for the adiabatic perfectly stirred reactor are developed, using one-step, first-order kinetics, to illustrate the basis for prediction of combustor performance characteristics such as flame temperature, extinction limits, combustion intensities, and combustion efficiencies, and the influence on these of firing rate and stoichiometry. The equations are also amplified with respect to increased engineering realism, by inclusion of the effects of nonadiabatic operation, heat recovery, dilution by recycle of tail gases, and higher order reactions. Mixing is introduced mainly as a qualitative phenomenon, but including a formal proof of the Bragg criterion, the "zero/one"-dimensional optimum mixing configuration of the PSR–PFR sequence (and also demonstrating the relation of the combustor equations to thermal explosion theory). The RTD and stirred reactor descriptions of mixing, as a basis for quantification, are also outlined. The discrete space (finite element) nature of stirred reactor theory is noted as a fidelity-reducing procedure that is still, as a space-modifying process, more acceptable than is the function-modifying process of more conventional methods of reducing complex equation constructs for solution.

I. INTRODUCTION

A combustion chamber may be described as a container into which fuels and oxidants are injected or otherwise supplied, and inside which the reactants are intended to burn completely, with a clean stable flame, with the objective of providing a heat or work output for some useful purpose. The design of combustion chambers and prediction of performance, however, still owes, in most instances, as much to art and past experience as it does to science. This situation is a consequence of formidable difficulties of design that are compounded by the widely varied applications of combustion, the different fuels in use, and the complexities of the combustion process itself. The difficulties of the problem, however, still have not deterred attempts to develop general, unified theories for predicting the performance of all combustors and furnaces; and one concept that has particular promise in this respect is that of the stirred reactor.

A stirred reactor is conceived as a device in which incoming reactants are assumed to disperse rapidly and reasonably uniformly throughout the reactor. The theory of such a device then shows that it is *potentially* possible to predict a variety of performance characteristics and parameters such as stable operating temperatures, ignition and extinction temperatures, concentrations and temperatures at the blow-off and extinction limits, combustion intensities, and overall combustion efficiencies. The theory can also be extended to include prediction of the effects of such additional factors as heat recovery, recirculation and recycle, dilution, and non-adiabatic operation due to wall loss. The theory is quite general and therefore applies in principle to all fuels: gas, liquid, or solid, when appropriate allowance is made for the different phases of reaction.

Prediction of combustor performance characteristics and parameters is not unique, of course, to the stirred reactor analysis; it is also possible from other theoretical approaches. However, the assumptions of the stirred reactor concept can lead to major simplifications, first, in the equations involved, and consequently, therefore, in the subsequent computations. Simplification in the computations therefore means that the potential of the stirred reactor theory is more readily realizable than is the potential of other theoretical approaches. With increasing complexity of the general theoretical background, valid simplifications are of increasing significance and value.

The initial appeal of the stirred reactor analysis therefore derives from its generality, relative simplicity, and predictive potential. However, there is a fourth reason for interest that is also substantially independent of the others. The consequences of stirred reactor theory indicate that inclusion of a stirred reactor section in a combustion chamber is not only a possible *theoretical* assumption but could be a desirable or even essential *physical* requirement for good flame holding with fast ignition at good turndown. There are also rules for sizing the stirred section to optimize the mixing pattern and, with it, the combustor performance. Acceptance of such results then inverts the point of view, and instead of developing a theory to match a combustor, one would develop a combustor to match the theory (if possible).

In spite of the evident appeal and potential of stirred reactor theory, however, its development since inception 20 or 30 years ago has been erratic, and most of the potential has yet to be realized. There has been a reluctance to accept the claims of generality made for the theory considering the relative simplicity of the assumptions and the extraordinarily varied and complex nature of the systems being analyzed. The central issue was the question of engineering realism; the analysis is valueless unless there are indeed stirred reactor sections in combustion chambers. This issue, however, is now being resolved in three ways. First, mixing studies, particularly over the last decade, have shown that virtually all industrial combustors of any importance have sections that can be identified as stirred, with recirculation or back-mix flows moving from downstream to upstream. Second, the original stirred reactor concepts were mainly developed in terms of the limiting case of the "perfectly" stirred reactor, and this certainly was partly responsible for the reluctance to accept the generality of the analysis or its applicability to an engineering as opposed to an ideal system. The objection is now being removed by extension of the analysis to imperfectly stirred or "well-stirred" systems with increasing engineering realism. Third, there is the appreciation of the value of designing the combustor to match the theory, or at least investigating the extent to which this can be done.

Currently, therefore, there is substantially increased interest in the application of stirred reactor theory to combustion chamber design; and the purpose of this chapter is to provide an introduction to the elements of the theory. Space prevents a full coverage of the ramifications of the theory which are increasing rapidly. Coverage is therefore confined to the salient aspects obtainable from the perfectly stirred theory with first-order kinetics, plus a brief qualitative survey of some of the recent conclusions from well-stirred theory. This, it is hoped, will provide sufficient basis for understanding by the reader of some of the more recent and expected future developments of the analysis.

II. ELEMENTS OF THE STIRRED REACTOR ANALYSIS

A. Objectives and Basis of Method

The objective of this, or of any other combustor design analysis is to predict required performance characteristics and parameters such as, for example, those listed in the Introduction. The basis of the method for predicting *some* of the characteristics is simple and well understood: It requires setting up and solving the steady-state equations for heat, mass, and momentum flows, with combustion, inside the reactor. Execution of such a program is extremely complex if the full range of spatial variations of concentrations, temperatures, reactions and reaction rates, and so forth, are included; and where the internal flow patterns are not even approximately known, solution is usually impossible. The particular value of the perfectly stirred reactor (PSR) analysis is the simplicity that is introduced. The simplicity derives from the perfect stirring assumption since, by definition, this eliminates any need to consider the effect of spatial variations. We only need to measure the exit conditions to know also the combustion conditions throughout the rest of the chamber. If the combustor is only well stirred, not perfectly stirred, the analysis becomes more complex, but the theory can be suitably modified to cover many well-stirred situations without losing its predictive potential or too much of its generality and relative simplicity.

Use of the PSR assumption reduces the theoretical requirements to an overall heat and mass balance on the reactor, plus a kinetics rate equation per unit of the reactor volume that is uniform throughout the reactor. Setting up and solving the equations, however, only provides information on the stable operating conditions (temperatures, concentrations, combustion intensities, and combustion efficiencies). Information on the so-called "criticality" conditions of ignition and extinction (temperatures, concentrations, residence times, etc.) requires a modified approach, based on the Semenov thermal explosion type of analysis, that is most easily introduced by way of the relevant equations as set out in the section following.

B. The PSR Equations for First-Order, One-Step Reaction

This is the simplest possible case; the analyses of more complex reactive systems, with multiple-order, multiple-step reactions in a well-stirred reactor, are derivatives of this simplest case. This analysis substantially follows that originally developed by Vulis (1961), with some minor changes in nomenclature. Analysis proceeds by development of mass and heat balances on the reactor.

1. Mass Balance

The system is illustrated in Fig. 1. A carrier stream flowing at J kg/hr brings into the combustor (of volume V_c m^3) a reactant of calorific value

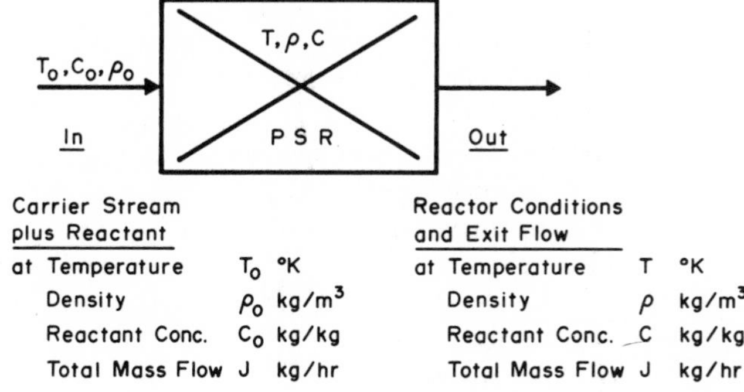

Fig. 1. Representation of an adiabatic perfectly stirred reactor in steady-state flow. Heat Generation: $\eta_{\text{I}} = 1 - (1 + \tau_{s/k}e^{-1/\theta})^{-1}$; heat loss: $\eta_{\text{II}} = (\theta - \theta_o)/\omega$. [See Eqs. (11) and (13).]

h_f kcal/kg at temperature T_0°K, and concentration c_0 kg reactant/kg of flowing mass. By the PSR assumption, the incoming material (carrier and reactant) is distributed instantaneously and uniformly throughout the reactor. In the process, it is mixed with and diluted by the material already present, a mixture of carrier, reactant, and reactant products. Uniformly throughout the reactor, therefore, the reactant is present at temperature T and concentration c (kg/kg). The reactant is being supplied at a rate Jc_0, and is being removed at a rate Jc. There is therefore an excess of supply over removal that is matched, at steady state, by the reaction rate summed over the combustion volume. If f is the volumetric concentration (kg/m³) the local rate of reaction is $(df/dt) = d(\rho c)/dt$, where ρ is the local density (kg/m³). Multiplication by the combustion chamber volume V_c sums the reaction rate over the volume, as required. At steady state, we then have

$$Jc_0[1 - (c/c_0)] = F\eta = (\rho V_c)(dc/dt), \tag{1}$$

where F is the rate of reactant supply or firing rate equal to Jc_0, and η is introduced as the *combustion efficiency*

$$\eta = 1 - (c/c_0). \tag{2}$$

Note: In Eq. (1), the density ρ appears outside the differential because ρ is invariant with time at steady state.

The assumption of first-order, one-step reaction is now introduced by writing (see Note on Units, Section II,D)

$$df/dt = kf = k\rho c = \rho(dc/dt), \tag{3}$$

where k is the velocity constant for the reaction, and we adopt the usual Arrhenius expression

$$k = k_0 e^{-E/RT}, \tag{4}$$

where k_0 is the frequency factor or preexponential constant (reciprocal time), and E is the activation energy (kcal/mole).

Equation (1) is implicit in two dependent variables, c and T. A unique solution for the steady state therefore requires an additional equation. This is supplied by the heat balance (Section II,B,2); but it is convenient to rearrange Eq. (1) in comparative thermal terms. If Eq. (1) is multiplied through by the calorific value, it is transformed into an expression for the heat generation rate $\dot{Q}_G$ (kcal/m³-hr). It is also convenient to relate this to the heat generation parameter used by Vulis: $\dot{Q}_I$ (kcal/kg-hr), where $\dot{Q}_G = (\rho V_c)\dot{Q}_I$. We can also substitute

$$J\tau_s = \rho V_c, \tag{5}$$

where τ_s is the mean residence time in the combustor. Equation (1) can then be written

$$\dot{Q}_G/(\rho V_c) = \dot{Q}_I = (c_0 h_f/\tau_s)\eta = k_0 c e^{-E/RT}. \tag{6}$$

2. Heat Balance

Since carrier and reactants flow into the reactor at T_0 and flow out at T there is a rate of loss of sensible heat, $\dot{Q}_L$ kcal/hr, that is proportional to the temperature difference between outlet and inlet, and to the mass flow rate J. Dividing $\dot{Q}_L$ by (ρV_c) yields $\dot{Q}_{II}$ kcal/kg-hr, and the expression for heat loss can be written [cf. (6)]

$$\dot{Q}_L/(\rho V_c) = \dot{Q}_{II} = (J/\rho V_c)C_p(T - T_0) = C_p(T - T_0)/\tau_s, \tag{7}$$

where C_p is the specific heat at constant pressure of the inlet reactants and outlet products, assumed equal and independent of temperature.

At *steady state* the heat supplied equals the heat lost

$$\dot{Q}_I = \dot{Q}_{II}. \tag{8}$$

Simultaneous solution of Eqs. (6) and (7) under the condition of (8) then yields the stable state temperature, concentration, and combustion efficiency. Typical results obtained for this ideal case of the one-step, first-order reaction are developed and discussed below. What can be emphasized here, however, is the generally representative nature of the analysis and the subsequent results.

3. Dimensionless Form of the PSR Equations

Equations (6) and (7) are more conveniently handled in dimensionless form. One dimensionless group, the combustion efficiency η is given by Eq. (2). Manipulation in Eq. (6) shows that η has an alternative interpretation. Multiplying through by $(\tau_s/h_f c_0)$, we find that η is also a dimensionless *rate of heat generation group*

$$\eta_I = \dot{Q}_I \tau_s / c_0 h_f, \tag{9}$$

where the subscript I to η is to distinguish from η_{II} [Eq. (13) below]. Introducing also the dimensionless *temperature group*

$$\theta = RT/E, \tag{10}$$

Eq. (6) can be written, after substituting and rearranging,

$$\eta_I = \tau_{s/k}/(\tau_{s/k} + e^{1/\theta}) = 1 - (1 + \tau_{s/k} e^{-1/\theta})^{-1}, \tag{11}$$

where $\tau_{s/k}$ is a dimensionless *time group*

$$\tau_{s/k} = k_0 \tau_s = \tau_{s/k}{}^0 (\theta_0/\theta)(P/P_0), \tag{12}$$

where $\tau_{s/k}{}^0$ is the dimensionless residence time at ambient temperature θ_0 and pressure P_0.

Equation (7) can be treated likewise by multiplying through also by $(\tau_s/h_f c_0)$. The heat loss rate $\dot{Q}_{\mathrm{II}}$ is transformed into η_{II}, giving

$$\eta_{\mathrm{II}} = (\theta - \theta_0)/\omega, \tag{13}$$

where ω is a dimensionless *initial concentration group*

$$\omega = c_0 R h_f / E C_p = (\theta_m - \theta_0), \tag{14}$$

where θ_m is the maximum or adiabatic flame temperature.

C. Interpretation of the PSR Equations

1. Functional Forms of the Equations

Equations (11) and (13) are both equations for the monotonic variation of the dimensionless rate of heat generation or loss η with the dimensionless temperature θ. In both cases, η has low and high limits of zero and one; θ has low and high limits of zero and infinity. Equation (11) is a sigmoid with a point of inflection; Eq. (13) is a straight line with slope $(1/\omega)$ and intercept at zero η of $\theta = \theta_0$.

The functional form of Eq. (11) is illustrated in Fig. 2 for a range of values of constant $\tau_{s/k}$ (solid lines). The meaning of the additional dashed and semidashed lines is explained below; they are related to the thermal ignition and extinction behavior.

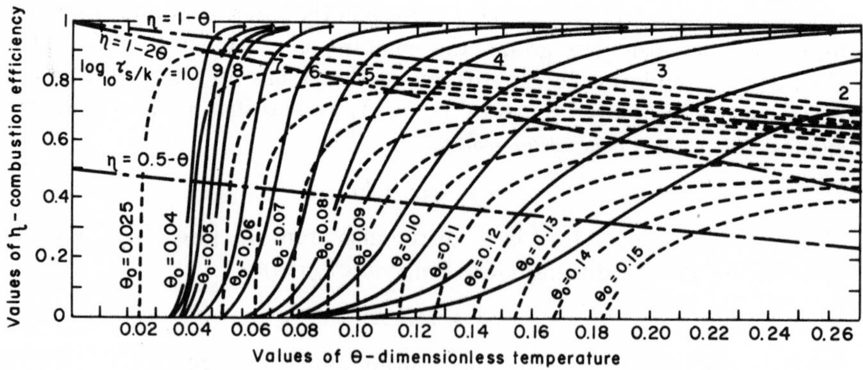

Fig. 2. Variation of η with θ, illustrating (i) Eq. (11) (solid curves), heat generation; (ii) Eq. (17) (dashed curves), criticality curves; (iii) Eqs. (18)–(20), semidashed curves. See text. Section II, C. [Adapted from "Thermal Regimes of Combustion" by L. A. Vulis. Copyright 1961, McGraw-Hill. Used with permission of McGraw-Hill Book Company.]

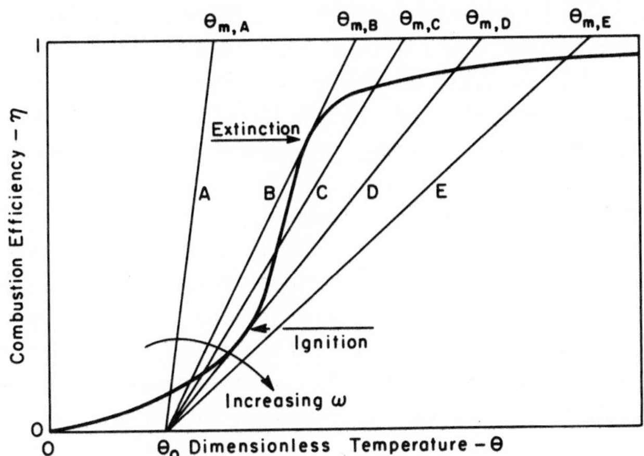

Fig. 3. Illustration of graphical solution of Eq. (15) to obtain stability conditions by equating η_{I} [Eq. (11)] and η_{II} [Eq. (13)], for five different values of ω. The sigmoid is η_{I} and the straight lines are for η_{II}. Lines B and D represent extinction and ignition, respectively.

2. Stability

Stable solutions that predict the operating temperature and exit concentration or combustion efficiency are obtained when Eq. (8) is satisfied. In dimensionless form this becomes

$$\eta_{\text{I}} = \eta_{\text{II}} = \tau_{s/k}/(\tau_{s/k} + e^{1/\theta}) = (\theta - \theta_0)/\omega. \tag{15}$$

This can be regarded as an equation in five variables: η, θ, $\tau_{s/k}$, ω, and θ_0, where the first two are dependent, and the last three are independent (externally controllable) variables. Solutions to Eq. (15) can be obtained graphically, by drawing straight lines through the sigmoids illustrated in Fig. 2. A typical result is illustrated in Fig. 3, in which it has been assumed that the dimensionless concentration has been steadily decreased. Excluding the condition of the two tangent lines (B and D) which are discussed below, it is evident that the solution of η with respect to θ can be either single valued (lines A and E) or triple valued (line C); that is to say, for given values of the independent variables, there can in some instances be three stable temperatures.

Where there are three stable temperatures, however, only two of the intersections represent *positively* stable values: These are the highest and lowest intersections on line C, and they can be formally shown to be positively stable by considering the effects of a slight displacement from the stable point and determining that the subsequent behavior is return to the original point before displacement. The middle intersection is a metastable point so that, on slight displacement, the system does not return to that point. The metastable point is not, therefore, a significant physical condition. The two positively stable points, however, do represent significant physical conditions: The upper point represents stable operation, with flame and substantial combustion efficiency (but never 100%, which is a point discussed in some detail below); the lower point represents passage of reactant through the combustor with theoretically finite, but practically negligible degree of combustion—the system could operate stably but has not been ignited. This is in contrast with the single intersection of line *A* where there is no combustion and the system cannot be ignited. Conversely, the intersection of line *E* represents combustion at significant temperature, under conditions such that extinction is (theoretically) impossible; the occurrence of such circumstances is somewhat rare: it is found with pyrophoric or hypergollic materials.

3. Criticality

The so-called critical conditions are those of ignition and extinction. Graphically, they are represented by the tangency between the heat generation and the heat loss lines (lines *B* and *D* in Fig. 3). Analytically, the condition is

$$d\eta_{\mathrm{I}}/d\theta = d\eta_{\mathrm{II}}/d\theta, \tag{16}$$

in addition, of course, to the stability condition expressed by Eq. (15). Differentiation and rearrangement yield an equation that can be written in a number of different forms; a convenient one is

$$\eta_{\mathrm{cr}} = 1 - [\theta^2/(\theta - \theta_0)]. \tag{17}$$

The critical conditions are obtained by the simultaneous solution of Eqs. (17) and (11) or (13). Such simultaneous solution yields equations for either of the two dependent variables, η or θ, as functions of any two of the other three *in*dependent variables, θ_0, ω, or $\tau_{s/k}$. The equations so obtained are quadratics in θ or η, showing that there are two solutions: One solution represents ignition and the other extinction. In the graphical representation of Fig. 3, line *D* is the condition for ignition, and line *B* is that for extinction.

Equation (17) has been plotted on Fig. 2 as the set of dashed lines. Intersection of the dashed and the solid lines yields a criticality point. It can be seen that all such points lie within a certain region of the graph. This region is bounded by the equation

$$\eta_h = 1 - \theta. \tag{18}$$

Only for values of η less than η_h for a given θ can ignition or extinction be defined by the tangency method illustrated in Fig. 3. Line A, for example, if translated across the diagram without changing slope, would never exhibit more than one intersection.

Two other lines of significance are also illustrated on Fig. 3. Identification of the extinction and ignition points shows that all ignition points lie below a given line, and all extinction points lie above it. The line is the equation of the point of inflection of the η_I sigmoids, and it has the form, when the ignition and extinction points coincide,

$$\eta_{i=e} = 0.5 - \theta. \tag{19}$$

One other line is the locus of the peak values of Eq. (17); it is

$$\eta = 1 - 2\theta. \tag{20}$$

These lines map out the regions of ignition, extinction, and noncriticality, in very general terms. From such graphs, or from the equations that the graphs represent, the plots can be translated into real terms. Figure 4 is an

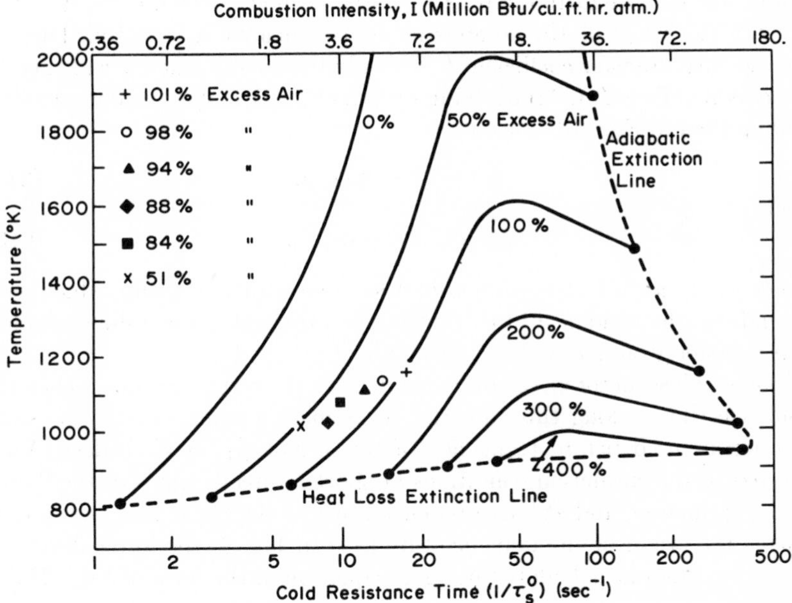

Fig. 4. Illustrates variation of reactor stability temperature (°K) with reciprocal of cold residence time at different levels of excess air for combustion of smoke. The figure shows two extinction lines. The experimental points are also indicated. [From Biswas and Essenhigh (1972).]

example for some experiments on smoke combustion (Biswas and Essenhigh, 1972) where a few experimental points were used as the basis of plots to determine the frequency factor and activation energy of the burning smoke. With the numbers so obtained, the stability curves for constant initial concentration c_0 were calculated for six different values of c_0 as a function of the cold residence time in the combustor (τ_s^0). The theory outlined above was also used to determine the extinction boundaries as marked. The practical objective was to determine the bounding conditions of operation under which the smoke flame would extinguish, and the smoke burning unit would emit smoke as a pollutant.

D. Other Factors

1. Nonadiabatic Operation

In the equations so far developed, adiabatic operation has been assumed. This is, of course, a substantial approximation except at very high firing rates (very short residence times) in well-insulated enclosures. The effect of wall loss may be introduced by adding a term $\dot{Q}_W$, which is the heat lost through the walls of the combustor across an area A, controlled by an average heat transfer coefficient h. The additional wall loss $\dot{Q}_W$ will then be given as $Ah(T - T_0)$. Multiplying by $(\tau_s/\rho V_c c_0 h_f)$ makes it dimensionless. Adding then to Eq. (13) yields

$$\eta_{II} = (\theta - \theta_0)/\omega_H, \tag{21a}$$

where

$$\omega_H = \omega/(1 + \beta\tau_{s/k}{}^0), \tag{21b}$$

where ω_H is an effective dimensionless concentration group which now depends on the residence time, that is, the firing rate; β is a dimensionless heat transfer group, $h(A/V_c)/(\rho_0 C_p k_0)$.

The heat loss factor η_{II} is now dependent on the residence time. When the residence time is long this can lead to extinction using exactly the same criterion as used before (tangency of the η_I and η_{II} lines). Indeed, Vulis referred to the extinction conditions at short residence times as the "adiabatic" extinction, and at long residence times as the "heat loss" extinction. This is the origin of the nomenclature used in Fig. 4; the lower, heat loss extinction line was calculated using an equation of the form of Eq. (21).

2. Heat Recovery

This plays an opposite function to wall loss, but the equations obtained have a similar form. If the combustor is followed by a heat exchanger of efficiency η_R, an equation of the same form as (21a) is obtained but with an effective $\omega = \omega_R$, given by

$$\omega_R = \omega/(1 - \eta_R). \tag{22}$$

The interesting point about this result is that the effect of heat recovery is to increase the initial concentration effectively, not, as one would intuitively expect, to increase the effective ambient temperature.

3. Dilution

Dilution by cooled, recycled tail gases is being increasingly used for the control of NO_x. With tail gases the combustion is assumed to be complete. Their effect is then to reduce both the inlet concentration and the residence time because of the greater mass throughput at constant firing rate. If the recycle ratio is r (kg of recycled gases/kg of fresh reactants as input), the residence time and inlet concentration become

$$\tau_s'/\tau_s = c_0'/c_0 = 1/(1 + r). \tag{23}$$

where the primed values are the values with recycle.

The reduction of residence time has the effect of reducing the combustion efficiency in Eq. (11). The effect can be expressed in terms of the combustion efficiency without recycle, η_I. If the modified combustion efficiency with recycle dilution is written $\eta_{I,D}$, we have

$$\eta_{I,D} = 1 - [r + 1/(1 - \eta_I)]^{-1}. \tag{24}$$

From this it is clear that the combustion efficiency falls as the recycle increases.

This is an effect of tail gases recycled outside the combustor with cooling. It should be noted that recirculation (which is internal recycle) inside the combustor has no influence on the PSR system since additional recirculation on the condition of perfect stirring can have no further effect. (This is contrary to the conclusion reached by Vulis.)

E. Note on Units

The units mainly used in this article for concentration are kilograms per kilogram, or kilograms per cubic meter. They are used in preference to mole fraction or moles per unit volume because a molar basis generally leads to mixed units, and because the molar concentration is often difficult or impossible to calculate with many practical fuels such as coal, commercial fuel oils, many natural gases, smokes, and combustible wastes. As reaction proceeds, the net molar flux across any cross section is very often changing when the mass flux (at steady state) is constant. Consequently, the mole fraction of an inert component can change even without reaction, when its mass fraction is constant.

If conversion to other units is required, there are seven different bases in common use, as follows:

(1) mass of component i per unit volume, f_i;
(2) mass of component i per unit mass of all components, c_i;
(3) moles of component i per unit volume, y_i;
(4) moles of component i per mole of total components, Y_i;
(5) fractional partial pressure of component i, $p_i = Y_i$;
(6) partial pressure of component i, $P_i = p_i P$;
(7) moles of component i per mass of all components, σ_i.

These units are related as follows:

$$\begin{aligned} c_i &= (1/\rho)f_i = (RT/P)(1/\bar{M})f_i \\ &= M_i\sigma_i \\ &= (RT/P)(M_i/\bar{M})y_i = (1/\rho)M_i y_i \\ &= (M_i/\bar{M})(P_i/P) \\ &= (M_i/\bar{M})p_i = (M_i/\bar{M})Y_i, \end{aligned} \tag{25}$$

where $\bar{M}$ is the mean molecular weight of the mixture, and M_i is the molecular weight of the component i.

The concentrations are all proportional to each other so that rates of reaction that are proportional to concentration or concentration products can be expressed in any of the units listed. A common simplification is to assume that $\bar{M}$ is constant throughout the reactor, and that $\bar{M}$ is approximately equal to M_i. However, the only consistent base in a flowing system is c (kg/kg) or the mixed base σ (moles of i/kg).

III. ELEMENTARY APPLICATIONS TO COMBUSTION CHAMBERS

A. The General Concepts

The material developed in the previous section provides quantitative descriptions of simple or ideal combustor behavior that can also be regarded as qualitative descriptions of combustor behavior in general. The quantitative agreement between the simple model as described and more complex real systems steadily diminishes, of course, with progressive departure from perfect stirring and with increasing complexity of the reaction systems. The next theoretical objective, therefore, is to identify and analyze the significant sources of difference between the simple ideal model and the real system, and to modify the simple model accordingly. This program has now been under way by numerous investigators for some years, and the modification of the simple model is a progressive and continuing process. This present section introduces two of the simplest extensions: The first relates to the mixing pattern and the second relates to the reaction order. Before outlining the two points mentioned, however, it is worthy of note that even the simple PSR model with first-order kinetics has practical value: first, as a general qualitative guide to operational behavior; and second, as a model by which departures from ideality can be quantified. It is also of interest that the effect of many of the model modifications would seem to be mainly to change the scales of such plots as those in Figs. 2–4. The graphs might be said to remain topologically invariant, which would seem to be a common consequence of essentially phenomenological analyses such as that developed in this article.

In outline, past attention was called to the problem of the mixing pattern in two ways. From the practical point of view, studies of jets showed the existence of recirculation patterns which led to skepticism that the jet region was even well stirred. From the theoretical point of view, even if the injection region *was* perfectly stirred or even well stirred, the theory showed that the combustion efficiency was less or substantially less than 100% (see parenthetical comment in Section II,C,2). It therefore followed that a stirred reactor region had to feed into a final burn-out region that Bragg (1953) first proposed had to be in plug flow (PFR). This prescription of a PSR followed by a PFR is usually referred to now as the Bragg criterion. This is the point regarding mixing to be expanded on in this section; note that it still adopts the view that the injection region is perfectly stirred. The question of nonperfect stirring in the injection region is considered in Section IV.

The other factor to be considered in this section is the more straightforward one of the effect of second- and multiple-order reactions. Both aspects are conveniently set in context by a prior consideration of combustion intensity.

B. Combustion Intensity

1. The Rosin Analysis

The first analysis of combustion intensity was carried out by Rosin (1925). The derivation related the combustion intensity, defined as the rate of heat release per unit volume and atmospheric pressure, to an average flame temperature (T_f) and combustion time (t_c). The expression obtained was

$$I = Fh_f/V_cP = K(T_0/T_f)/t_c, \tag{26}$$

where

$$K = \rho_0 h_f/(1 + G_s(1 + E\%/100)), \tag{27}$$

where G_s is the stoichiometric ratio between the air and the fuel (kg air/kg fuel); and $E\%$ is the percentage of excess air. $E\%$ is related to the equivalence ratio ϕ, defined as the actual fuel–air ratio divided by the stoichiometric fuel–air ratio, thus:

$$\phi = 1/(1 + E\%/100). \tag{28}$$

Note that $E\%$ can take negative as well as positive values.

The Rosin analysis provides a context in which all systems can be compared. In the first place, the range of values for the constant K [Eq. (27)] is surprisingly limited for all fossil fuels (coal, oil, and gas). Statistically, fossil fuels require about 1 kg of air per 500 kcal released by combustion; this means that the group ($\rho_0 h_f/G_s$) in Eq. (27) is about 500. Since G_s for fossil fuels is 10 or greater, the constant K reduces to $500/(1 + E\%/100)$, to an error that is generally less than 10%. Because of the statistical approximation of the air required per unit of heat released, the adiabatic flame temperatures of the fuels are also very close. When the weight of air substantially outweighs the weight of fuel, the adiabatic combustion of fossil fuels would put 500 kcal into about 1 kg of air. The adiabatic flame temperature would therefore be about 2300°K. Actual values for hydrocarbon fuels mostly fall in the range 2100–2400°K. Flame temperatures in industrial furnaces and combustors are rather lower because of excess air and thermal loads. The temperature T_f in Eq. (26) will generally lie in the range 1500–2100°K. The temperature ratio T_0/T_f therefore lies in the range $\frac{1}{5}-\frac{1}{8}$. Excess air likewise varies over a relatively small range, in most instances between 5 and 25%.

The parameters in Eq. (26) are therefore relatively invariant, with the exception of the combustion time t_c. The combustion time is the sum of entry delays, dispersion (i.e., mixing delay τ_D, if any), evaporation (if relevant), and chemical reaction τ_r. Upper limits to combustion are therefore determined by reaction time alone. Table I lists ranges of reaction times for industrial fuels in the three phases. As can be seen, they range over five decades. Inserting values appropriately into Eq. (26) yields an upper limit, in a detonation, of about 10^{11} kcal/m^3-hr-atm, and in the reaction zone of a premixed flame of about 10^{10} (Palmer, 1968). Table II lists other values for a variety of different fuels and conditions.

TABLE I

RANGES OF REACTION TIMES FOR HYDROCARBON FOSSIL FUELS AT NORMAL PRESSURE

Fuel	Conditions of reaction	Time (msec)
Solids (coal)	Normal pulverized coal combustion	10^3 (1 sec)
	Possible limit in explosion flame	10^2
Fuel oils	Heavy fuel oils (carbon forming)	10^2
	Light fuel oils: large drops	10
	Light fuel oils: small drops	1
Gases	Premixed flames (industrial)	1
	Adiabatic flat flame combustion	10^{-1}
	Detonation or shock combustion	10^{-2}

The Rosin equation brings out very clearly [see Essenhigh (1967b)] the extent to which combustion intensity is influenced primarily by the mixing and reaction kinetics (determining, with temperature, the combustion time) and to a lesser extent by heat transfer which helps to determine the flame temperature. The equation also brings out the relative importance of chemical reaction and mixing for the different fuels. Calculation of the combustion intensities from the reaction times listed in Table I yields values close to those found in practice for coals, but values one, two, or three orders of magnitude higher than practice for gases, with oil intermediate. The conclusions are that mixing dominates most industrial gas combustion systems; for oil, mixing and chemical reaction seem to be of roughly equal importance; and reaction dominates in coal combustion.

2. PSR Prediction of Combustion Intensity

Prediction of combustion intensity from PSR theory is the first step in amplifying the Rosin equation. In particular, Eq. (26) shows I as a function of two related parameters, T_f and t_c, but the Rosin analysis leaves the relation implicit (Essenhigh, 1967b); one objective is to make the relationship explicit.

In the definition of combustion intensity in Eq. (26), the product Fh_f is, of course, the total rate of heat release for complete combustion, that is to say, the value of $\dot{Q}_G$ in Eq. (6) at $\eta = 1$. At $\eta = 1$, that equation therefore yields the maximum combustion intensity for the given conditions, I_m:

$$I_m = \dot{Q}_G/PV_c = (\rho/\tau_s)(c_0 h_f/P) \tag{29}$$

TABLE II

TABLE OF COMPARISONS OF COMBUSTION INTENSITIES OBTAINED WITH DIFFERENT FUELS[a]

Combustion intensity (Btu/hr ft³-atm)	Fuel type: Gas	Fuel type: Liquid	Fuel type: Solid	Combustion intensity (kcal/hr m³-atm)
10^{11}				10^{12}
—	Palmer upper limit (theoretical)			
10^{10}				10^{11}
4×10^{9}	Mullins theoretical upper limit			
10^{9}				10^{10}
	Longwell bomb (80% comb. special research reactor)			
10^{8}	Wingaersheek torch	Liquid fuel rockets		10^{9}
	Countervortex burner (PSU)			

10^7	Premixed industrial gas burners (intensity defined on *flame* volume)	Ram jet	Solid fuel rockets	10^8
		Gas turbines with pressure jet burners		
10^6	Premixed or turbulent diffusion gas flames with intensity defined on furnace volume	Industrial furnaces (light oil)	Pulverized coal MHD experimental also fluid bed also cyclone excluding radiant chamber	10^7
		Medium fuel oils (pressure jet and air atomized)		
10^5				10^6
		Heavy fuel oils (air and steam atomized)	Pf and stoker firing (industrial)	
10^4	Ceramic (batch furnaces)			10^5
		Household oil burners (No. 2 oil)		
10^3			Coal (domestic)	10^4
10^2	*All fuels*—for drying and baking ovens			10^3

[a] *Note:* The industrial furnace operations are generally at atmospheric pressure. Gas turbines and rockets are normally pressurized. (1 Btu/hr-ft^3-atm = 9 Kcal/hr-m^3-atm.)

(and $\rho/\tau_s = \rho_0/\tau_s^0$). For incomplete combustion in the reactor, we have

$$I/I_{\mathrm{m}} = \eta = \tau_{s/k}(1 - \eta)e^{-1/\theta}. \tag{30}$$

Since η is given by Eq. (11), the ratio I/I_{m} can be determined as a function of η alone or of θ alone, when the stability condition has been determined by equating η_{I} and η_{II}.

For the variation of I/I_{m} with θ, this is given, of course, by Fig. 2, for different *constant* values of residence time $\tau_{s/k}$. If, however, we search for an optimum combustor intensity by changing the firing rate, this changes the residence time if the stoichiometry is kept fixed, and $\tau_{s/k}$ then becomes a *variable* of the system. The variation of I with η is then modified by the variation of I_{m} with τ, according to Eq. (29). This is clear if we write out the relation

$$I = I_{\mathrm{m}}\eta = (\rho c_0 h_f k_0/P)(\eta/\tau_{s/k}) = (\rho_0 c_0 h_f k_0/P)(\eta/\tau_{s/k}^0). \tag{31}$$

We are therefore led to define a parameter, the "specific intensity" χ, proportional to I, given by

$$\chi = \eta/\tau_{s/k} = (1 - \eta)e^{-1/\theta}. \tag{32}$$

If the firing rate is progressively increased in steps, we are conceptually looking at a series of stable intersections of η_{I} and η_{II}. We can therefore use Eqs. (13) and (14) to replace θ in Eq. (32), yielding

$$\chi = (1 - \eta)\exp\{-1/[\eta(\theta_{\mathrm{m}} - \theta_0) + \theta_0]\}. \tag{33}$$

The typical curve obtained from this equation is illustrated in Fig. 5. The

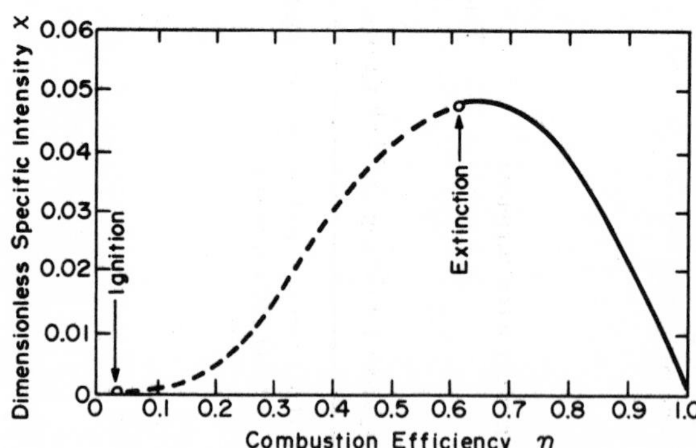

Fig. 5. Variation of specific intensity (χ) with combustion efficiency (η) illustrating optimum intensity (at minimum residence time) at 60–70% combustion efficiency. Illustration of Eq. (33). [Adapted from "Thermal Regimes of Combustion" by L. A. Vulis. Copyright 1961, McGraw-Hill. Used with permission of McGraw-Hill Book Company.]

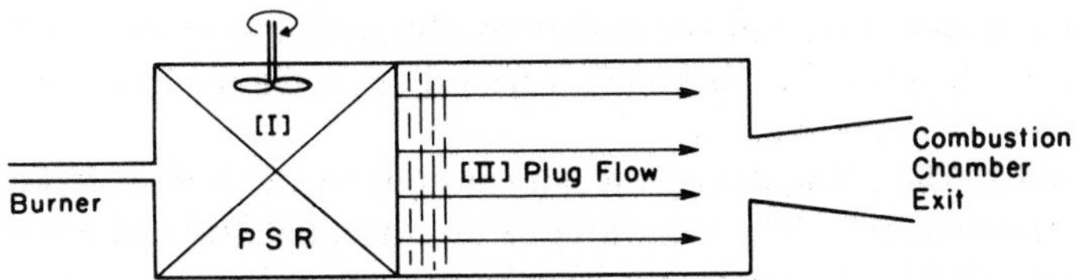

Fig. 6. Bragg's idealized combustion chamber illustrating the Bragg criterion. [I] Perfectly stirred reactor section; [II] plug-flow section. [See Bragg (1953).]

most significant point is the existence of an optimum combustion intensity or specific intensity at some combustion efficiency that generally lies between 0.5 and 1.0, and typically will more likely be 0.6 to 0.7.

Because of the optimum in Fig. 6, this figure is one of the most important results of PSR theory relating to combustion chamber design. It provides part of the basis for the relative sizing of the PSR and PFR sections in practical combustors designed according to the Bragg criterion (Sections III,A and C).

C. The Bragg Criterion

The essential proposition of the Bragg criterion is that the total volume of a combustion chamber should be divided, as already mentioned, into a PSR section for ignition and flame holding, followed by a PFR section for final burn-out. The arrangement is illustrated in Fig. 6. The need for a final burn-out section follows from Fig. 5 which shows that a PSR section at optimum or near optimum combustion intensity achieves only 60 or 70% combustion efficiency. A general proof that the final burn-out section should be in plug flow can be developed through consideration of the one-dimensional or plane flame. It is not intuitively obvious that this should be so; it was a result that appeared (Essenhigh, 1967a) as an extension to a previous analysis of the plane flame by application of PSR theory (Essenhigh, 1966) following a lead from Vulis (1961).

1. Plane Flame Analysis by PSR Theory

In the classical analysis of the plane flame, the flame is divided into elemental volumes on which heat and material balances are set up, and the equations thus developed are taken to the limit of the appropriate differential equations for the system. In the PSR analysis, the procedure is identical, except for the final step of forming the limit of the differential equations. The elemental volumes are not taken to the zero limit; they are left as finite elements, and the equations obtained appear in a finite difference form. The procedure is known variously as the finite element analysis, or the discrete space analysis.

Applied to the plane flame, we can consider two variants of two related sets of circumstances. The two variants are the adiabatic and nonadiabatic conditions, and in this context it is sufficient to consider only the adiabatic case. The two different sets of circumstances are the conditions of: *no* thermal flux upstream against the flow direction; and *with* upstream thermal flux. The two situations are illustrated as PSR sequences in Figs. 7 and 8. The first of these two cases (Fig. 7) represents the classical thermal explosion system that led to the Semenov analysis: Conceptually, we think of each PSR cell as the same cell at different moments in time so that there can be no thermal feedback to an earlier moment in time. The second case is the same as the thermal explosion system but with the addition of heat transfer upstream.

Derivation of the equations is out of place here—the derivations are given in the original article (Essenhigh, 1966)—but the essence of the method is to set up first the η_{I} and η_{II} equations for a single cell (the ith cell), following the principle of the analysis developed in Section II. The equation for η for the nth cell is then obtained by summing over all cells. The results are as follows:

(a) *For the thermal explosion system*:

$$\eta_{\mathrm{I},n} = 1 - \prod_{i=1}^{n} [1 + \tau_{s/k} e^{-1/\theta_i}]^{-1}, \tag{34}$$

$$\eta_{\mathrm{II},n} = (\theta_n - \theta_0)/\omega. \tag{35}$$

A first point of interest is the product form of the equation for η_{I}. For any cell n the variation of η with θ is a sigmoid as in Figs. 2 and 3; however, each sigmoid starts, at $\theta = 0$, with a value of η equal to the stable condition value of the previous cell, $(n - 1)$. The second point of interest is that η_{II} is the same as for a single cell [cf. Eq. (13)].

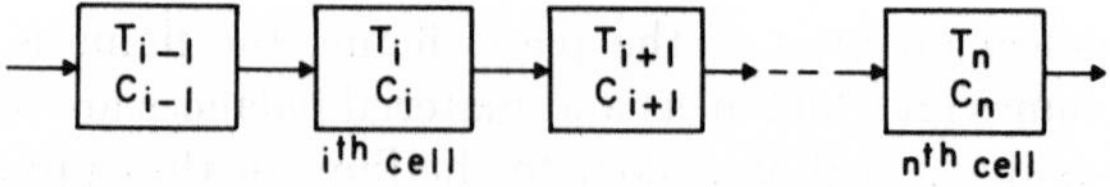

Fig. 7. PSR sequence representation of thermal explosion, equivalent to plane flame with no thermal feedback upstream. Heat generation: $\phi_{n,\mathrm{I}} = 1 - \prod_i[1 + \tau_{s/k}e^{-1/\theta_i}]^{-1}$; heat loss: $\phi_{n,\mathrm{II}} = (\theta - \theta_0)/\omega$. [From Essenhigh (1966).]

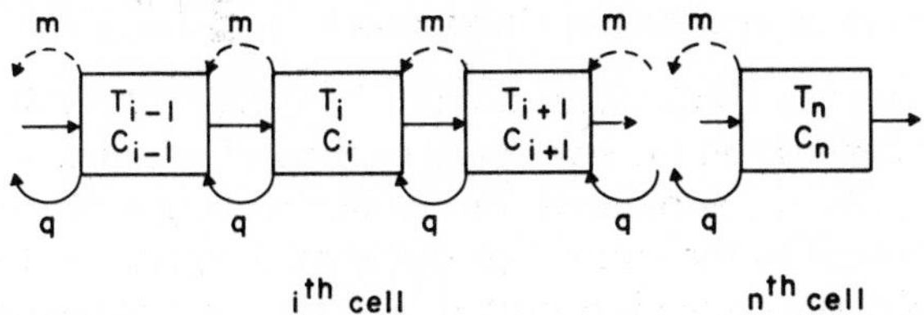

Fig. 8. PSR sequence representation of plane flame, with thermal feedback upstream. Heat generation: $\phi_{n,\mathrm{I}} = 1 - \prod_i [1 + \tau_{s/k} e^{-1/\theta_i}]^{-1}$; heat loss: $\phi_{n,\mathrm{II}} = \{\theta_n - [\theta_0 + \gamma(\theta_{n+1} - \theta_n)]\}/\omega$ (dashed lines: diffusion of active species). [From Essenhigh (1966).]

(b) *For the plane flame system:*

$$\eta_{\mathrm{I},n} \equiv \text{Eq. (34)},$$

$$\eta_{\mathrm{II},n} = (\theta_n - \theta_0')/\omega, \tag{36a}$$

where

$$\theta_0' = \theta_0 + \gamma(\theta_{n+1} - \theta_n) \tag{36b}$$

and γ is a dimensionless heat exchange coefficient for heat transfer inside the flame, given by

$$\gamma = (h_i A / V_c \rho C_p)\tau_s = h_i / J_A C_p, \tag{37}$$

where h_i is the intercell heat exchange coefficient (that can be related to the thermal conductivity in the case of pure conduction), A is the area of the heat exchange interface, and J_A is the mass flow rate into any cell per unit of area of cross section (introducing J_A eliminates the cell residence time τ_s).

Equation (36) shows that the effect of the upstream heat transfer is to introduce an effective and variable surroundings temperature θ_0' whose value depends on the temperature of the cell beyond the nominal point of computation. The result is to create ignition earlier than would be the case for the thermal explosion system under otherwise identical conditions.

In the original article (Essenhigh, 1966) a number of consequences of the equations were explored, including the demonstration that they lead to flame temperature profiles that can be calculated by appropriate iterative computer techniques. These are out of place here except for the qualitative statement of a quantitative result: that the equations show the existence of an ignition cell in the middle of the flame, defined by the tangency criterion of Eq. (16) (thus enabling definition of ignition without need for recourse to arbitrary criteria such as point of inflection on the temperature profile, and so forth). The results show that ignition in the flame occurs at or before the point of inflection on the temperature profile. This is a result of pertinence to the Bragg criterion proof to follow.

2. Derivation of the Bragg Criterion

In computing the temperature profiles from Eqs. (34)–(36), the shape of the profile depends on the size of cell or element selected, which is determined by $\tau_{s/k}$. As $\tau_{s/k}$ is reduced, the profile converges to that computed from the solutions to the equivalent differential equations. Conversely, if $\tau_{s/k}$ is increased, the profile is distorted, and the temperature gradient may be either greater or less than the correct value; that is, the temperature is rising faster or slower than accurate prediction would indicate. To the extent that this is a deviation from accurate prediction, the calculation is in error, and is valueless. However, if a large $\tau_{s/k}$ represents a realizable *physical* condition, to that extent the deviation from the differential equation solution is meaningful and possibly important. Since in some combustors, at least, there is a stirred section or an approximation thereto, we have therefore some circumstances in which $\tau_{s/k}$ is large for physical reasons. We can also conceive of utilizing turbulence to adjust the size of the finite element cells to suit some theoretical requirement. The question at issue, therefore, is what conclusions we can draw from the one-dimensional flame analysis regarding the desirability of large or small cell size.

The problem may be set in the following terms. In a one-dimensional flame, treated as a sequence of cells as in Fig. 8, the temperature is rising monotonically, and cell size, or residence time $\tau_{s/k}$, affects the rate of temperature rise. Consider, therefore, two adjacent cells i and $i + 1$. Between entry to cell i and exit, the temperature rises from θ_{i-1} to θ_i; and likewise for the $(i + 1)$th cell, the temperature rise is from θ_i to θ_{i+1}. Let these differences be $\Delta\theta$ and $\Delta\theta'$. We now replace the two cells by a single cell, and we adjust the residence time, $\tau_{s/k}'$, so that the temperature across the cell jumps from θ_{i-1} to θ_{i+1}. We now can construct a criterion to determine whether the temperature rise is faster or slower with the one cell than with the two cells; it depends on whether $\tau_{s/k}'$ is greater or less than twice $\tau_{s/k}$.

From Eq. (36), and dropping the subscripts I and II because $\eta_{\mathrm{I}} = \eta_{\mathrm{II}}$, we can write, for the ith cell,

$$\Delta\theta = \omega(\eta_i - \eta_{i-1}) + \gamma(\theta_{i+1} - \theta_i) - \gamma(\theta_i - \theta_{i-1}), \tag{38}$$

and likewise for the $(i + 1)$th cell with the subscripts increased by one unit. If γ is less than one, and/or the temperature gradient is close to linear over the small temperature range involved in Eq. (38), the last two terms on the RHS will be approximately equal and will cancel. This leaves

$$\Delta\theta/\Delta\theta' = (\eta_i - \eta_{i-1})/(\eta_{i+1} - \eta_i). \tag{39}$$

From the predecessor equation to Eq. (34) we have, for the ith cell,

$$\eta_i - \eta_{i-1} = (1 - \eta_{i-1})\tau_{s/k}(\tau_{s/k} + e^{1/\theta_i})^{-1}, \tag{40}$$

and similarly, both for the $(i + 1)$th cell, with the subscripts increased by units, but also for the replacement cell, with τ' in place of τ and the combustion efficiency difference across the cell as $(\eta_{i+1} - \eta_{i-1})$. Manipulation of the equations yields the separate result

$$(\eta_i - \eta_{i-1})/(\eta_{i+1} - \eta_i) = (\tau_{s/k}'/\tau_{s/k}) - 1 \tag{41a}$$

$$= \Delta\theta/\Delta\theta'. \tag{41b}$$

This equation shows that τ' is more than twice τ if the temperature increments along the profile are progressively *de*creasing, that is, the rate of rise is slower in the larger cell; and the converse holds where the temperature increments are progressively *in*creasing along the profile. Consequently, on a sigmoid temperature profile the rate of temperature increase is enhanced either by having a few large cells where the profile is concave upward or by having many small cells beyond the point of inflection where the profile is concave downward. The limit of this is contraction of all cells prior to the inflection point into a single (PSR) cell, reverting to an infinite number of infinitely small cells (PFR) beyond the inflection point. This configuration describes the Bragg criterion.

D. Multiple-Order Reactions

1. Scope of the Problem

In actual combustion chambers, particularly using industrial fuels, the one-step, first-order reaction is rare. In most instances, reactions are multiple order with multiple steps. Jenkins *et al.* (1967), for example, solved an equation set for the hydrogen–oxygen reaction using nine elementary reaction steps (using the mixed units of gm mole/gm mixture, σ, as the computation base). Many others have extended such calculations to hydrocarbons, and for these it is common to use 10–100 elementary reaction steps. In a PSR such computation is tedious rather than difficult but, if mixing is included, the size and complexity of the calculation escalate rapidly. One solution to the problem was devised by Edelman and Fortune (1969) who introduced a "quasi-global" kinetic scheme in which the initial partial oxidation of the hydrocarbon molecule was assumed to occur with great rapidity, reducing to hydrogen and CO, with detailed kinetic steps only used for the final hydrogen and CO oxidation. This reduced the overall reaction to a set of only 10 "elementary" reaction steps. With this limited set, Edelman *et al.* (1972) recently extended their computations to a heterogeneous droplet combustion system that included coupling between reaction and flow, and the calculations have met with substantial success.

Even the quasi-global schemes, however, may not always be necessary, or even desirable when complexities are added from other sources. Few programs, for example, contain internal radiative heat transfer, which is of major importance in furnace combustion studies and which must be included in due course. More limited kinetic schemes have been successfully used, particularly in application to gas turbine combustors (Kretschmer and Odgers, 1972; Odgers and Carrier, 1972). The limitation imposed is to consider only a one-step reaction scheme, using phenomenological kinetics [e.g., Penner (1957)]. Further discussion in this section is therefore limited to multiple-component, one-step reactions.

2. One-Step, Two-Component Reactions

The one-step, two-component reaction was originally introduced as a means of treating the reaction of fuel with oxidant. It is convenient to retain this connection by using f_R and f_{ox} for the reactant (fuel) and oxidant concentrations, respectively. Generality is increased by assuming an overall reaction order n that can be different from 2, with the reactant order r, and the oxidant order $n - r$. The reaction rate per unit volume is then assumed to be proportional to $f_R{}^r f_{ox}{}^{(n-r)}$, with the velocity constant k as the proportionality constant. Multiplying by V_c, the combustion chamber volume, yields the total reaction rate $F\eta$ (kg/hr), where η is based on the fuel consumption (when oxidant is in excess overall). Converting to mass fraction concentrations c_f and c_{ox}, the equivalent form of the rate equation (1) can be written

$$Jc_R{}^0\eta = F\eta = (\rho V_c)(\rho^{n-1}k_0)c_R{}^r c_{ox}{}^{n-r}e^{-E/RT}. \tag{42}$$

The ρ term has been split for reasons to be developed [cf. Eq. (6)].

Equation (42) can now be reduced in one of two ways. First, multiply by $(h_f/\rho V_c)$ and nondimensionalize by introducing

$$\phi\eta = 1 - c_{ox}/c_{ox}{}^0, \tag{43}$$

yielding

$$\eta/(1-\eta)^r(1-\phi\eta)^{n-r} = (c_R{}^0)^{r-1}(c_{ox}{}^0)^{n-r}\tau_{s/m}e^{-1/\theta} \tag{44}$$

where

$$\tau_{s/m} = \tau_s k_m{}^0 = \tau_s(\rho^{n-1}k_0). \tag{45}$$

The term ρ^{n-1} is paired with k_0 to convert the units of the pair to reciprocal time (k_m is used to denote a *mass*-based kinetic constant, also with units of reciprocal time, but then dependent on pressure and temperature). An equation of the form of (44) was used by Biswas and Essenhigh (1972), first, to evaluate kinetic data on smoke combustion, and then to predict the combustion map of Fig. 4. If $n = r = 1$, Eq. (44) reduces to (11).

A common alternative treatment is to expand the density term in temperature and pressure and rearrange to form a "loading" parameter (J/V_cP^n) which has been found [e.g., by Herbert (1960)] to be a significant correlative parameter for stability and blow-off for varied combustor designs. Writing the loading parameter as ψ,

$$\psi\eta = (\rho_0^{n-1}k_0)(\rho_0/P_0^n)c_R^r c_{ox}^{n-r}(T_0/T)^n e^{-E/RT}. \tag{46}$$

Clarke and Odgers with associates (Clarke *et al.*, 1959, 1962, 1965; Kretschmer and Odgers, 1972; Odgers and Carrier, 1972) in particular, and also other workers, have applied Eq. (46) to gas turbines. The latest results (Kretschmer and Odgers, 1972) are of particular interest because of the clear correlations found between the reaction order and the stoichiometry; the activation energy was also found to vary with stoichiometry. The possibility of such a variation was first proposed by Hottel *et al.* (1957) but Odgers' results were the first to provide any quantification. They concluded that E rose with decreasing ϕ; and on reaction order, best fit was obtained with $n = 2\phi$ for lean mixtures, and $n = 2/\phi$ for rich mixtures. Of particular interest is the apparent change of overall reaction order from second near the stoichiometric to first as ϕ goes to either the lean or the rich limit. This is a correlative result, with no mechanistic explanation yet advanced—it could be due to changes in mixing pattern, or to changes in operating temperature as ϕ changes, or to real changes in the dominant reaction steps—but in the meantime, the developed equations, regarded as operational, are being successfully used for modeling gas turbine combustors and for prediction of performance (Odgers and Carrier, 1972).

3. Other Results

Most stirred reactor studies have been carried out in small laboratory reactors, of which the Longwell bomb (Longwell and Weiss, 1955) was the prototype. Table III lists some of the more readily accessible results. Analysis was generally based, overall, on Eq. (42) or its equivalent, using moles per unit volume or mole fraction as the concentration basis.

It is also appropriate to mention at this point some of the historical origins of stirred reactor theory. The principal origins are generally attributed to four sources, Bragg (1953), Avery and Hart (1953), Longwell *et al.* (1953), and Vulis (1961) in a book published in Russia in 1950 and republished in translation in 1961. Bragg was concerned with jet engines; Avery and Hart were concerned with ram jets. Additional input about that time was derived from van Heerden (1953) dealing with thermal theory of chemical reactors, and from DeZubay and Woodward (1955) using a

TABLE III

EXPERIMENTAL RESULTS FROM STIRRED REACTOR INVESTIGATIONS

Investigator	Fuel	Reaction order				Activation energy kcal/mole	Preexponential factor	
		Fuel	O_2	H_2O	n		Volume basis (k_o)	Mass basis ($k_m{}^\circ$) (sec^{-1})
Longwell and Weiss (1953)	Heptane	—	—	—	1.8	42	—	—
	Isooctane	—	—	—	1.8			
DeZubay and Woodward (1955)	Propane	—	—	—	2	30.5	4.07×10^{14} (liters/mole) sec^{-1}	1.82×10^{13}
Hottel *et al.* (1957)	Mixtures A and B	—	—	—	1.3 ± 0.3	30 (A)	1.6×10^{11} (liters/mole)$^{0.3}$ sec^{-1}	6.3×10^{10}
						50 (B)	9.1×10^{14} (liters/mole)$^{0.3}$ sec^{-1}	3.58×10^{14}
Clarke *et al.* (1959)	Propane	0.8	1	—	1.8	42	1.3×10^{10} (liters/mole)$^{0.8}$ sec^{-1} (°K)$^{-0.5}$	1.78×10^{10}
Blichner (1962)	Propane	—	—	—	1.8	—	—	—
Clarke *et al.* (1962)	Propane	0.92	0.88	—	1.8	22.6	2.8×10^{8} (liters/mole)$^{0.8}$ (°K)$^{-0.5}$ sec^{-1}	3.84×10^{8}
Herbert (1962)	Kerosine	1	1	—	2	42	1.9×10^{14} (ft^3/lb) sec^{-1} (°K)$^{-0.5}$	2.51×10^{14}
Clarke *et al.* (1965)	Propane	1	1	—	2	28	1.8×10^{12} (cm^3/mole) sec^{-1} (°K)$^{-0.5}$	1.33×10^{9}
Hottel *et al.* (1965)	CO	1	0.3	0.5	1.8	16	1.2×10^{11} (cm^3/mole)$^{0.8}$ sec^{-1}	9.95×10^{9}
	Propane	1	0.35	0.4	1.75	15	2.9×10^{10} (cm^3/mole)$^{0.75}$ sec^{-1}	2.81×10^{9}

Kydd and Foss (1965)	CH_4,	—	—	—	—	57	—	—
	C_3H_8,	—	—	—	—	61	—	—
	C_2H_4	—	—	—	—	46	—	—
	H_2	—	—	—	—	48	—	—
	CO	—	—	—	—	38	—	—
Jenkins *et al.* (1967)	H_2/O_2	(Multiple order, multiple set of reactions, basis of evaluation)						—
Williams *et al.* (1969)	CO	1	0.5	0.5	2	25	1.8×10^{10} (liter/mole) sec^{-1}	8.05×10^8
	CH_4	1	0.5	0.5	2	57	5.3×10^{15} (liter/mole) sec^{-1}	2.37×10^{14}
Pratt and Starkman (1969)	NH_3	0.75	1.2	—	1.95	35	4.3×10^{13} $(cm^3/mole)^{0.95}$ sec^{-1}	3.12×10^9
Miles (1971)	Propane	—	—	—	1.8	—	—	—
Biswas and Essenhigh (1972)	Smoke	1	1	—	2	16.5	4.6×10^9 (cm^3/gm) sec^{-1}	5.98×10^6
Kretschmer and Odgers (1972)	C_3H_8, C_4H_{10}, C_5H_{12}, $C_{20}H_{42}$	ϕ	ϕ	—	2ϕ lean; $2/\phi$ rich	26–75	1.3×10^{10} $(liter/mole)^{n-1}$ $(°K)^{-0.5}$ sec^{-1}	—
Kuwata and Essenhigh (1972)	CH_4	(pseudo 1)		—	—	20	10^6 sec^{-1}	10^6
Shieh and Essenhigh (1972)	Cellulose volatiles	1	1	—	2	20.1	7×10^9 (ft^3/lb) hr^{-1}	1.55×10^5

ceramic-walled reaction chamber with reverse flow. Since then, additional investigations have proliferated, as indicated by Table III, mostly using variants of the Longwell bomb (Longwell and Weiss, 1955), with some later ones based on the jet mixed reactor introduced by Spalding (Jenkins *et al.*, 1967).

IV. A QUALITATIVE OUTLINE OF THE MIXING PROBLEM

A. Nature of the Problem

The essence of the mixing problem can be presented in two ways. If mixing is not perfect, we may ask either: (1) how can this be demonstrated, characterized, and translated into modifications of the mathematical model of the reactor; or (2) how can we modify a combustion chamber so that the mixing can be improved to within an acceptable level of the ideal?

1. Mixing as a Phenomenon

Mixing imperfections have been demonstrated in a number of different ways. Apart from visual inspection and flame probing, one of the earliest was by direct tracer in a flame; for example, Nicholson and Field (1949) used high-speed photography of bluff-body stabilized flames containing finely divided sodium salt as the tracer. More recently there has been much use of tracers in cold model simulations, with analysis based on residence time distribution measurements. In particular, three early sets of measurements (Bartok *et al.*, 1960; Beér and Lee, 1965; Drake and Hubbard, 1966) provided support for the PSR–PFR mixing pattern required for the Bragg configuration in both experimental reactors and in vortex flames (single-swirl oil flames). Later, more detailed investigations (Rao *et al.*, 1970; Rao and Essenhigh, 1971; Zeinalov *et al.*, 1972) modified the simpler view: The first results (Rao *et al.*, 1970; Rao and Essenhigh, 1971) seemed to support the proposition that the back-mix stirred region could be represented by a plug-flow section to describe mixing delay, followed by a PSR, and the combination would represent a well-stirred reactor. By measuring a mixing delay time (τ_D) from tracer experiments, a stirring factor (W) was defined as

$$W = \tau_s/(\tau_D + \tau_s), \tag{47}$$

where τ_s is the residence time of the PSR section. Unfortunately, the model required the existence of a definite margin between the mixing delay section and the PSR, and search for the boundary between the two was unavailing. Nevertheless, the model still has some conceptual value, and possibly some realistic value by reinterpretation, particularly as Zeinalov *et al.* (1972) have used a more complex model to show a relation between the stirring factor W and the Peclet number for mixing in the system.

A less direct approach to mixing is provided by air loading correlations [e.g., Herbert (1960)]. The argument is that, if reaction rate is controlled by the chemistry of the process—as it should be in the PSR or its close equivalent—then the combustor performance should be insensitive to or independent of changes in mixing, with changes assumed to be monitored by Reynolds number. Conversely, sensitivity to Reynolds number would denote the importance of mixing. Writing

$$\psi \propto \mathrm{Re}^m, \tag{48}$$

Herbert (1960) quotes values of m less than 0.2 as indicative of approach to the Longwell bomb level of mixing, and values of 0.5 or higher as indicative of the mixing influence being unusually large. Gas turbine combustor cans generally have values of m lying in the region of 0.2, with homogeneity further dependent on the scale of the recirculation inside the can, being more homogeneous the smaller the scale, as intuition would suggest. In contrast, one would predicate a value for m of 0.5 or higher for a single, unswirled turbulent diffusion jet of the type used in glass tanks and open hearth furnaces. The value of m has never been measured for such systems but appropriate research in the last two decades has indeed shown very clearly the influence of mixing, with *jet momentum* as a dominant factor. In the extreme cases of gas combustion in turbulent diffusion flames, combustion is evidently determined almost entirely by mixing, in line with the combustion engineer's rule of thumb "When it has mixed it has burned." Such mixing patterns were characterized as "heterogeneous" by Rao and Essenhigh (1971), in contrast to the homogeneous stirring of the PSR or its close equivalent.

Inhomogeneity is also apparent in many cases by direct visual examination and by measuring thermal and concentration maps inside a combustor by probing. However, although the evidence for adequate or inadequate mixing is usually unambiguous, this is not always the case, and it is desirable to rely rather on tracers and measurements that are not part of the combustion process itself. Even so, flame measurements are good indicators in the absence of other independent measurements. For example, Biswas and Essenhigh (1972) relied on temperature and oxygen concentrations in the smoke combustion investigations mentioned earlier (see Fig. 4).

2. Mixing as a Mechanism

In describing mixing as a mechanism we now embark on the beginnings of a long, and still uncompleted study, with many controversial points still being argued. Nevertheless, some consensus is now developing about the broad structure of mixing patterns and, without suggesting that the following picture is necessarily acceptable to all, it seems to have appreciable support.

Measurements have now established that mixing can be regarded as due to a combination of turbulence and convection. It is a decidedly arguable point whether the convective flow exists in its own right or whether it should be regarded as part of a much larger turbulence vortex. However, since laminar flow (which by definition is nonturbulent) can result in mixing—generally referred to then as blending—convective flows can clearly exist independently of turbulence. The separation of turbulence and convection is not therefore an unreasonable concept, and it is as a concept that the separation is presented here, and not necessarily as a "real" description. It would also seem to be intuitively reasonable to base the separation of the convection from the turbulence by an isotropic definition. Any identifiable flow stream inside a combustor can then be thought of as a convective flow with a net (nonisotropic) translational direction that *on average* will result in transport of material always in the same direction at the same velocity. Turbulence is then superimposed on the convective stream as a random process that is isotropic with respect to axes traveling in the convective stream at the stream velocity. The turbulence can then have any appropriate scale or intensity, where the Prandtl mixing length concept has a rough equivalence to both combined.

On this description, mixing is then the result of both convection and turbulence, with convection as the main distribution process throughout the reactor, and with cross mixing between adjacent convective streams due to turbulence. If the cross mixing is to be effective, the turbulence vortices must have sufficient range (mixing length) that they decay inside the adjacent convective stream (which thus determines the turbulence scale required: the mixing length should be comparable to the convective stream width). When the turbulence vortices do decay, mixing is completed with assistance from molecular diffusion. Therefore, two mixing scales are usually identified, as follows:

(a) *Coarse* or macromixing is attributed jointly to convection flows in the reactor aided by turbulent *dispersion*.

(b) *Fine* or micromixing is attributed jointly to turbulent *decay* aided by molecular diffusion.

On this basis we can now conceive of four main mixing patterns to model mathematically:

(1) The *PSR*, as already described, with homogeneous mixing at both the coarse and the fine level.

(2) The *well-stirred reactor*, with homogeneous mixing at the fine level, but nonhomogeneously mixed at the coarse level. An appropriate instrumental investigation would detect appropriate variations in the average local temperatures and compositions throughout the reactor, but no variations within a small locality.

(3) The *well-mixed reactor*, with homogeneous mixing at the coarse level, but not at the fine. Instruments would detect local variations, but their average would be unchanged at macroscopically different locations throughout the reactor.

(4) The *operational reactor*, with nonhomogeneous mixing at both the coarse and the fine level.

B. Basis for Quantification

Turning the qualitative model just described into mathematical form is a still uncompleted task. What follows here is merely a sketch of the two principal approaches being developed, outside, of course, the relatively more mechanistic approaches deriving from studies of the Navier–Stokes equations with reaction added [e.g., Beér and Chigier (1972)].

1. The Residence Time Distribution Description

This has a long background history deriving largely from chemical engineering [e.g., Levenspiel (1962), Aris (1965)]. The RTD description provides the basis for the analysis of tracer experiments, and the background is the age distribution functions (see Levenspiel or Aris) of molecules or particles injected into a reactor at some certain time, and then monitored either inside the reactor or as they escape.

There are four standard response curves, to which a fifth is added here for reasons to be given: I is the internal age distribution function; E is the exit age distribution function; the C curve is the exit response from a pulse (δ function) input; and F is the exit response from a step function start-up input. To this set is added here a G_∞ curve, the exit response from a step function cutoff input.

These functions can also be related to probability distribution functions $P(\varepsilon, \theta_i)$, where P is the probability of finding a molecule injected some time between θ_i and $\theta_i + d\theta_i$ earlier in a normalized volume element $d\varepsilon(x, y, z)$

inside the reactor, and θ is the normalized time, that is, $\theta = t/\tau_s$. The probability of finding the molecule in the exit volume, from which it can only leave the reactor, is $P(e, \theta_i)$. We then have the following relationships between the different functions:

$$\begin{aligned} I(\theta_i) &= 1 - F(\theta_i) = G_\infty(\theta_i) \\ &= \int_\varepsilon P(\varepsilon, \theta_i)\, d\varepsilon = \int_0^{\theta_i} E(\theta)\, d\theta = \int_0^{\theta_i} C(\theta)\, d\theta \\ &= 1 - \int_0^{\theta_i} P(e, \theta)\, d\theta = \int_{\theta_i}^{\infty} P(e, \theta)\, d\theta. \end{aligned} \tag{49}$$

For instrumental reasons, one of the easiest experiments to carry out is the step function cutoff. This was the technique used by Rao and associates (Rao *et al.*, 1970; Rao and Essenhigh, 1971). The response is then compared with the expected behavior from a PSR condition. The PSR response is obtained from the equations above by assuming that the probability of finding a molecule is equal at all points, including the exit volume. Inserting this condition yields

$$G_\infty = e^{-t/\tau_s}. \tag{50}$$

In practice, the response generally includes an initial period of constant concentration (the mixing delay period τ_D) before the exponential decay sets in. Conceptual models have been developed to explain this delay behavior, and primarily it appears to be due to the finite transit time through the reactor (Zeinalov *et al.*, 1972). However, the problem is complex and still under investigation.

A point of importance, however, that the exponential decay brings out is the distribution of residence times that exist in a stirred reactor that is frequently ignored in the combustion equations by assuming an average residence time. It has indeed been suggested (Kuwata and Essenhigh, 1972) that such a distribution can be responsible for pollutant emissions, and recent measurements are in support of this view (Kuwata and Essenhigh, 1973). If turbulence is very coarse, being possibly required for mixing across large convective streams, the turbulence lifetimes are correspondingly long. If the residence time distribution lies mainly inside the vortex lifetimes, then the turbulence vortices will mostly travel through the reactor without mixing with the background material. The shortest lifetime vortices can then travel through unburned, and be the principal source of unburned hydrocarbons; and the long-life vortices will either burn nearly adiabatically if the fuel and oxidant are premixed, and generate exothermic centers that can be sources of NO_x, or they will enable fuel-rich volumes to crack, thus being sources of smoke and soot.

Further study of residence time distributions is therefore likely to have increasing priority in the future.

2. The Stirred Reactor Description

This description seeks to present mixing in terms of sets of PSR's and PFR's, appropriately linked and cross coupled. It is essentially an enlargement of the concept used by Bragg as basis for the description of a PSR in series with a PFR; and the attempted model by Rao *et al.* (1970) using a PFR for mixing delay is a typical example of the procedure in a simple case. This is a procedure that again has a long history in chemical engineering (Levenspiel, 1962). In developing such constructs, however, care must be taken to distinguish between the *concept* that can be used to account for observed results, and the idealized *model* of a real physical situation. The background issue is that of establishing necessity of the construct as well as sufficiency. Equation (49) may help to clarify the point. For any given internal distribution function, there can be many probability functions that can yield the given I function on integration (sufficiency); the problem is always to determine the *unique* probability function for a given case (necessity). For that reason, constructs that are developed without test of validity are of interest but cannot be considered too seriously.

Zeinalov *et al.* (1972) therefore approached the problem from the point of view of reducing the options on possible constructs by prior experimental investigation. Three standard conditions were investigated, the unswirled, single-swirled, and double-swirled jet. In all three cases it was possible to identify a "flow tube" along which there was both forward and back-mix convective flow. A model was set up as a variant of the plane flame system of Fig. 8. The variant is shown in Fig. 9. Two outcomes can be

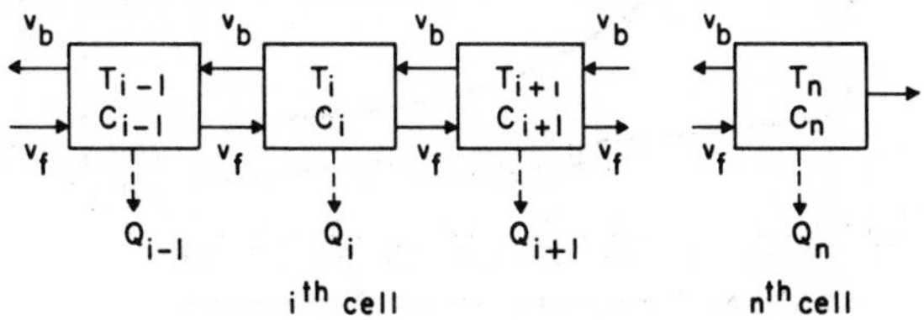

Fig. 9. PSR sequence representation of combustor flame along a flow tube with back-mix flow. Cell size characterized turbulence level. Notation used as as follows: c_i, fuel concentration of ith cell; i, ith cell; N, total number of cells (PSR); Q_i, heat loss from ith cell; v, net flow velocity $= v_f - v_b$, where v_b is convective back-mix flow velocity and v_f is convective forward flow velocity. [From Kuwata and Essenhigh (1972).]

mentioned. First, that the model was able to establish the significance of a definable Peclet number in terms of the back-mix ratio v_b/v_f and the number of cells used in the model N taken to be an index of the turbulence level, and to show that the Peclet number was a function of the stirring factor W:

$$\mathrm{Pe} = (vL/D_s) = -\ln[(v_b/v_f)^N] = fn(W), \tag{51}$$

where $v = v_f - v_b$, L is a characteristic dimension (length of a flow tube), and D_s is the dispersion coefficient.

The second outcome of the model is the demonstration of the influence of mixing on combustion behavior. Figure 10 illustrates a comparison of prediction with some experiments in a double-swirled combustor.

In spite of the relative success of the model so far, however, it is still deficient in respect to the mixing level of each PSR in the sequence (Fig. 9). The model describes the well-stirred reactor, with local homogeneity assumed. Neverthelcss, a model for the well-mixed reactor already exists

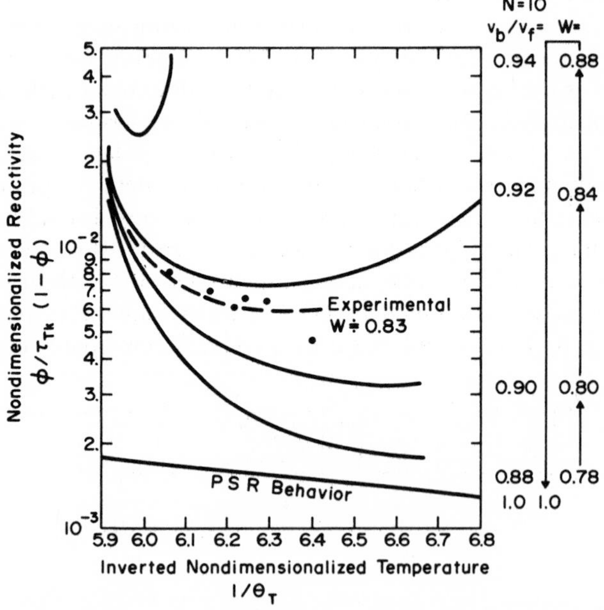

Fig. 10. Variation of reactivity group $\eta/(1-\eta)\,\tau_{T/k}$ with dimensionless temperature θ_T for different stirring factors W on an Arrhenius plot. Comparison of prediction and experiment. Values for activation energy E and frequency factor k_o adopted for plotting experimental data were 20 kcal/mole and 10^6 sec^{-1}, respectively. [From Kuwata and Essenhigh (1972).]

in a proposal by Vulis (1961). Vulis assumed that the effect of imperfect mixing would be to create concentration variations through the reactor, and then he proposed that the concentration differences would dissipate at a rate proportional to the concentration difference and to a turbulent mixing time. Relating this to a mixing scale, Vulis obtained the relation

$$\tau_D' \approx d^2/D_T \approx d/u', \qquad (52)$$

where τ_D' is the turbulent mixing time (or dissipation time), d is a characteristic (turbulence vortex) dimension, D_T is the turbulent diffusion coefficient, and u' is the maximum value of the turbulent velocity fluctuations. Recently, Swithenbank *et al.* (1972) have adopted this suggestion with tentative identifications in terms of the Kolmogorov scales. The consequences of the influence of mixing time have been developed by Vulis (1961) in some detail.

V. CONCLUSIONS

Combustion is a complex coupling of reaction kinetics, mixing, and heat transfer, with complexity further compounded in many instances by two- or even three-phase reaction, and aggravated by the widely varying properties and qualities among many fuels of the same phase. Fuels can be coal, oil, or gas, with increasing interest in solid waste, hydrogen, and denatured rocket fuels. Applications of combustion also vary widely: Examples from the thermal process industries include steam raising in boilers, roasting or reducing ores, firing brick, melting and fining glass, refining steel, heat treating metals and nonmetals, distilling petroleum or coal, drying, sintering; and examples from the power industries include reciprocating engines, gas turbines, jet engines, rockets, electricity production. Furnace and combustor development in the different industries has also involved very different time scales, being a few decades for gas turbines, up to millenia for some ceramic process furnaces.

With such widely varied antecedents, development time scales, fuels, combustion phases, and applications, it is no surprise to find that the theoretical development of combustion chamber design has been somewhat fragmented. Even so, the common basis in flame chemistry, combustion aerodynamics, and heat transfer has long been recognized; what has also been long recognized has been the difficulty of unifying the common fundamentals in a general and practically useful theory. General theories apparently were not useful; and useful theories were not general. The point about the stirred-reactor analysis is not that it is necessarily particularly useful in its very simplest form—except to the extent that it provides a simple basis for qualitative understanding of very complex phenomena (and possibly in a "topological" form)—but that it is capable of enlargement of complexity in relatively small increments, and not necessarily with the addition of all increments simultaneously, with the prospect of gaining utility thereby. It is true that this must reduce generality, but as shown by the gas turbine applications (Section III,D) utility was obtained in that instance with relatively little, if any, loss of generality, by the addition of a modest quantity of complexity.

This is not to say that stirred reactor theory will necessarily prove to be the solution to all possible combustion problems, but it is reasonable to conceive that most could reduce to a limited number of subsets of subgeneral problems that can be adequately handled by different expansions of the simple theory; for example, some systems may approximate to the well-stirred reactor condition, and others to the well-mixed condition; and of course, some others again may just demand the combination of well stirred and well mixed. What would determine selection of the modifications would be the degree of sensitivity of the result to the assumptions and approximations made, and to the accuracy required in the result. Spalding (1972) recently stressed the importance of a *balance of complexity* in a mathematical model, pointing out the absurdity of a highly sophisticated reaction scheme in a primitive flow model. The same principle can be applied to a part of a model: There is no point in including a factor that will have a tenth order of magnitude influence on the result when all factors responsible for fourth to ninth order of magnitude effects, *in that particular context,* have been omitted (and a factor important in one combustor configuration may be totally irrelevant in another). The answer does not therefore (necessarily) lie in construction of a "kitchen sink" theory in which *all* conceivable and many inconceivable factors have been included. The answer can lie in selection of parameters such that all are in balance in complexity, and the "accuracy" of the answer does not grossly exceed the accuracy of experimental determination and test. A good example of the positive appreciation of this point is the Edelman quasi-global reaction scheme with many very fast reactions reduced to one equivalent (Edelman and Fortune, 1969).

A current example of the need for different degrees of complexity for different objectives is to be found in the problem of NO_x production and emission from combustors. The relat vely modest variants of the simple stirred reactor theory that are relatively successful for gas turbine combustor performance have been found to be quite inadequate for NO_x prediction. It would seem to be a general rule that pollutant emissions, being usually present only in trace quantities, require very fine-grained theories for their prediction. We therefore seem to be faced with the need for greatly enlarging the scope of all combustor theories, and in such essential detail that it will be necessary to construct virtually a new theory for each new combustor. It is possible, however, to look at the problem from another point of view. In the first place, we do not need to know the absolute concentration of NO_x formed or emitted, to some small or large degree of error; we only need to know that it will not *exceed* some given value, and this removes some degree of constraint on the theoretical requirements. In the second place, we now know that NO_x is formed mainly in the hottest region of the combustor. It may, therefore, suffice if we are able to use a much less complex theory to isolate the thermal center of the flame and to specify at the same time, in necessary and sufficient detail, the effective boundary conditions (inputs and outputs, thermal exchanges, etc.) of that thermal center to be able then to use much more detailed theory on a less complex, more precisely defined zone to calculate the expected upper NO_x limits. One advantage of this procedure can be speed. Construction of a fully detailed theory for the full combustor may be more intellectually satisfying, but it does not have too much point if it is not completed in time to be of use.

To recapitulate, therefore, the stirred reactor analysis has the advantages of generality with relative simplicity, with total scope nevertheless for detailed amplification to the point of convergence with any other valid theory. The generality derives from the phenomenological basis of the theory, in contrast to more specific mechanistic theories. If accurately constructed, of course, mechanistic theories will in principle have greater fidelity of representation of the combustor (except in the limit when the physical combustor is a PSR). However, as Preisendorfer (1965) has pointed out in comparing the utility of phenomenological in contrast to mechanistic theories, any equation system of appreciable complexity in the mechanistic construct, that is so intractable that simplifying assumptions are required, is thereby involved in a fidelity-reducing procedure, and sometimes to a degree that is not fully explored. Concession to complexity to achieve a solution generally modifies the structure of the equations themselves, and therefore the functions they govern, which means modification of the physical processes the functions are supposed to represent. A finite element formulation, such as a stirred reactor representation, is also a fidelity-reducing procedure; but, as Preisendorfer emphasized, this is a space-modifying rather than a function-modifying procedure, and as such it is generally a better compromise between fidelity to the physical model and the needs for realistic computational schemes. There is therefore a good basis for the belief that increasing use of the stirred reactor approach to combustion chamber design can be expected in the future.

List of Symbols

A	Area of combustor wall surface (m^2)
c	Concentration, mass basis (kg/kg)
c_0	Initial concentration (kg/kg)
c_0'	Initial concentration with recycle (kg/kg)
c_R	Reactant concentration (kg/kg)
c_{ox}	Oxidant concentration (kg/kg)
$C(\theta)$	Output response curve to δ function input
C_p	Constant pressure specific heat (kcal/kg °K)
D_s	Dispersion coefficient (m^2/hr)
D_T	Turbulent diffusion coefficient (m^2/hr)
E	Activation energy (kcal/mole)
$E(\theta)$	Exit age distribution function
$E\%$	Excess air
f	Concentration, volume basis (kg/m^3)
f_R	Reactant concentration (kg/m^3)
f_{ox}	Oxidant concentration (kg/m^3)
F	Firing rate (kg/hr)
$F(\theta)$	Exit response function to an input step function start-up
G_s	Stoichiometric oxidant–fuel ratio (kg/kg)
$G_\infty(\theta)$	Exit response function to an input step function cutoff
h_f	Calorific value of the fuel (heat of combustion) (kcal/kg)
h	Wall heat transfer coefficient (kcal/m^2-hr °K)
i	ith cell of PSR sequence; ith component
I	Combustion intensity (kcal/m^3-hr-atm)
I_m	Maximum combustion intensity (kcal/m^3-hr-atm)
$I(\theta)$	Internal age distribution function
J	Total mass flow rate (kg/hr)
J_A	Mass flow rate per unit area (kg/m^2-hr)
k_0	Preexponential constant [(m^3/mole)$^{n-1}$/hr]
k_m	Preexponential constant, mass basis ($\rho_0^{n-1}k_0$) (hr^{-1})
K	Rosin combustion intensity constant defined by Eq. (27)
m	Index in Eq. (48)
$\bar{M}$	Mean molecular weight of mixture (kg/mole)
M_i	Molecular weight of component i (kg/mole)
n	Overall reaction order
N	Number of cells in back-mix PSR sequence
p_i	Fractional partial pressure of component $i = Y_i$
P_i	Partial pressure of component i (atm)
P	Pressure (atm)
P_0	Ambient pressure (atm)
Pe	Peclet number (vL/D_s)
$P(\varepsilon, \theta_i)$	Probability distribution function of molecules in reactor
$P(e, \theta_i)$	Probability distribution function of molecules in exit volume
$\dot{Q}_G$	Heat generation rate in combustor (kcal/hr)
$\dot{Q}_L$	Heat loss rate in combustor (kcal/hr)
$\dot{Q}_W$	Heat loss rate through combustor wall (kcal/hr)
$\dot{Q}_I$	Heat generation rate, mass basis (kcal/kg-hr)
$\dot{Q}_{II}$	Heat loss rate, mass basis (kcal/kg-hr)
r	Reactant reaction order; recycle ratio (kg/kg)
R	Gas constant
t	Time (hr)
t_c	Combustion time (hr)

T Temperature (°K)
T_0 Ambient temperature (°K)
T_f Mean flame temperature (°K)
u' Turbulent velocity fluctuation (m/hr)
v Resultant forward flow velocity (m/hr)
v_b Back-mix flow velocity (m/hr)
v_f Forward flow velocity (m/hr)
V_c Combustor chamber volume (m³)
W Stirring factor [Eq. (47)]
y_i Concentration (moles of component i/m³)
Y_i Mole fraction of component i (mole/mole)
β Dimensionless heat transfer coefficient for wall loss
γ Dimensionless heat transfer coefficient inside flame
η Combustion efficiency: dimensionless heat rate
η_I Dimensionless heat generation rate
η_{II} Dimensionless heat loss rate
$\eta_{I,D}$ Value of η_I with recycle dilution
η_{cr} Ignition or extinction value of η
η_h Boundary between critical and noncritical region
η_i Combustion efficiency of ith cell
$\eta_{I,n}$ Value of η_I for nth cell in PSR sequence
$\eta_{II,n}$ Value of η_{II} for nth cell in PSR sequence
η_R Heat exchanger efficiency
θ Dimensionless temperature (RT/E)
θ_0 Ambient temperature
θ_m Maximum or abiadatic temperature
θ_i Dimensionless temperature in ith cell of PSR sequence
$\theta_{(i)}$ Dimensionless time in residence time distribution functions
ρ Fluid density (kg/m³)
ρ_0 Ambient fluid density (kg/m³)
σ_i Concentration, mixed units (moles of i/kg mixture)
τ_D Mixing delay time (hr)
τ_D' Turbulent mixing time (hr)
τ_r Chemical reaction time (hr)
τ_s Combustor residence time (hr)
τ_s^0 Cold residence time (hr)
τ_s' Residence time with recycle (hr)
$\tau_{s/k}$ Dimensionless residence time ($\tau_s k_0$)
$\tau_{s/k}^0$ Cold dimensionless residence time
$\tau_{s/m}$ Dimensionless residence time for nth order reaction ($\tau_s \rho^{n-1} k_0$)
ϕ Equivalence ratio
χ Specific intensity [Eq. (33)]
ψ Loading parameter [Eq. (46): $J/V_c P^n$] (kg/m³-hr-atm^n)
ω Dimensionless initial concentration group ($c_0 R h_f / E C_p$)
ω_H Value of ω with heat loss [Eq. (21)]
ω_R Value of ω with heat recovery [Eq. (22)]

References

Aris, R. (1965). "Introduction to the Analysis of Chemical Reactors." Prentice-Hall, Englewood Cliffs, New Jersey.

Avery, W. H., and Hart, R. W. (1953). *Ind. Eng. Chem.* **45,** 1634.

Bartok, W., Heath, C. E., and Weiss, M. A. (1960). *J. AIChE* **6,** 685.

Beér, J. M., and Chigier, N. A. (1972). "Combustion Aerodynamics." Appl. Sci. Publ., London.
Beér, J. M., and Lee, K. B. (1965). *Proc. Int. Symp. Combust., 10th*, pp. 1187–1202. Combust. Inst., Pittsburgh, Pennsylvania.
Biswas, B. K., and Essenhigh, R. H. (1972). *In* "Air Pollution and Its Control" (R. W. Couglin, A. F. Sarofim, and N. J. Weinstein, eds.), AIChE Symp. Ser. No. 126, Vol. 68, pp. 207–215. Amer. Inst. Chem. Eng., New York.
Blichner, O. (1962). *Proc. Int. Symp. Combust., 8th*, pp. 995–1002. Williams & Wilkins, Baltimore, Maryland.
Bragg, S. L. (1953). Brit. Aeronaut. Res. Council Paper No. 16170 (CF. 272).
Clarke, A. E., Harrison, A. J., and Odgers, J. (1959). *Proc. Int. Symp. Combust., 7th*, pp. 664–673. Butterworths, London.
Clarke, A. E., Odgers, J., and Ryan, P. (1962). *Proc. Int. Symp. Combust., 8th*, pp. 982–994. Williams & Wilkins, Baltimore, Maryland.
Clarke, A. E., Odgers, J., Stringer, F. W., and Harrison, A. J. (1965). *Proc. Int. Symp. Combust., 10th*, pp. 1151–1166. Combust. Inst., Pittsburgh, Pennsylvania.
DeZubay, E. A., and Woodward, E. C. (1955). *Proc. Int. Symp. Combust., 5th*, pp. 329–335. Combust. Inst., Pittsburgh, Pennsylvania.
Drake, P. F., and Hubbard, E. H. (1966). *J. Inst. Fuel* **39**, 98.
Edelman, R. B., and Fortune, O. (1969). AIAA Paper 69–86.
Edelman, R. B., Fortune, O., and Weilerstein, G. (1972). *In* "Emissions from Continuous Combustion Systems" (W. Cornelius and W. A. Agnew, eds.), pp. 55–87. Plenum, New York.
Essenhigh, R. H. (1966). Western States Sect. of the Combust. Inst., Paper No. WSCI-66-28 (ONR Tech. Rep. No. FS66-3(u)).
Essenhigh, R. H. (1967a). Central States Sect. of the Combust. Inst. Spring Meeting, Cleveland, Ohio, Paper No. III. 2 (ONR Tech. Rep. No. FS67-1(u)).
Essenhigh, R. H. (1967b). *Ind. Eng. Chem.* **59**, 52.
Herbert, M. V. (1960). *Progr. Combust. Sci. Technol.* **1**, 61.
Herbert, M. V. (1962). *Proc. Int. Symp. Combust., 8th*, pp. 970–982. Williams & Wilkins, Baltimore, Maryland.
Hottel, H. C., Williams, G. C., and Baker, M. L. (1957). *Proc. Int. Symp. Combust., 6th*, pp. 398–411. Van Nostrand-Reinhold, Princeton, New Jersey.
Hottel, H. C., Williams, G. C. Nerheim, N. M., and Schneider, G. R. (1965). *Proc. Int. Symp. Combust., 10th*, pp. 111–121. Combust. Inst., Pittsburgh, Pennsylvania.
Hottel, H. C., Williams, G. C., and Miles, G. A. (1967). *Proc. Int. Symp. Combust., 11th*, pp. 771–778. Combust. Inst., Pittsburgh, Pennsylvania.
Jenkins, D. R., Yumlu, V. S., and Spalding, D. B. (1967). *Proc. Int. Symp. Combust., 11th*, pp. 779–790. Combust. Inst., Pittsburgh, Pennsylvania.
Kretschmer, D., and Odgers, J. (1972). *ASME J. Eng. Power* **192,** 173.
Kuwata, M., and Essenhigh, R. H. (1972). *Proc. AGA/IGT Conf. Natur. Gas Res. Technol., 2nd*. Paper No. IV.3. AGA, Washington, D.C. and IGT, Chicago, Illinois.
Kuwata, M., and Essenhigh, R. H. (1973). *Combust. Flame* **20,** *437*.
Kydd, P. H., and Foss, W. I. (1965). *Proc. Int. Symp. Combust., 10th*, pp. 101–110. Combust. Inst., Pittsburgh, Pennsylvania.
Levenspiel, O. (1962). "Chemical Reaction Engineering," Wiley, New York.
Longwell, J. P., and Weiss, M. A. (1955). *Ind. Eng. Chem.* **47,** 1634.
Longwell, J. P., Frost, E. E., and Weiss, M. A. (1953). *Ind. Eng. Chem.* **45,** 1629.
Miles, G. A. (1971). *Proc. Int. Symp. Combust., 13th*, pp. 483–487. Combust. Inst., Pittsburgh, Pennsylvania.

Nicholson, H. M., and Field, J. P. (1949). *Proc. Int. Symp. Combust., 3rd,* pp. 44–68. Williams & Wilkins, Baltimore, Maryland.

Odgers, J., and Carrier, C. (1972). ASME, Paper No. 72-WA/GT-5.

Palmer, H. B. (1968). *Ind. Eng. Chem.* **60,** 79.

Penner, S. S. (1957). "Chemistry Problems in Jet Propulsion," Chapter 17. Pergamon, Oxford.

Pratt, D. T., and Starkman, E. S. (1969). *Proc. Int. Symp. Combust., 12th,* pp. 891–899. Combust. Inst., Pittsburgh, Pennsylvania.

Preisendorfer, R. W. (1965). "Radiative Transfer on Discrete Spaces," Chapter 1. Pergamon, Oxford.

Rao, S. T. R., Kuo, T.J., and Essenhigh, R. H. (1970). *Proc. ASME Nat. Incinerat. Conf., 4th,* pp. 314–326. Amer. Soc. Mech. Eng., New York.

Rao, S. T. R., and Essenhigh, R. H. (1971). *Proc. Int. Symp. Combust., 13th,* pp. 603–615. Combust. Inst., Pittsburgh, Pennsylvania.

Rosin, P. O. (1925). *Braunkohole* **24,** 241. *Proc. Int. Conf. Bituminous Coal* **1,** 838.

Shieh, W., and Essenhigh, R. H. (1972). *Proc. ASME Nat. Incinerator Conf., 5th,* pp. 120–134. Amer. Soc. Mech. Eng., New York.

Spalding, D. B. (1972). *In* "Emissions from Continuous Combustion Systems" (W. Cornelius and W. G. Agnew, eds.), p. 20. Plenum, New York.

Swithenbank, J., Poll. I., Wright, D. D., and Vincent, M. W. (1972). *Proc. Int. Symp. Combust., 14th.* pp. 627–638. Combust. Inst., Pittsburgh, Pennsylvania.

van Heerden, C. (1953). *Ind. Eng. Chem.* **45,** 1242.

Vulis, L. A. (1961). "Thermal Regimes of Combustion." McGraw-Hill, New York.

Williams, G. C., Hottel, H. C., and Morgan, A. C. (1969). *Proc. Int. Symp. Combust., 12th,* pp. 913–925. Combust. Inst., Pittsburgh, Pennsylvania.

Zeinalov, M. A. O., Kuwata, M., and Essenhigh, R. H. (1972). *Proc. Int. Symp. Combust., 14th.* pp. 575–583. Combust. Inst., Pittsburgh, Pennsylvania.

XV

Recent Research and Development in Residential Oil Burners

DAVID W. LOCKLIN

BATTELLE MEMORIAL INSTITUTE
COLUMBUS, OHIO

I. INTRODUCTION

Combustion research has stimulated new developments in small oil-burning equipment, particularly as applied to residential space heating (Bolt and Locklin, 1967). This chapter reviews some of the research, new developments, and design trends in both Europe and the United States.

Much of this research effort in the past few years has been the result of the American Petroleum Institute's program (Locklin, 1960; Livingstone, 1961) to stimulate and coordinate industry research on domestic oil burners for No. 2 heating oil. Work under the API program included API-sponsored projects conducted at research institutes and independently financed research conducted by refiners. The results of the API program, plus related research carried out in Europe and the U.S., have been published in the proceedings of the six annual API Research Conferences on Distillate Fuel Combustion held from 1961 to 1966 (Proceedings, 1961–1966).

More recently, development programs on residential oil-burning equipment have been initiated as part of industry-sponsored oil burner development programs administered by the National Oil Fuel Institute, Inc. as an extension of the API program (Proceedings, 1968).

In the following sections, significant trends in conventional atomizing-type gun burners are reviewed, and other new developments are reviewed as they are applied to other basic burner types.

II. TRENDS IN HIGH-PRESSURE, GUN-TYPE BURNERS

The high-pressure atomizing gun-type burner is the most common type used for domestic heating with distillate fuel oils. Even though basic principles and component functions have generally remained the same, marked advances have been made recently in gun burners (Olson, 1965; Lord, 1968; Howe, 1964).

These improvements have resulted in more compact burners, having improved efficiency, flame stability, and serviceability. Many are operable at low firing rates. Major factors in these developments are

(1) higher speed fans and motors,
(2) improved combustion heads having higher pressure drops for better mixing,
(3) lower nozzle temperatures,
(4) integrated equipment designs,
(5) improved nozzles, pumps, and controls,
(6) more compact components.

The use of two-pole, high-speed, electric motors for blower drive has permitted higher pressure drops at the burner head for improved fuel–air mixing. Smaller fans and pumps have led to more compact burners. The trend toward two-pole motors, initially prevalent in Europe (Howe, 1964), has spread to the U.S. (Lord, 1968).

Since World War II the trend to smaller and well-insulated homes in the U.S., plus a wider use of oil-fired storage-type water heaters, has resulted in a greater demand for the development of burners capable of lower firing rates. Most small oil burners prior to 1940 were not capable of reliable operation below 1.0 gph. Today many operate well at 0.75 gph and some at 0.5 gph.

The ability to operate high-pressure burners reliably at lower firing rates is enhanced by the trend toward completely engineered heating units. Designers of domestic boilers and warm-air furnaces have incorporated features to improve fuel–air mixing and to minimize nozzle temperatures. In addition, lightweight ceramic-fiber refractory materials are being used for combustion chambers, reducing the heat radiated to the nozzle after shutdown. Pumps have been improved to give rapid fuel cutoff at shutdown. Along with these equipment improvements, an upgrading of fuel quality has contributed to the success of low-firing-rate operation (Olson, 1965).

Another recent trend in domestic oil burners is the use of photoelectric, burner-mounted combustion safety controls, particularly using cadmium-sulfide cells which give quicker response in case of flame failure. With the flame-sensing cell mounted in the air tube of the burner, there is less chance of improper installation than with a stack-mounted control. A number of solid-state-electronic circuits are being introduced in combustion safety controls (Evans, 1968).

New ignition transformers, some of pulse discharge type (Hazard and Stutz, 1962), are now more compact than previous transformers. New controls combine the functions of ignition transformer and combustion safety control, making possible further compactness in design (Rheaume, 1968; Florio, 1968).

Flame-retention combustion heads are growing in popularity (Olson, 1965). Generally characterized by multiple swirl vanes, these heads provide fuel–air mixing to give compact, stable flames that are quieter than those in conventional burners where the flame front is allowed to "float." Excellent mixing is provided in the vortex induced by the swirl vanes. Improved methods of fuel–air mixing in gun-burner operation have received considerable attention in the U.S. (Lang, 1961; Beach and Siegmund, 1962) and some oil marketers have taken steps toward upgrading existing burners by installing improved mixing devices on burners in the field (Anon., 1963).

III. DEVELOPMENTS IN OTHER TYPES OF BURNERS

New developments are discussed here as they apply to basic burner types, which are classified here for convenience as (A) atomizing burners of other types, (B) thermal-vaporizing burners, (C) combination atomizing–vaporizing burners, and (D) recirculation-type blue-flame burners.

A. Atomizing-Type Burners

In atomizing-type burners, the fuel is atomized into small droplets to accelerate its vaporization and its penetration into the fuel–air mixing zone of the combustion chamber. In research aimed at achieving a better understanding of droplet combustion, the burning rates of individual droplets under different conditions were investigated as part of the API program (Rosser *et al.*, 1963).

In domestic oil burners, high-pressure atomization is the most common method of fuel preparation and is used in most gun-type burners (as discussed in the previous section). A typical 0.8-gph nozzle operating at 100-psi oil pressure produces droplets ranging from about 10 to 150 μ, with a mass-median diameter of approximately 70 μ (Olson and Tate, 1962). Because the swirl slot and orifices in the nozzle are small and tend to clog, high-pressure atomizing nozzles designed for fuel rates below about 0.75 gph are not always reliable. Therefore, there is an incentive to develop new means of atomization suitable for firing rates below 0.75 gph.

As a guide to development of new types of atomizers, an analytical study was undertaken under the API program to provide a greater understanding of the fundamental mechanism of atomization processes. The stability and breakup of liquid surfaces were analyzed in detail, and the theory was applied to several types of oil atomizers (Peskin and Raco, 1966). In another API study, means of atomization were evaluated with respect to their applicability to domestic oil burners capable of reliable operation at low firing rates (Doyle *et al.*, 1966); ultrasonic atomization was judged to be the most promising new method.

1. Ultrasonic Atomization

When a film of oil is spread over a vibrating surface, capillary waves form on the film, and the crests of these waves break and form droplets (Lang, 1962). The higher the frequency, the smaller the droplets. This concept is attractive for atomizers that operate at low firing rates because relatively large oil-feed passages can be used in the atomizer.

Figure 1 shows an ultrasonic atomizer developed as part of the API research program (McCullough *et al.*, 1966), with the benefit of additional refiner background and experience (Lang *et al.*, 1962; Lang, 1963). Ultrasonic vibrations at 56 kHz, generated with a piezoelectric ceramic transducer, are increased in amplitude by an aluminum stepped-horn amplifier. Fuel is fed to the tip of the cylindrical horn where the amplitude is the

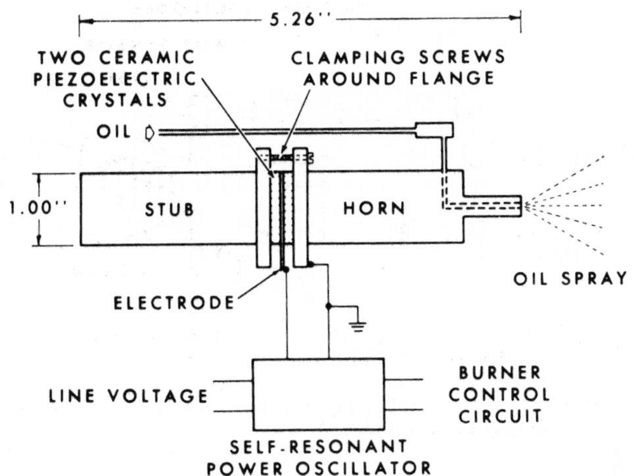

Fig. 1. Schematic representation of the API ultrasonic atomizer—56-kHz frequency. [From Bolt and Locklin (1967).]

highest. In long-term tests, such units operated while atomizing for more than 15,000 hr without failure (McCullough *et al.*, 1966).

Other designs of ultrasonic atomizers have been reported. A miniature atomizer (Hazard and Hunter, 1966), designed to operate at 85 kHz with a transistorized power supply using only 4 W from a battery, was developed to fire a portable multifueled, 100-W thermoelectric generator. The atomizer and burner operate satisfactorily on fuels ranging from gasoline to No. 2 heating oil, and the atomizer appears suitable for a variety of small burners with firing rates up to 0.3 gph.

A recent modification of this atomizer for military applications uses a

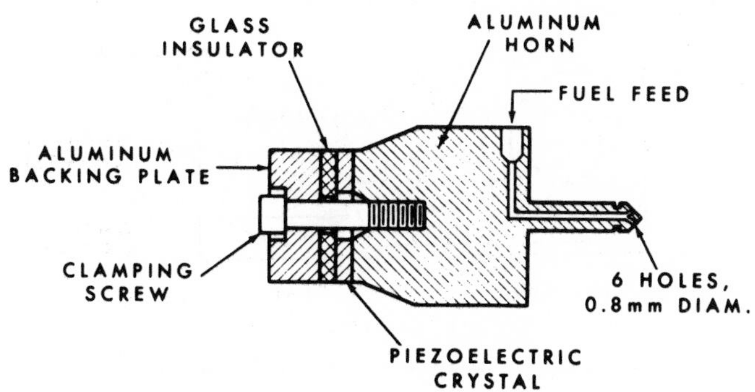

Fig. 2. British ultrasonic atomizer—40-kHz frequency. [From Bolt and Locklin (1967).]

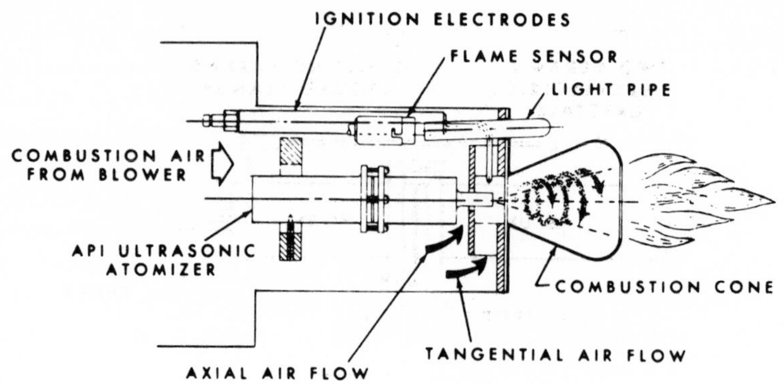

Fig. 3. Swirl-type burner prototype developed to demonstrate API ultrasonic atomizer. [From Bolt and Locklin (1967).]

power supply with a printed circuit board and fixed electronic components; higher firing rates have been achieved with the higher power levels of this system (Angelo, 1969).

Figure 2 illustrates an alternative method of clamping the piezoelectric crystal, as used in an atomizer developed in England for capacities up to 0.8 gph (Bradbury, 1966).

The spray-droplet velocity from an ultrasonic atomizer is low compared with that of a conventional high-pressure spray, so the droplet penetration into the airstream is less. Thus, for good combustion, air must be introduced in such a way as to provide proper droplet transport and fuel–air mixing. Swirl- or vortex-type burners have been employed successfully to burn sprays from ultrasonic atomizers.

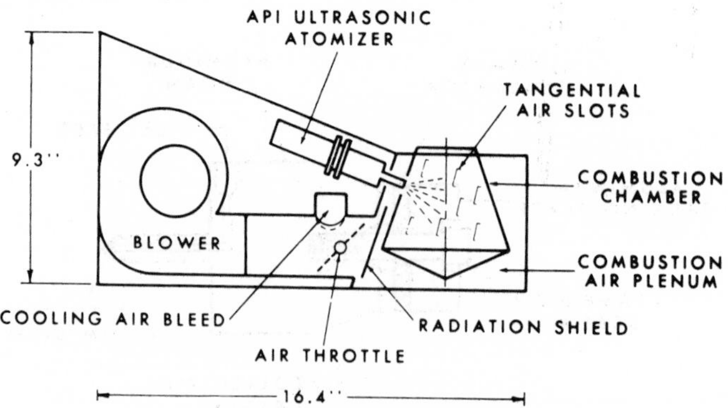

Fig. 4. API ultrasonic atomizer applied to double-vortex type combustion chamber. [From Bolt and Locklin (1967).]

Figure 3 shows an axially fired, swirl-type burner developed to match the characteristics of the API 56-kHz ultrasonic atomizer at 0.4 gph firing rate. Combustion performance was excellent in long term tests with these prototypes (McCullough *et al.*, 1966). An ultrasonic atomizer has also been successfully demonstrated with an axially fired burner that gave a conventional "sunflower" air pattern (Lang, 1963). Another approach is to fire the ultrasonic atomizer with the fuel spray perpendicular to the axis of a vortex in the burner, as shown in Fig. 4 (Dysart, 1966; MacCracken and McLendon, 1964); this arrangement is similar to that of the burner developed for the miniature atomizer (Hazard and Hunter, 1966).

2. Air Atomization

While not new to the art, air atomization is receiving renewed attention for low rates. In air atomization for domestic burners, shear forces produced by compressed air (generally at 3–15 psi) are used to induce liquid instability to form fine droplets. Air required for atomization is generally less than 5% of stoichiometric. Air-atomizing nozzles, because of their relatively large orifices, operate reliably at extremely low firing rates.

Various combustion systems have been demonstrated using siphon-type nozzles, in which the atomizing air supplied at 3 to 4 psi creates a suction in the swirl chamber, so that fuel can be lifted into the nozzle (Walsh, 1961, 1962, 1964; Briggs *et al.*, 1965). One example of this concept is a portable space heater which lifts oil from an integral tank (Briggs *et al.*, 1965). Figure 5 shows an experimental burner that relies on the momen-

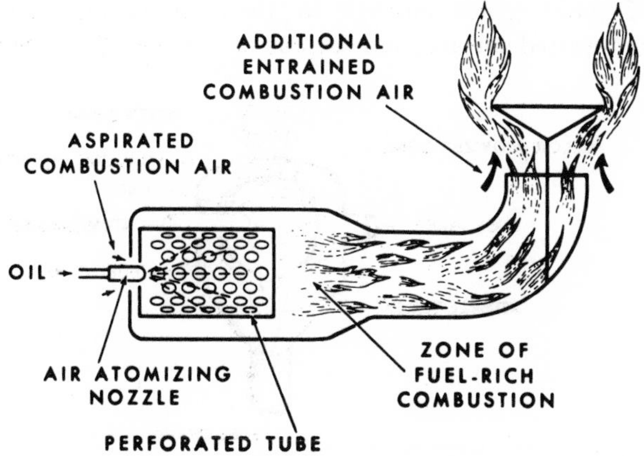

Fig .5. Concept of an air-atomizing, natural-draft burner. [From Bolt and Locklin (1967).]

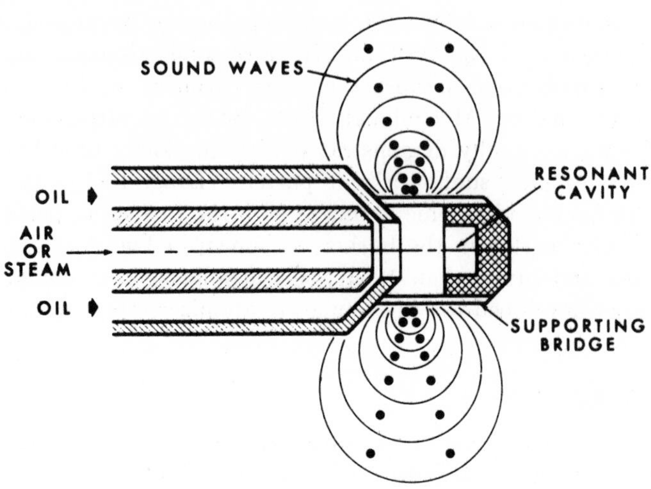

Fig. 6. An acoustic-type air-atomizing nozzle. [From Bolt and Locklin (1967).]

tum of the atomizing air and fuel spray to aspirate much of the secondary combustion air; the remainder of the secondary air is induced by means of natural draft.

Figure 6 illustrates a Hartmann-whistle (Hartmann and Trolle, 1926–27, 1931) type air-atomizing nozzle which uses acoustic energy for atomization. This design, with a central resonant cavity has been applied to a wide range of burners from residential to industrial sizes (Hughes *et al.*, 1965; Bell and Korn, 1968).

Figure 7 shows a new concept in which atomizing air issues from a slot in a fuel-wetted surface, and surplus oil is recirculated. Smaller quanti-

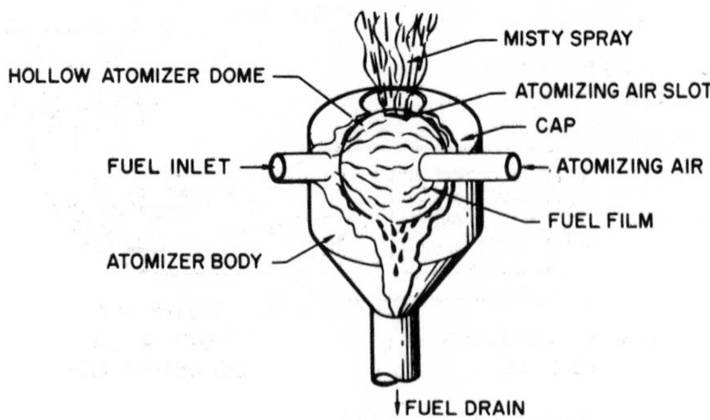

Fig. 7. New atomizing concept using an air-discharge slot in a fuel-wetted surface. [From Velie (1968).]

ties of air are required than for conventional air-atomizing nozzles, and relatively large ports can be used. A small burner using this concept has been successfully demonstrated (Velie, 1968).

B. Thermal-Vaporizing Burners

One of the oldest and simplest ways to prepare distillate fuel oils for combustion is to vaporize the fuel directly from the bulk liquid. Vaporizing or "pot-type" burners have been used with kerosene and No. 1 heating oil for many years, and mechanical-draft versions are now popular in the U.K. and in Europe. However, with No. 2 oil, vaporization deposits are formed —sometimes as much as 750 gm of deposits per thousand gallons of oil (Rakowsky, 1962). Nevertheless, a vaporizing burner for No. 2 oil is of considerable interest because of its potential advantages of simplicity and quiet, low-capacity operation.

A combination of API-sponsored and in-house oil-company research (Morgenthaler, 1962; Hein and Weller, 1963; Tuttle, 1961; Rakowsky and Meguerian, 1963; Hein *et al.*, 1966) established conditions for reducing deposits to an acceptable level. The requirements are rapid vaporization from a hot surface and periodic removal of trace deposits from the hot surface by air oxidation.

Figure 8 summarizes an API study of deposit formation from No. 2

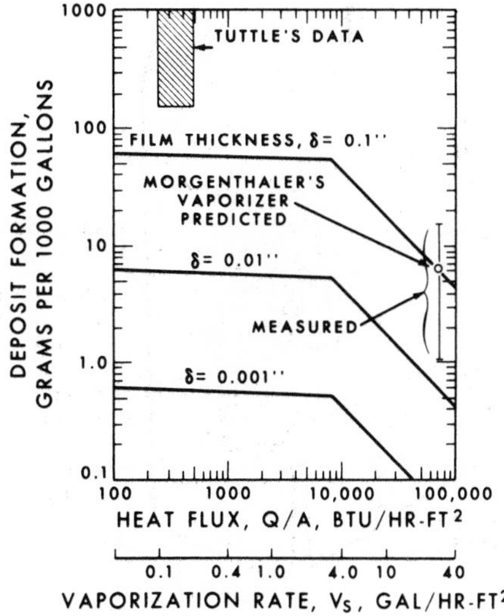

Fig. 8. Conditions affecting deposits during vaporization of No. 2 heating oil. [From Bolt and Locklin (1967).]

oil as a function of heat flux to the oil-vaporizing surface and of oil-film thickness, both of which determine oil residence time (Hein and Weller, 1963). To minimize deposits, residence time must be limited, continuous fractionation must be prevented, and oxidative fuel degradation must be limited. These conditions were incorporated in several demonstration prototype burners developed as part of the API program (Hein *et al.*, 1966).

1. API Prototype Vaporizing Burner

Figure 9 shows one of these prototype vaporizing burners which is unlike conventional pot-type burners in that the vaporizing chamber is separate from the combustion chamber and operates at a high temperature (above 800°F). Oil is introduced into the hot cast steel vaporizing chamber where it vaporizes quickly. The vapor is conducted to a combustion chamber, where it is mixed with air in a swirl-type burner. Heat is radiated from the refractory to the vaporizer through a small opening in the combustion chamber. The heavy vaporizer casting is extensively insulated except for the radiation-receiving surface and thus retains heat between firing cycles. However, when needed, the vaporizer is heated with an electric cartridge element, thermostatically controlled to main-

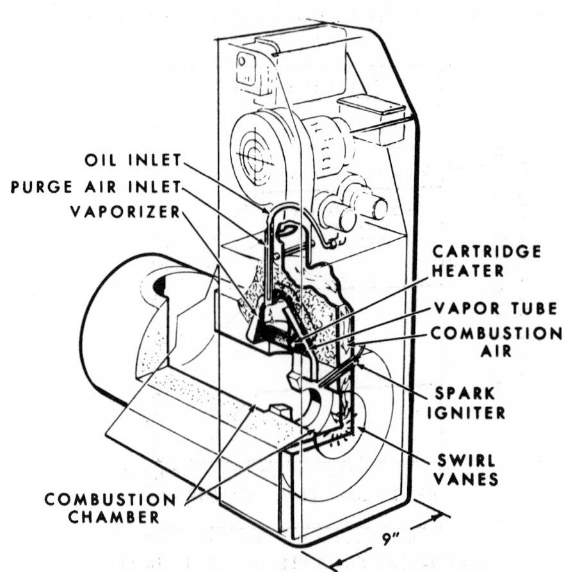

Fig. 9. Cutaway view showing API demonstration prototype vaporizing burner. [From Bolt and Locklin (1967).]

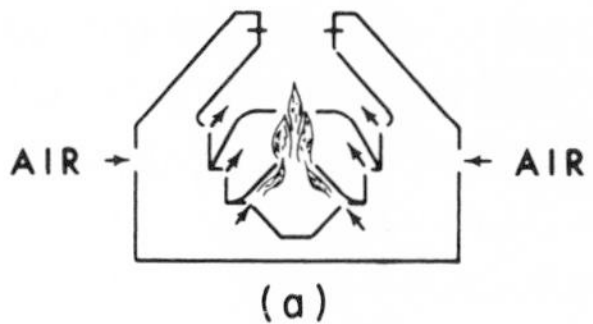

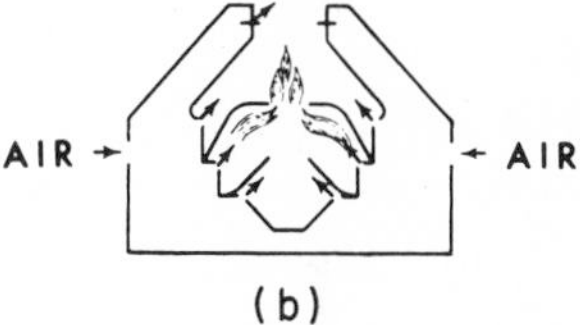

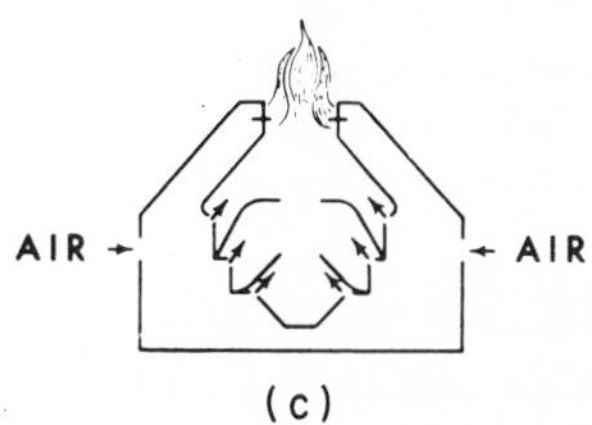

Fig. 10. Triple-stage vaporizing burner, showing flame position at various burning rates: (a) 0.058 gph; (b) 0.1 gph; (c) 0.29 gph. [From Bolt and Locklin (1967).]

tain it in a ready condition during off periods. Purge air, about 1% of the combustion air, is introduced through the vaporizer to prevent condensation of vapor in the oil inlet and to limit oil temperature. During off periods, the flow of purge air is maintained to oxidize deposits.

Such burners have performed satisfactorily on typical No. 2 oil for periods equivalent to a heating season, firing 0.5 gph on an intermittent basis. Installed in a storage-type water heater, one of these units logged over 1000 hr firing time (Hein *et al.*, 1966).

2. Triple-Stage Vaporizing Burner

Figure 10 illustrates another design approach used in a burner developed in Holland (Howe, 1964). This triple-stage burner represents a marked refinement of the conventional pot-type unit. The burner consists of a series of three vertically arranged chambers. Figure 10 shows a cross section of the chambers with flame positions at various firing rates. Kerosene or heavier distillate is introduced into the bottom of the center chamber and is vaporized along the length of the burner with heat from a bluish flame by radiation and recirculation. Oil vapor, below the cracking temperature, mixes with combustion air entering the lower row of ports.

At low fire, the fuel–air mixture burns partly under the baffle and partly in the aperture of the baffle; combustion gases in the aperture form a shield over the lower vaporizing and mixing chamber. As the firing rate is increased, the vapor mixes with primary air in the lower chamber and with secondary air in the middle chamber, so the flame burns from the middle chamber. At still higher oil-feed rates, the flame is lifted to the air holes of the upper chamber. The maximum capacity of a 16-in.-long burner is 0.35 gph.

C. Combination Atomizing–Vaporizing Burners

A promising concept for domestic burners is the combination atomizing–vaporizing burner. In this concept, fuel is atomized into relatively large droplets which are vaporized either while in flight in hot gases or after impinging upon a hot surface.

One example of this principle is in the vertical rotary wall-flame burner, formerly popular in the U.S. and recently introduced in Europe. A rotating distributor on a vertical shaft throws oil against a wall rim where the oil is vaporized. After mixing with air, the vapor burns with a bluish flame. Because of quiet, efficient operation at low firing rates, the wall-flame burner is used extensively to fire domestic boilers for central heating in the U.K. (Howe, 1964). Two significant improvements have recently been introduced in European wall-flame burners (Bullock, 1964) to reduce aldehyde odors on startup and shutdown: (a) the flame rim is electrically preheated on startup for rapid vaporization, and (b) an electric "brake" is used to stop rotation of the distributor to ensure quick oil cutoff. A similar rotary burner type being introduced in the U.K. uses forced recirculation of combustion products.

Figure 11 shows an example of an experimental low-capacity atomizing–vaporizing burner (Will *et al.*, 1964; Carl *et al.*, 1966) Fuel burns with a bright luminous flame in a much smaller volume than in the wall-flame burner. The distributor throws oil droplets against a ceramic-fiber wick to promote vaporization. Primary air is introduced around the distributor shaft and secondary air through the ceramic-fiber wick. Some recirculation of combustion gas occurs, but less than in wall-flame burners.

Several new atomizing–vaporizing burners use conventional pressure-atomizing nozzles but operate at reduced pressure to take advantage of larger nozzle orifices for low firing rates. One such experimental burner operates at 0.5 gph using 35-psi atomizing pressure with a conventional 0.8-gph high-pressure nozzle (Mueller and Schrader, 1966). On start-up, oil droplets impinge on a steel target until vaporizing temperatures are achieved. Combustion air introduced to the primary chamber is staged

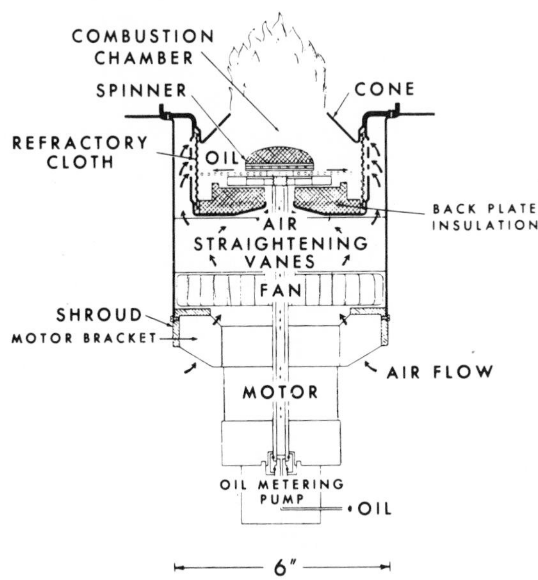

Fig. 11. Luminous-flame, vertical rotary atomizing-vaporizing burner. [From Bolt and Locklin (1967).]

much the same as in conventional pot-type burners. The flame is predominantly yellow, and the operation is extremely quiet.

D. Recirculation-Type Blue-Flame Burners

Quiet and clean combustion has also been achieved by recirculating a portion of the combustion products back to the primary combustion zone of a burner system (Dunn, 1961; Saxton and Lane, 1961; Hedley and Jackson, 1966). The effects of recirculation and fuel–air mixing were studied on a quantitative basis as part of the API research program (Cooper *et al.*, 1964).

Figure 12 illustrates representative results with the API experimental apparatus (Cooper *et al.*, 1964). This variation of smoke with excess air is typical of pressure burners having no special provision for recirculation; however, the tests with recirculation at stoichiometric conditions show a marked transition from high smoke levels to smokeless operation when the quantity of combustion products recirculated back to the burner inlet is about 45% of the stoichiometric air requirement. At still higher levels of recirculation, a transition is observed from a yellow flame to a blue flame with very quiet combustion.

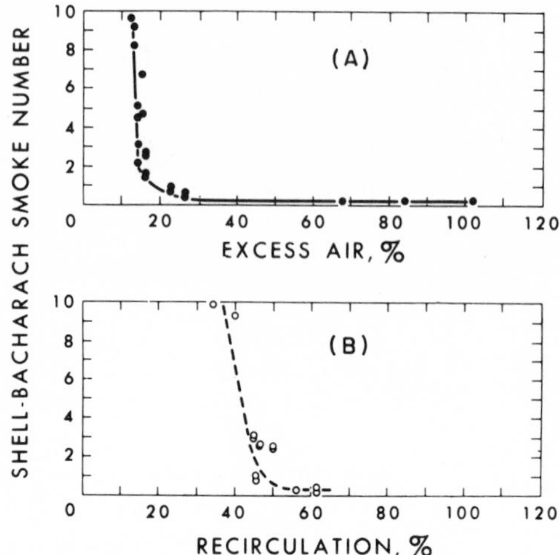

Fig. 12. Effect of excess air and recirculation of combustion products on smoke density. (A) Smoke at zero recirculation. (B) Smoke at zero excess air. [From Bolt and Locklin (1967).]

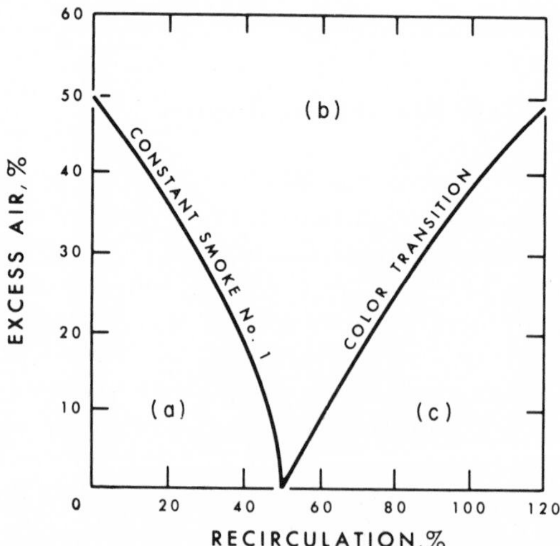

Fig. 13. Combined effects of excess air and recirculation showing transitions between yellow- and blue-flame burning. (a) Region of smoky yellow-flame combustion; (b) region of clean, yellow-flame combustion; (c) region of clean blue-flame quiet combustion. [From Bolt and Locklin (1967).]

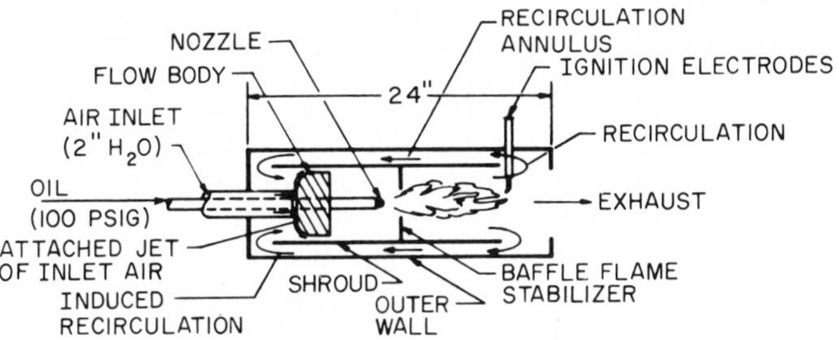

Fig. 14. Blue-flame recirculating-type oil burner utilizing attached-jet entrainment. [From Bolt and Locklin (1967).]

Figure 13 shows the combined effects of excess air and recirculation on the location of these transitions in the API experimental apparatus. They form boundaries between three regions: (a) one of smoky yellow-flame combustion, (b) another of clean yellow-flame combustion, and (c) a third of clean blue-flame quiet combustion.

Figure 14 shows a burner concept that applies the foregoing findings to achieve blue-flame combustion (Cooper and Marek, 1965). Combustion products are entrained by the inlet air jet, which follows the surface of a curved solid body by the attached-jet effect. In tests at rates from 0.65 to 1.1 gph, clean blue-flame burning was obtained with air pressures as low as 2 in. of water.

Figure 15 illustrates the concept used in another recirculation-type blue-flame burner (Saxton and Lane, 1961). This design depends on recirculation of gases entrained into a central vaporizing tube where some fuel droplets impinge and vaporize in a fuel-rich zone. The final combus-

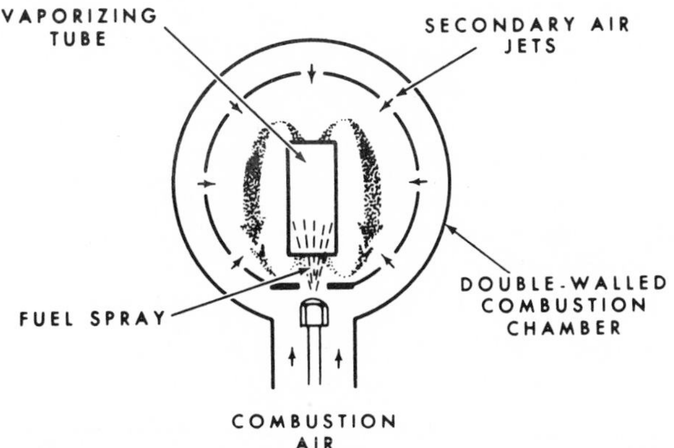

Fig.15. Recirculation pattern in one concept of a blue-flame oil burner. [From Bolt and Locklin (1967).]

tion air is added through ports in the outer double-walled chamber where small blue-flame jets appear. Measurements of flame radiation spectra (Janssen and Torborg, 1965, 1966) indicate the flame characteristics to be similar for the prototypes shown in Figs. 14 and 15.

A number of other approaches to recirculation-type burners have been reported, some using atomizing-air momentum for inducing recirculation (Bailey, 1968), and others using recirculation induced by swirl-type burners (Ranze and Schindler, 1965).

The effectiveness of recirculation is enhanced if the combustion products can be cooled before reentering the flame zone. This suggests the design advantage of integrating blue-flame burners with the heat exchanger portion of the equipment (Paterson, 1968). The cold-starting transient condition presents a problem with any recirculating-type burner, but further developments are expected because of the advantages of clean and quiet combustion in the blue-flame regime.

IV. OTHER CONSIDERATIONS IN BURNER DEVELOPMENT

Quieter operation is being emphasized in burner development, particularly where equipment is installed in the living spaces (Kennedy, 1965). This is important in mobile homes, which are growing in popularity. It is also a significant factor in the U.K. where basements are rare (Howe, 1964).

Another noteworthy trend, though not a new burner development, is the use of central oil-storage and distribution systems for groups of homes, with flowmeters at each home for billing purposes. These are increasing both in the U.S. and in Europe.

Performance levels of oil-heating equipment are being significantly raised by upgrading of acceptance standards (Peoples, 1965). Another major factor in upgrading is the equipment selectivity in performance and reliability being exercised by those companies that market both heating oils and equipment (Hunt, 1964).

A new factor in burner development is that imposed by air pollution considerations. While particulate emissions and CO have long been factors considered by burner designers, greater attention is now being focused on oxides of nitrogen and other emissions (Wasser, 1968; Andrews *et al.*, 1968; Howekamp, 1970). Some measure of emission control can be achieved in the design of combustion systems, and new research is being directed to establishing design criteria (Barrett *et al.*, 1971).

V. EXPECTED FUTURE DEVELOPMENTS IN SMALL BURNERS

The following trends and emphasis in burner development are forecast over the next few years:

1. Recirculation-type burners operating in the blue-flame region will be developed in response to consumer desire for burner quietness. Burners with intense swirl patterns show promise in this respect. It is likely that boilers and warm-air furnaces will incorporate recirculation passages as an integral part of the heat exchanger and burner system.
2. Higher speed fans will be used with conventional gun-type burners to improve fuel–air mixing and thus produce more stable and compact flames.
3. Air atomization will find increasing application at low rates, primarily with mechanical-draft burners; reliable low-cost air compressors may continue to be a limitation. Ultrasonic atomization will be introduced first in special applications; its use may become widespread for low capacities if costs of the atomizer and electronic power supply continue to be reduced.
4. Forced-draft vaporizing burners and combination atomizing–vaporizing burners will be further developed and will find special appeal where low-firing-rate capability and quietness are desired. Some burners will be designed especially for operation with central fuel distribution systems.
5. Air pollution considerations will have more influence upon the design of oil burners, especially in larger sizes. Concepts such as flue gas recirculation and two-stage combustion will be utilized to control nitrogen oxides and other emissions.

Acknowledgment

The author acknowledges the information and comments supplied by representatives of the petroleum and equipment-manufacturing industries.

References

Andrews, R. L., Levine, D. G., and Siegmund, C. W. (1968). Effect of Flue Gas Recirculation on Emissions from Heating Oil Combustion. APCA Paper No. 68-21.

Angelo, J. P. (1969). Static, Silent, Thermoelectric Power Sources. Presented at the *Power Sources Conf., 23rd, Atlantic City, New Jersey,* May 20–22, 1969.

Anonymous (1963). New Shell 'Burner-Pak' Modernizes Existing Burners. *Fueloil Oil Heat* **22**, 62.

Bailey, F. W. (1968). OOHA Blue Flame-Burner Systems. Paper No. 68-WS-106 *in Proceedings* (1968) (*q.v.*).

Barrett, R. E., Locklin, D. W., and Weller, A. E. (1971). The Federal R&D Plan for Air Pollution Control by Combustion Modification. Final Rep. by Battelle, Columbus Lab. to the Environmental Protection Agency, Contract CPA 22-69-147, PB No. 198066.

Beach, W. A., and Siegmund, C. W. (1962). Flame Studies Leading to Improved Combustion in Domestic Burners. Paper CP62-14 *in Proceedings* (1962) (*q.v.*).

Bell, F., and Korn, J. (1968). Developments in Sonic Atomization for Domestic Oil Burner Applications. Paper No. 68-WS-113 *in Proceedings* (1968) (*q.v.*).

Bolt, J. A., and Locklin, D. W. (1967). Recent Developments in Oil Burners for Space Heating and Industrial Applications. *Proc. World Pet. Congr., 7th,* Vol. 7, Sect. on Application and New Uses, Part 1, p. 119. Elsevier, Amsterdam.

Bradbury, C. H. (1966). The Ultrasonic Atomization of Liquid Fuels—Some British Developments. Paper CP66-4 *in Proceedings* (1966) (*q.v.*).

Briggs, E. C., Farkas, R., and Wilson, R. D. (1965). Air-Atomizing Oil Burner for Low-Firing-Rate Applications. Paper SP65-J *in Proceedings* (1965) (*q.v.*).

Bullock, G. J. (1964). Developments in Wallflame Burners. *Plumber J. Heating,* September (1964). p. 705.

Carl, R. W., Hunt, R. A., Jr., and Martz, R. G. (1966). Field Tests of a Rotary-Vaporizing Burner. Paper CP66-16 *in Proceedings* (1966) (*q.v.*).

Cooper, P. W., and Marek, C. J. (1965). Design of Blue-Flame Oil Burners Utilizing Vortex Flow or Attached-Jet Entrainment. API Publ. 1723-A.

Cooper, P. W., *et al.* (1964). Recirculation and Fuel-Air Mixing as Related to Oil-Burner Design. API Publ. 1723.

Doyle, A. W., Perron, R. R., and Shanley, E. S. (1966). A Study of New Means for Atomization of Distillate Fuel Oil. API Publ. 1725.

Dunn, F. R., Jr. (1961). Improved Oil-Burner Performance with Recirculation of Combustion Gases. Paper CP61-8 *in Proceedings* (1961) (*q.v.*).

Dysart, W. D. (1966). Application of the API Ultrasonic Atomizer to a Vortex Combustor System. Paper CP66-2 *in Proceedings* (1966) (*q.v.*).

Evans, C. R. (1968). Honeywell's Approach to Solid State on Oil-Fired Equipment. Paper No. 68-WS 109 *in Proceedings* (1968) (*q.v.*).

Florio, S. A. (1968). Solid State Electronic Ignition and Control System for Domestic and Small Commercial Oil Burners. Paper No. 68-WS 102 *in Proceedings* (1968) (*q.v.*).

Hartmann, J., and Trolle, B. (1926–1927). A New Acoustic Generator: The Air-Jet Generator. *J. Sci. Instrum.* **4**, 101.

Hartmann, J., and Trolle, B. (1931). *Phil. Mag.* **11**, 926.

Hazard, H. R., and Hunter, H. H. (1966). A Miniature Ultrasonic Burner for a Multi-fueled Thermoelectric Generator. Paper CP66-3 *in Proceedings* (1966) (*q.v.*).

Hazard, H. R., and Stutz, D. E. (1962). A Survey of Ignition Technology for Residential Oil Burners. API Publ. 1721.

Hedley, A. B., and Jackson, E. W. (1966). The Influence of Recirculation in Combustion Processes. Paper CP66-8 *in Proceedings* (1966) (*q.v.*).

Hein, G. M., and Weller, A. E. (1963). Feasibility of a Vaporizing Burner for No. 2 Heating Oil. Paper CP63-5 *in Proceedings* (1963) (*q.v.*).

Hein, G. M., Weller, A. E., and Locklin, D. W. (1966). Vaporization and Combustion of No. 2 Heating Oil. API Publ. 1726.

Howe, E. L. (1964). Domestic Heating with Oil in North and Western Europe. Paper SP64-C *in Proceedings* (1964) (*q.v.*).

Howekamp, D. P. (1970). Flame Retention-Effects of Combustion-Improving Devices on Air Pollutant Emissions from Residential Oil-Fired Furnaces. APCA Paper No. 70-45.

Hughes, N., Schurig, R. E., and Hommel, D. P. (1965). Application of Air-Powered Sonic Energy in Atomization for Domestic Oil Burners. Paper SP65-I *in Proceedings* (1965) (*q.v.*).

Hunt, R. A., Jr. (1964). Evaluation of Oil-Heating Equipment. Paper SP64-G *in Proceedings* (1964) (*q.v.*).

Janssen, J. E., and Torborg, R. H. (1965). Flame Radiation in Domestic Oil Burners. Paper CP65-3 *in Proceedings* (1965) (*q.v.*).

Janssen, J. E., and Torborg, R. H. (1966). Performance and Radiation Studies with the API Prototype Blue-Flame Burner. Paper CP66-9 *in Proceedings* (1966) (*q.v.*).

Kennedy, D. R. (1965). Domestic Oil Burner Noise. Paper SP65-C *in Proceedings* (1965) (*q.v.*).

Lang, R. J. (1961). Air-Oil Pattern Matching in Gun Burners. Paper CP61-7 *in Proceedings* (1961) (*q.v.*).

Lang, R. J. (1962). Ultrasonic Atomization of Liquid. *J. Acoust. Soc. Amer.* **34**, 6.

Lang, R. J. (1963). Optimization Studies on the Ultrasonic Burner. Paper CP63-2 *in Proceedings* (1963) (*q.v.*).

Lang, R. J., Wilson, J. A., and Young, J. C. O'C. (1962). An Ultrasonic Oil Burner. Paper CP62-7 *in Proceedings* (1962) (*q.v.*).

Livingstone, C. J. (1961). Objectives and Scope of the New API Research Program. Paper CP61-1 *in Proceedings* (1961) (*q.v.*).

Locklin, D. W. (1960). Recommendations for an Industry-Wide Oil-Burner Research Program. API Publ. 1700.

Lord, W. (1968). Design Modifications of High Pressure Oil Burner. Paper No. 68-WS-101 *in Proceedings* (1968) (*q.v.*).

MacCracken, C. D., and McLendon, H. B. (1964). A Double-Vortex Recirculating Burner. Paper SP64-F *in Proceedings* (1964) (*q.v.*).

McCullough, J. E., Perron, R. R., and Shanley, E. S. (1966). The Development of Ultrasonic Atomizers for Domestic Oil Burners. API Publ. 1725-A.

Morgenthaler, J. H. (1962). Surface Vaporization Studies with No. 2 Fuel Oil. Paper CP62-8 *in Proceedings* (1962) (*q.v.*).

Mueller, R. H., and Schrader, G. F. (1966). A New Domestic Oil Burner—Low Capacity, Pressure, Noise, and Cost. Paper CP66-15 *in Proceedings* (1966) (*q.v.*).

Olson, E. O. (1965). Recent Progress in Conventional Oil-Fired Equipment. Paper SP65-A *in Proceedings* (1965) (*q.v.*).

Olson, E. O., and Tate, R. W. (1962). Spray Droplet Size of Pressure-Atomizing Burner Nozzles. *ASHRAE J.* **4**, 39.

Paterson, R. E. (1968). Some Design Considerations for Heat Exchangers. Paper No. 68-WS-111 *in Proceedings* (1968) (*q.v.*).

Peoples, G. (1965). Upgrading Oil-Burner Performance Standards. Paper SP65-H *in Proceedings* (1965) (*q.v.*).

Peskin, R. L., and Raco, R. J. (1966). Analytical Studies on Mechanisms of Fuel Oil Atomization. API Publ. 1727.

Proceedings (1961). *API Res. Conf. Distillate Fuel Combust., 1st, March 14–15, 1961.* API Publ. 1541.

Proceedings (1962). *API Res. Conf. Distillate Fuel Combust., 2nd, June 19–20, 1962.* API Publ. 1701.

Proceedings (1963). *API Res. Conf. Distillate Fuel Combust., 3rd, June 18–19, 1963.* API Publ. 1702.

Proceedings (1964). *API Res. Conf. Distillate Fuel Combust., 4th, June 2–4, 1964.* API Publ. 1703.

Proceedings (1965). *API Res. Conf. Distillate Fuel Combust., 5th, June 1–3, 1965.* API Publ. 1704.

Proceedings (1966). *API Res. Conf. Distillate Fuel Combust., 6th, June 13–15, 1966.* API Publ. 1705.

Proceedings (1968). *New and Improved Oil Burner Equipment Workshop, Nat. Oil Fuel Inst., September 17–18, 1968.* NOFI Tech. Publ. 106 ED.

Rakowsky, F. W. (1962). The Influence of Surface on Vaporization of No. 2 Heating Oil. Paper CP62-9 *in Proceedings* (1962) (*q.v.*).

Rakowsky, F. W., and Meguerian, G. H. (1963). Precombustion Deposits. Paper CP63-6 *in Proceedings* (1963) (*q.v.*).

Ranze, W. E., and Schindler, R. E. (1965). Recirculation in a Vortex-Stabilized Oil Flame. Paper CP65-1 *in Proceedings* (1965) (*q.v.*).

Rheaume, R. (1968). Pulse Ignition and Control System for Oil Burning Equipment. Paper No. 68-WS-105 *in Proceedings* (1968) (*q.v.*).

Rosser, W. A., Jr., Wise, H., and Wood, B. J. (1963). A Study of Ignition and Combustion of Fuel Droplets. API Publ. 1722.

Saxton, D. J., and Lane, E. R. (1961). The Ventres Blue-Flame Burner. Paper CP61-10 *in Proceedings* (1961) (*q.v.*).

Tuttle, J. R. (1961). Vaporization of Distillate Fuels in an Idealized System. Paper CP61-3 *in Proceedings* (1961) (*q.v.*).

Velie, W. W. (1968). New Concept for Domestic Oil Burners—Una-Spray Oil Burner Development Progress. Paper No. 68-WS-116 *in Proceedings* (1968) (*q.v.*).

Walsh, B. R. (1961). Air-Aspirated Oil Burners. Paper CP61-11 *in Proceedings* (1961) (*q.v.*).

Walsh, B. R. (1962). Burner Research Related to the API Program. Paper CP62-13 *in Proceedings* (1962) (*q.v.*).

Walsh, B. R. (1964). Aspiration in Fuel-Oil Combustion. Paper CP64-6 *in Proceedings* (1964) (*q.v.*).

Wasser, J. H. (1968). Effects of Combustion Gas Residence Time on Air Pollutant Emissions from an Oil-Fired Test Furnace. Paper No. 68-WS-110 *in Proceedings* (1968) (*q.v.*).

Will, R. G., Hunt, R. A., Jr., Martz, R. G., and Downs, E. S. (1964). A New Rotary-Vaporizing Burner. Paper SP64-A *in Proceedings* (1964) (*q.v.*).

Author Index

Numbers in italics refer to the pages on which the complete references are listed.

H

L

M

N

O

P

Q

R

S

T

Y

Z

Subject Index

D

E

F

T

U

V

W

Y

Z

A 4
B 5
C 6
D 7
E 8
F 9
G 0
H 1
I 2
J 3